Making *Nature*

Making *Nature*

The History of a Scientific Journal

MELINDA BALDWIN

The University of Chicago Press

CHICAGO AND LONDON

MELINDA BALDWIN is a lecturer in the Department of the History of Science at Harvard University.

The University of Chicago Press, Chicago 60637
The University of Chicago Press, Ltd., London
© 2015 by The University of Chicago
All rights reserved. Published 2015.
Printed in the United States of America

24 23 22 21 20 19 18 17 16 15   1 2 3 4 5

ISBN-13: 978-0-226-26145-4 (cloth)
ISBN-13: 978-0-226-26159-1 (e-book)
DOI: 10.7208/chicago/9780226261591.001.0001

Library of Congress Cataloging-in-Publication Data

Baldwin, Melinda Clare, 1981– author.
Making Nature : the history of a scientific journal / Melinda Baldwin.
pages cm
Includes bibliographical references and index.
ISBN 978-0-226-26145-4 (cloth : alkaline paper) — ISBN 0-226-26145-x (cloth : alkaline paper) — ISBN 978-0-226-26159-1 (e-book) — ISBN 0-226-26159-x (e-book) 1. Nature (London, England) 2. Science—Periodicals—History. 3. Science publishing—History. I. Title.
Q1.N23B35 2015
505—dc23
2014040048

♾ This paper meets the requirements of
ANSI/NISO Z39.48-1992 (Permanence of Paper).

Contents

Citations and Abbreviations • vii
A Note to the Reader • 1

INTRODUCTION • 4
Who Is a "Scientist"?

CHAPTER ONE • 21
Nature's Shifting Audience, 1869–1875

CHAPTER TWO • 48
Nature's Contributors and the Changing of
Britain's Scientific Guard, 1872–1895

CHAPTER THREE • 74
Defining the "Man of Science" in *Nature*

CHAPTER FOUR • 100
Scientific Internationalism and Scientific Nationalism

CHAPTER FIVE • 123
Nature, Interwar Politics, and Intellectual Freedom

CHAPTER SIX • 145
"It Almost Came Out on Its Own":
Nature under L. J. F. Brimble and A. J. V. Gale

CHAPTER SEVEN • 170
Nature, the Cold War, and the Rise of the United States

CHAPTER EIGHT • 200
"Disorderly Publication": *Nature* and
Scientific Self-Policing in the 1980s

CONCLUSION • 228

Acknowledgments • 243
Notes • 247
Bibliography • 285
Index • 301

Citations and Abbreviations

PRIMARY AND SECONDARY SOURCE CITATIONS

Several hundred *Nature* articles are mentioned or discussed in this book, rendering a primary source bibliography unmanageably long. Therefore, all primary sources will be cited in full in the endnotes and will not have a separate bibliographic entry. Secondary sources are cited in short form in the endnotes and in their full form in the bibliography.

ARCHIVAL ABBREVIATIONS

ARWP: Alfred Russel Wallace Papers, British Library
MP:BL: Macmillan Papers, British Library
MP:UR: Macmillan Papers, Special Collections, University of Reading Library
NLP: Norman Lockyer Papers, Special Collections, University of Exeter Library
SRGP: Sir Richard Gregory Papers, Special Collections, University of Sussex Library
SRGP:SA: Sir Richard Gregory Papers, Addition from the University of Sheffield, Special Collections, University of Sussex Library
THHC: Thomas Henry Huxley Collection, Records and Archives, Imperial College London Library

A Note to the Reader

If you are reading this book in the hope that it will reveal the secret to getting your article published in *Nature*, I should warn you that you will be disappointed. I have been surprised at how often people ask whether I know any tips or tricks for getting published in *Nature*—but perhaps I should have anticipated the question. Today, *Nature* is unquestionably one of the world's most prestigious and selective scientific journals. Even as an undergraduate chemistry major, it was clear to me that a *Nature* article was one of the most impressive publishing achievements a scientist could have on her CV. In the history of science, too, *Nature* was everywhere. *Nature* was the journal that first described the neutron, the journal that published groundbreaking work about plate tectonics, the journal where the "liquid drop" model of fission was published.

The goal of this book, however, is not simply to list all of the important papers *Nature* has published (although there are many discussions of famous—and infamous—*Nature* papers). Instead, this book examines how *Nature* has changed over time, what role it has played within the scientific community at different points in history, and how *Nature* both responded to and influenced changes in science.

Most interesting to me as a historian, *Nature* was founded at a time when many of the scientific practices we take for granted did not yet exist. A modern reader who picks up a nineteenth-century copy of *Nature* will see a lot that is familiar—research articles, book reviews, letters to the editor, and plenty of graphs and illustrations. A reader with an 1875 or a 1905 or even a 1955 issue of *Nature* will also encounter a scientific community very different from the one we know today. In 2015, no researcher would engage in a serious scientific debate with someone whose "expertise" was based solely on reading others' accounts of coral reefs or solar phenomena—but in the nineteenth century, the line between "layman" and "expert" was significantly blurrier than we consider it now. Similarly, if an important new paper from an elite laboratory was published without peer review today, there would be an uproar and accusations of favoritism—but the seminal 1953 Watson and Crick paper describing the structure of DNA was not peer reviewed before it was published, and few people seemed to mind. To observers at the time it was not obvious that people such as the Duke of Argyll or Richard Proctor could not be true "men of science," or that all *Nature* papers should go through external peer review before publication. One reason that I wrote this book about *Nature* instead of a single-discipline journal is that *Nature* was a major site where scientific practitioners proposed, argued over, and ultimately established many of the scientific norms that we take for granted today.

Nature's history reveals a lot about how the rules that govern science have evolved over time. It also reveals how scientific practitioners have thought about their place in society. *Nature* was founded in 1869 in London, a place and time that presents a fascinating science-and-society paradox. Victorian Britain produced household scientific names such as Charles Darwin and William Thomson (Lord Kelvin), but many of Darwin and Thomson's contemporaries felt deep anxiety about the social and intellectual status of "men of science" (a term that was quite purposely gendered). Science gained a tremendous amount of cultural and social authority over the course of the nineteenth and twentieth centuries, a change that is arguably one of the most significant developments in modern history. *Nature*'s editors and contributors continued to think and write about science's role in society even after they had gained the social respectability their Victorian predecessors had desired. When the National Socialist government fired Jewish scientists and claimed that Aryan superiority was a biological fact, *Nature*'s contributors called attention to the Nazi government's actions and insisted that science did not support the Nazi social order. At the end of the twentieth century, when two chemists announced that they had produced nuclear fusion at room temperature, *Nature*'s contributors took them to task for making the announcement prematurely and threatening public trust in science. As we will see throughout this book, *Nature*'s editorial staff, readers, and scientific contributors were deeply aware of science's relationship with politics, culture, and the social order.

I wrote this book with several audiences in mind, and one group I hope will pick up this book is regular readers of *Nature:* practicing scientists, science journalists, and others who love science. That brings me to a brief word about this book's methodology for readers who may be less familiar with how academic historians tend to approach the history of science.

In order to understand the rich historical story of science and its development, historians of science work to put ourselves in our actors' shoes. To avoid anachronism, historians of science tend to take a fairly neutral tone when discussing historical debates that might seem to have obvious "winners" according to modern science. Some readers might find this disconcerting. It can sometimes seem as if historians are criticizing current scientific ideas when we describe older ones, or that we are claiming that the rules that govern the modern scientific community are fraudulent "constructs" when we explore their origins.

Neither of these is my intent. To say that science used to operate differently is not to criticize the way things are now; giving careful consideration to both sides of a historical debate is not an implicit criticism of the opinion that won out. But to understand the development of scientific knowledge, we cannot write about the past as if the standards of modern science—things such as the appropriate qualifications for a "scientist," acceptable publication procedures, or proper experimental techniques—are now the standards because they were immediately agreed on by everyone as soon as they were suggested. We also cannot assume that in a debate, one view "won" simply because it was right; figuring out what is "right" is not a simple process in 2015, and it was no easier in 1869.

One of the most challenging and rewarding things about studying the history of science is putting aside what we know and letting the familiar become unfamiliar and vice versa. In doing so, we are not rejecting the present, but trying to see the past as clearly as possible.

INTRODUCTION

Who Is a "Scientist"?

On 29 November 1924, the British scientific journal *Nature*—then in its fifty-fifth year of publication—printed a letter to the editor from the respected physicist Norman R. Campbell. Campbell had a simple request: that *Nature* discard the terms *man of science* and *scientific worker* and, in their place, begin using the word *scientist*.

"There is a prejudice against this word," Campbell's letter began. The word *scientist*, he said, had been coined at a time "when scientists were in some trouble about their style" and "were accused, with some truth, of being slovenly." Campbell argued, however, that such questions of "style" were no longer a concern—the scientist had now secured social respect. Furthermore, said Campbell, the alternatives were old fashioned and unsuitable: *scientific worker* was cumbersome and *man of science* was outright offensive to the increasing number of women in science. Campbell closed with a direct appeal to *Nature*'s editor, Sir Richard Gregory:

> Let me therefore plead with you, Sir, who have done so much to raise the standard of scientific literature, and with all others who have striven to show that scientific and linguistic precision are not incompatible, to give us a lead in this matter. If you will not have "scientist," at least provide us with some other single word.[1]

Gregory, a man who was not above being pleased at a little flattery, responded in a note following Campbell's letter. The editor announced that he and the *Nature* staff had "invited a number of authorities on good English, including distinguished men of science, to favour us with their opinions" on the use of *scientist*. He promised that replies "of a critical or constructive kind" would be printed in future issues of *Nature*.[2] The Letters to the Editor column then continued its usual business with letters on "Cell-Wall Formation," "Nitrogen and Uranium," and "Edible Earth from Travancore."[3]

Modern readers might be surprised to see a debate about the word *scientist* nearly a hundred years after the English academic William Whewell first put the term in print, but as Campbell's letter suggested, the word had a surprisingly fraught history among English-speaking scientific practitioners.[4] *Nature* was far from the only British scientific institution that had refused to use the word *scientist* in an official capacity. As Gregory would observe in a later editorial on the matter, the Royal Society of London, the British Association for the Advancement of Science (BA), the Royal Institution, and the Cambridge University Press all rejected *scientist*.[5] And yet Norman Campbell directed his plea to *Nature* when he sought to establish *scientist* as a respectable term. *Nature* was not just the forum for the discussion; the journal's editor, contributors, and readers were also the arbiters of the dispute, the best authorities on whether the term should be used. Why *Nature*? Why did Campbell and others think that this weekly, for-profit journal of general science would be a decisive voice in the acceptance or rejection of the word *scientist*?

Today, modern readers know *Nature* as perhaps the most prestigious scientific journal in the world, a publication that scientists (as we call them now) from every country in the world read and where many scientists hope to publish their work. *Nature* has printed some of the most celebrated articles in the history of science, and famous figures from Charles Darwin to Stephen Hawking have contributed their scientific work to its pages. But as the debate over the word *scientist* suggests, *Nature* was not simply a vessel for important research articles: it was a site where scientific practitioners could debate about how to define themselves and discuss their place within a wider society.

This book will explore the history of *Nature* from its foundation in 1869 to the present day. I might justify such close focus on a single journal by pointing to *Nature*'s longstanding importance, but my goal is not to produce a hagiography. Over the course of 146 years of publication, *Nature*'s editors and contributors have engaged in a process of defining and redefining membership in their scientific community and have used *Nature* to promote both their own work and their visions of what science and its practitioners should be like. In short, *Nature* is an important publication not only because of the famous papers printed in its pages but also because of its significance as a place where scientific practitioners have worked to define what science is and what it means to be a scientist.

SCIENTIFIC COMMUNICATION AND SCIENTIFIC COMMUNITIES

As we shall see, *Nature*'s founder, Norman Lockyer, initially intended for his publication to be a place where British men of science would tell British laymen about the latest scientific advances. The publication quickly lost this focus because *Nature*'s contributors found the journal more useful for another purpose—namely, communicating with their fellow researchers. The 1924 debate about the word *scientist* provides useful insight into the role *Nature* came to play in the British scientific community. The week following Campbell's letter, ten letters on the subject appeared in *Nature*'s Letters to the Editor column. The physician Clifford Allbutt, author of a book on the composition of scientific papers, wrote that *scientist* "is quite a normal word, such as artist, economist, etc.," and he endorsed its use.[6] The renowned physicist Oliver Lodge also wrote in favor of using the word, albeit less than enthusiastically. He admitted that he personally disliked *scientist* but felt that "the public has forced the word upon us, and that we must succumb."[7]

But other voices pushed back against Campbell's proposal. Sir D'Arcy Wentworth Thompson, a zoologist, argued that *scientist* was a tainted term used "by people who have no great respect either for science or the 'scientist.'"[8] The eminent naturalist E. Ray Lankester, who had been a *Nature* contributor since its foundation, expressed the most vehement sentiments against the word, protesting that any "Barney Bunkum" might be able to lay claim to such a vague title:

> I hope NATURE will continue to refuse to use the word "scientist." Its formation can be defended, it is true, as parallel to that of "artist." But the example of the word "artist" gives us no encouragement, for it is the most vague and ill-used word in our language. All sorts of mysterious qualities are claimed for "the artist," and any imposter can defend his claim to be "an artist," and to worship art with a big A. We shall have others saying they "stand for" science with a big S and calling each other "Scientist." The eminent scientist Barney Bunkum is already flourishing in the United States and in English newspapers. I think *we* must be content to be anatomists, zoologists, geologists, electricians, engineers, mathematicians, naturalists ... "scientist" has acquired—perhaps unjustly—the significance of a charlatan's device.[9]

A handful of additional letters trickled in over the next two months. The physicist Herbert Dingle declared that the English language needed a word like *scientist* and said that he could not see "any justification" for refusing to use the word.[10] A contributor writing under the pseudonym of "A Chemist" called the word "cacophonous," but wrote, "You cannot help it; scientist is an established term. . . . But if *Nature* maintains its policy on the subject, I am sure we shall not grudge the Editor this little tyranny."[11] The chemist, educational reformer, and noted curmudgeon Henry Armstrong announced that he hated the term and proposed another: "The real men, those who do things—bakers, butchers, builders, boxers, grocers, even green-grocers— all have names ending in *er*. . . . Of late, I have often used sciencer, and I like it."[12]

Gregory and the *Nature* staff returned to the discussion on 21 February 1925 with an editorial titled "Words, Meanings, and Styles, I." Gregory wrote that *Nature* would not forbid authors to use the word *scientist* but that the *Nature* staff would continue its practice of avoiding the word. Gregory argued that *scientist* was "too comprehensive in its meaning. . . . The fact is that, in these days of specialized scientific investigation, no one presumes to be 'a cultivator of science in general.'"[13]

The next week Gregory continued with the second part of his editorial on "Words, Meanings, and Styles," offering more ruminations on writing, from elegant phrasing to the importance of quoting sources accurately to the use of plurals and capital letters. Gregory scoffed at "the common belief that writers on scientific subjects compare unfavourably with workers in other intellectual fields in the capacity to express themselves in suitable words," noting that most men of science spoke one or two foreign languages. He further argued that "the vocabulary of a man of science is probably more extensive than that of a man of letters."[14] If only the educated British public would make an effort to learn about science, implied Gregory, they would see that scientific writing was not deficient in the slightest.

Several rich threads emerge from the discussion about the word *scientist*. Although some contributors indicated that they had no objections to being called "scientists," other writers' discomfort with the term is revealing. There were concerns over the word's linguistic suitability—was it a clumsy hybrid of Latin and Greek? Or worse, an Americanism? Might a term such as *sciencer* more accurately portray what a scientific worker did in his day-to-day life? Significantly, however, the primary objections to *scientist* were not about the form of the word but about what the word implied.

Lankester and Gregory both fretted over whether the term was exclusive enough. Would any "Barney Bunkum" be able to designate himself a "scientist," thereby cheapening the accomplishments of naturalists, zoologists, chemists, and physicists who had built their reputations through advanced degrees and painstaking research? Was the term actually an insult, meant to portray the "scientist" as slovenly or ungenteel? Finally, there was a subtle but clear sense of rivalry between men of science and men of letters. Although Gregory had solicited contributions from both scientific practitioners and linguistic experts about the term *scientist*, he aligned *Nature* with what he claimed was the prevailing dislike for the term among men of science. Gregory also expressed frustration at the way literary men scorned scientific writing, although he considered scientific prose at least as accomplished as any other kind of writing. The difficulty, he suggested, was due to the public's ignorance of scientific terms. Anxiety over science's social status and intellectual respectability pervaded the "scientist" debate.

In many ways, the 1924 debate about the word *scientist* is a microcosm of the journal's history. *Nature* served as a site where its contributors could explore what it meant to be a scientist—or a man of science, or a scientific worker, or even a sciencer—in Great Britain and in the world. What made *Nature* unique was, in large part, its ability to act as a venue for such discussions via its correspondence columns and its weekly publication schedule. Historians of early modern and Enlightenment intellectual life have written a great deal about the Republic of Letters, the correspondence networks through which philosophers communicated to colleagues in other cities and nations, exchanging information and ideas and establishing intellectual credibility.[15] The ideal behind the Republic of Letters was that philosophers were part of a wider intellectual community and had ties with their fellow philosophers even if they never met in person. Participation in the Republic of Letters meant connections with other philosophers; communication among them made them an intellectual community. As historian Robert Mayhew evocatively puts it, "Scientists were a community held together by ink, both on the printed page and in the written letter."[16] In the nineteenth and twentieth centuries, *Nature* fulfilled a similar role with the added benefit of making letters and observations available to many readers at the same time. *Nature* grew into a forum where individuals interested in the advancement of scientific knowledge could talk to one another and discuss the intellectual and social issues affecting scientific work—in other words, *Nature* came to define a scientific community.[17]

But it is important to recognize that *Nature*'s contributors were not deliberately trying to build a monolithic "scientific community" with clearly defined membership qualifications that they all agreed on. The boundaries of *Nature*'s scientific community were constantly shifting, constantly being renegotiated and redefined. At different times, laymen were either included or excluded, political issues were acceptable or unacceptable subjects of discussion, and foreigners could be either a rare curiosity or an integral part of the journal's community. Both the editors' and the contributors' interests shaped the journal's development and the community it defined. It is an obvious but nonetheless important point that the journal itself had no agency beyond what the editors and contributors put into it. I have tried to avoid using *"Nature"* as the subject of a sentence—for example, *"Nature"* argued this"—but when I do, it should be taken as shorthand for *"Nature*'s editors and contributors."

When we think about communities, we must consider not only the community's members but the people who are excluded from its ranks. Unfortunately, Macmillan and Company and the *Nature* offices did not preserve much official correspondence before 1990.[18] The absence of an official editorial archive makes it difficult to determine whose contributions were rejected and why, but there are many indications in personal correspondence and in the journal itself that *Nature*'s contributors had strong ideas about whose voices ought to count in their community. Initially Lockyer intended to include laymen among *Nature*'s readers; the journal quickly shifted to a more technical publication written by and for men of science. *Nature*'s contributors often used the journal to set forth a vision of science as a specialized, expert discipline based on original investigations. By the 1880s, the astronomer and writer Richard Proctor felt that *Nature* was a bastion of scientific elitism that excluded popularizers and informed laymen from scientific discussions. He was so annoyed at *Nature*'s exclusionary tendencies that he founded a competing journal called *Knowledge*.[19] *Nature* continued to be a place where scientists discussed the rules and boundaries of their community well into the twentieth century.

A final aspect of *Nature*'s community is one that readers might find surprising: up until the mid-twentieth century, it was primarily a community of British scientists. Many historians have observed the profound influence national context exercised over research methods and scientific styles in the nineteenth and early twentieth centuries. Journals, including *Nature*, appear to fall under a wider pattern of scientific nationalism during this period.[20] Although about half of the contributions to the Letters to the Editor column came from outside Britain's borders by the 1930s, *Nature*'s editorials, articles, and book reviews were the products of a journal focused on the needs of a British readership. Editorials urged the British government to take more advantage of scientific knowledge in order to enhance the quality of life in the empire; correspondents discussed how scientists should best involve themselves in the seemingly inevitable war between Britain and Germany. It was not until the mid-1960s that *Nature* began to position itself as a spokesman for an international scientific community, not just a British one. That might seem counterintuitive, given the rise of the United States as the world's scientific powerhouse, but it was precisely because Britain's scientific influence was on the decline that *Nature* could broaden its contributor base and the range of international scientific issues that *Nature*'s staff could comment on without alienating subscribers in other countries.

Because *Nature*'s development was so intimately tied to the development of science in Britain, this book contributes to several wider discussions in the historiography of British science, including the question of the "professionalization" of science in nineteenth- and early twentieth-century Britain.[21] Although this book will not spend much time on the issue of language, it is worth noting that an obvious corollary of *Nature*'s Britishness is that it was published in English. In the nineteenth century, when French and German were far more dominant intellectual languages, an English-language periodical could hardly expect a large international readership. But as English became the most widely used scientific language in the twentieth century, *Nature* gained a significant advantage in the international publishing community.[22]

NATURE, SCIENCE, AND THE SCIENTIFIC JOURNAL

As the previous section suggests, social and cultural historians, especially historians of the book, will likely find this book interesting as an account of how print communication influenced a community's identity. Social and cultural historians will also find this book valuable as a history of an institution—*Nature*—that rose to prominence within the scientific community during a time when science made significant gains in social and cultural authority. In the nineteenth and twentieth centuries, few groups experienced a more dramatic change in their social and cultural status than scientific researchers. These centuries saw increased belief in the uniqueness of scientific knowledge and increased trust in scientific research to produce reliable knowledge claims. Trust in science is not and has never been absolute, of course, but it is clear that science held far greater social and cultural authority at the end of the twentieth century than it did in 1869, the year *Nature* was founded in London.[23]

As science gained more cultural authority, its practitioners became increasingly concerned with placing limits on who could claim to be a "scientist," as the 1924 *Nature* debate suggests. Scientific practitioners worked to establish their discipline as a specialized and expert pursuit that required particular qualifications. One of these qualifications became publishing scientific knowledge claims in a single approved format: the specialist scientific journal.

Nature's early history is intimately tied with the rise of the specialist scientific journal, and it is worth saying a few words on the journal's history and place in current scholarship. The world's first scientific periodicals, the British *Philosophical Transactions of the Royal Society* and the French *Journal des Sçavans*, printed their first issues in 1665. Other scientific periodicals, such as the *Acta Eruditorum* in Germany, were founded in the decades that followed.[24] These examples may tempt us to assume that the scientific journal we know today was fully formed by the beginning of the eighteenth century, but that assumption would be incorrect.[25] As A. J. Meadows observes in *Communication in Science*, seventeenth- and early eighteenth-century journals largely served to record presentations at meetings of scientific societies or to reprint valuable foreign papers. While periodicals were considered useful for sharing results, most readers expected that important scientific work would eventually be published as a book.[26] Famous examples include Isaac Newton's 1687 *Principia Mathematica* and 1704 *Opticks* or Linnaeus's 1735 *Systema Naturae*. Monographs continued to be an important form of scientific communication well into the nineteenth century, the obvious example being Charles Darwin's 1859 *On the Origin of Species*, arguably the most significant scientific work of the nineteenth century.

As Alex Csiszar's recent work has shown, however, over the course of the nineteenth century, the modern scientific journal began to emerge as the "principal institutional site for the representation, certification, and registration of scientific knowledge."[27] *Nature* was founded at the very moment when the scientific community was solidifying the specialist journal's status as the "embodiment of authoritative scientific knowledge."[28] By the early twentieth century, journal articles were the overwhelmingly dominant form of sharing one's research with the scientific community.

As James Secord recently observed, historians of science have tended to read specialist journals as historical sources rather than as historical phenomena in their own right, often taking the existence of journals for granted rather than viewing them as objects whose existence requires explanation.[29] This is beginning to change. Recently, there has been a great deal of excellent scholarship on scientific periodicals and the periodical press more generally in Victorian Britain.[30] However, the existing scholarship on scientific periodicals has usually focused on serials with a general or "popular" readership. Few historians have examined scientific writing intended for an audience of scientists or considered the substructure of the world of specialist periodicals.[31] Some historians have written about changes in the language of the scientific article, but these studies are not interested in journals as much as in the prose form of individual articles.[32] Furthermore, journals have played almost no role in the literature on scientific internationalism. In his book on international genetics congresses, Nikolai Krementsov writes that scholars have studied internationalism by looking at "international associations, research facilities, philanthropies, and societies"—a list from which journals are notably absent.[33]

Despite *Nature*'s influence, only a few scholars have written articles about *Nature*'s history. R. M. Macleod and Gary Werskey wrote a series of short historical pieces for *Nature*'s centenary issue.[34] Another account of the journal's first decades by Sir John Maddox (*Nature*'s editor from 1966 to 1973 and 1980 to 1995) is available as an introduction to a reprint edition of *Nature*'s first ten years of issues.[35] These brief essays provide a valuable overview of several interesting episodes in *Nature*'s history and remain the most widely read sources of scholarly information on *Nature*'s development. More recently, Ruth Barton and Peter Kjærgaard have both written extremely useful articles about scientific controversies in nineteenth-century *Nature*.[36] The literary historian David Roos also wrote a frequently cited piece about *Nature*'s earliest years of publication for a 1981 collection of essays on Victorian science.[37] The broader questions this book poses, however, remain unaddressed in the current literature.

This book approaches the study of *Nature* from a different angle than most previous works about specialist journals. Rather than focusing on a journal's linguistic or rhetorical characteristics or considering *Nature* as evidence for the development of a single scientific discipline, this book will address broader questions about *Nature*'s place in British and international science and its function in the world of science publishing. *Nature* is an ideal publication through which to study the evolution of specialist journals because of its long history and administrative stability. Not only has *Nature* been published every week since 1869, it has been published by the same publishing group since its foundation. This consistency enables us to trace the development of a scientific periodical from the nineteenth-century transition to journal publishing up to present concerns about open access and online publication.

The most obvious characters in *Nature*'s story are its editors in chief, the men (and so far, it has only been men) who have been responsible for guiding *Nature* into print. In nearly 150 years of publication, *Nature* has had only seven editors in chief (table 1): Sir Norman Lockyer, Sir Richard Gregory, A. J. V. Gale, L. J. F. "Jack" Brimble, David Davies, Sir John Maddox, and Philip Campbell. The intellectual and career ambitions of the different editors shaped the journal in particular ways, and it is always important to bear in mind who was at *Nature*'s helm when evaluating its development over the years.

A book about an individual journal might reasonably be expected to analyze decisions about which pieces were accepted and rejected. Unfortunately, without an editorial archive it is impossible to do a systematic study of how many or what kinds of articles *Nature* rejected. There are a handful of letters in the editors' personal archives that refer to particular editorial decisions and that give us a few insights into the editors' choices, but these hardly provide a complete picture. This book makes use of the personal papers of *Nature*'s editors, several contributors, and of Macmillan's existing archives, but the most important archival source for this study is *Nature* itself. The journal's issues provide a record of the accepted pieces, and we can determine a great deal about the motivations behind the journal by examining the journal's contents. The topics of the lead editorials indicate what issues the editors felt affected—or should affect—their readership; the research articles in the journal can indicate which research areas were particularly active at a given time; the "Letters to the Editor" (and later, the "Correspondence" column) frequently serve as a bellwether for readers' thoughts on controversial theoretical, political, and social questions affecting science and its practitioners.

Readers are another key component of the history of publishing.[38] What about the men and women who subscribed to *Nature* or regularly read a library copy of the weekly? Why did they read *Nature*? How did they incorporate it into their daily conversations, their scientific work, or their understanding of the current state of science? As with questions about editorial decisions, the absence of an official archive makes answering questions about readers more difficult—we have little in the way of subscription data or archives of letters from readers. However, it is possible to gain information about readers' uses of *Nature* through personal correspondence, interviews, and occasionally articles in the popular press. For example, Charles Kingsley, an early reader of *Nature*, was bold enough to tell Norman Lockyer that his magazine was becoming highly technical and difficult to under-

TABLE 1. *Nature*'s editors

Name	Previous career	Years as editor
Sir Norman Lockyer	Astronomer, science writer	1869–1919
Sir Richard Gregory	Science writer, member of *Nature* staff	1919–1939
A. J. V. Gale	Agronomist, member of *Nature* staff	1939–1962
L. J. F. Brimble	Botanist, member of *Nature* staff	1939–1965
Sir John Maddox	Physicist, science journalist	1966–1973
David Davies	Geophysicist, journal editor	1973–1980
Sir John Maddox	Physicist, science journalist	1980–1995
Philip Campbell	Physicist, *Nature* subeditor	1995–present

stand. Later, the physicist Bertram Boltwood considered *Nature* important enough to his scientific work that he would wait to read the latest issue before telling his colleague, Ernest Rutherford, about his recent findings. At the end of the twentieth century, *Nature* had become so influential that when editor John Maddox announced that cold fusion was "over" as a field of serious research, many science reporters took that as the final word on the topic.

Perhaps the most important characters in this story are *Nature*'s contributors, the men and the increasing number of women who wrote articles for the journal. What drew them to submit their work to *Nature*? Did they submit to *Nature* in addition to publishing in other journals, or did they prefer *Nature* over other publications? How did publishing in *Nature* serve their ends? Did articles in *Nature* fulfill a specific purpose for their authors, or was *Nature* simply one of a number of journals where scientific workers could publish their results? By analyzing the publication patterns of some of *Nature*'s key contributors—from Charles Darwin and George J. Romanes to James Watson and Francis Crick—we see that *Nature* was a unique publication that usually served a specific goal. In the nineteenth century, contributors began using *Nature* and its weekly turnaround time to debate scientific questions and to give abstracts of longer forthcoming papers in monthly or quarterly journals. In the early twentieth century, some contributors began employing a new strategy and used *Nature* for the immediate publication of interesting results before a paper was prepared or submitted elsewhere. This approach to *Nature* would reshape the journal's identity and transform it from Britain's most important scientific periodical to a major international venue for announcing—and debating—new scientific results. By the late twentieth century, *Nature* was renowned for printing some of the most exciting and important scientific results in a range of fields. Publication in the journal became so prestigious that, in the words of one *Nature* contributor, it seemed as if some scientists might "donate a kidney or something to get a paper into it."[39]

CHAPTER OUTLINE

This book contains eight chapters presented in roughly chronological order. In order to organize a potentially vast amount of material, the book uses both major international events (such as World War I) and changes in *Nature*'s editorship to divide the chapters. Chapter 1 tells the story of *Nature*'s foundation and its first few years of publication. Norman Lockyer, the astronomer who founded the journal with the financial backing of the publishing house Macmillan and Company, originally intended to produce a popular science magazine that would be read by both laymen and scientific researchers. However, the respected men of science whom Lockyer wanted as his contributors preferred to write for an audience of their scientific peers. Lockyer encountered further difficulties with his plan to direct the journal at laymen when his editorial policies clashed with the wishes of the X Club, a group of influential men of science who were also prominent science popularizers. As a result, Lockyer lost the support and the pens of the very group of people who would have been most likely to write the kinds of pieces he wanted. By the mid-1870s, even scientifically sophisticated laymen found it difficult to understand much of *Nature*'s content.

Chapter 2 explores how and why a younger generation of British scientists adopted *Nature* as a forum of scientific communication in the 1870s and 1880s. Although men of science Lockyer's age and older certainly contributed to the journal, they did not see *Nature* as a particularly desirable place to conduct scientific debates. The older generation preferred more established venues such as literary periodicals or the annual meeting of the BA. In contrast, younger men of science—those born in the 1840s and later—adopted *Nature* as the primary forum where they could debate the most important scientific questions of the day before a knowledgeable readership. The contributions of this younger generation established *Nature* as essential reading for British men of science in the final decades of the nineteenth century.

Chapter 3 focuses on the identity of the "man of science" in the late nineteenth and early twentieth centuries. By the end of the nineteenth century, British scientific workers viewed their discipline as a demanding pursuit that required complete devotion in order to attain expertise. They used

Nature to set forth their vision of the man of science as an individual who made original contributions to scientific knowledge and exhibited a commitment to scientific truth. In order to establish these standards, *Nature* contributors identified outsiders who were attempting to comment on scientific matters—such as literary critics who critiqued scientific theories and the anti-Darwinian politician George Douglas Campbell (the Duke of Argyll)— and took to *Nature*'s pages to explain why these writers were not qualified to participate in discussions about science. Chapter 3 also introduces Richard Gregory, a science journalist who joined the *Nature* staff in 1893 after a brief stint as Lockyer's observatory assistant, and discusses the difficulties surrounding his assumption of the editorship in 1919. Lockyer was concerned that Gregory might not be a suitable successor because he was not a Fellow of the Royal Society and was not a researcher—concerns that nicely illustrate the qualifications that *Nature*'s readers and contributors considered necessary for a true man of science. However, Gregory managed to create a niche for himself within this research-centric community by acting as a spokesman for their vision of science.

Chapter 4 focuses on the relationship between scientific publishing and scientific internationalism at the turn of the twentieth century. *Nature*'s speed of publication made the journal an invaluable resource for scientists working in the rapidly advancing field of radioactivity, and the Nobel Prize–winning physicist Ernest Rutherford was instrumental in establishing the Letters to the Editor column as a venue for announcing new and exciting scientific findings even before a complete scientific paper had been written. The frequent contributions from physicists such as J. J. Thomson, Frederick Soddy, and most importantly Rutherford made *Nature* essential reading not just for British scientists but for anyone interested in the most recent advances in physics. Radioactivity was an international field, with important contributions coming from several cities in Europe and the Americas, and *Nature* began to draw contributions on the subject from foreign scientists such as Otto Hahn in Germany, Bertram Boltwood in the United States, and scientists elsewhere in the British Empire, such as Rutherford in Canada. Despite the growth in international contributions from physicists, however, *Nature* remained firmly grounded in its British roots. Other growing disciplines, such as genetics, did not attract nearly as many non-British contributors as radioactivity, and the journal remained focused on science and scientific issues in Great Britain.

Chapter 5 centers on *Nature*'s conflict with Germany's National Socialist Party, a clash that led to the Nazi government banning *Nature* from German universities and libraries in 1937. Following the First World War, *Nature*'s editor and contributors shifted from portraying Britain as a nation that lagged behind other countries in its support of science to portraying Britain as a nation with superior respect for intellectual freedom. This transition was likely due in part to Richard Gregory assuming the editorship in 1919, but it also had a great deal to do with the journal's difficult relationship with German science and scientists during and after the First World War. *Nature*'s contributors had once held Germany up as the shining example of government support for science that Britain ought to emulate, but the journal's editors and contributors vehemently decried the behavior of German scientists during the war. Even after the war the journal did not resume its former envious tone with respect to Germany—or any other nation. Instead, in discussions ranging from the Scopes trial in the United States to the academic policies of the Soviet and National Socialist governments in Europe, *Nature* proudly praised Britain's commitment to intellectual freedom. Gregory and the editorial staff were also pleased to point out the number of contributions from foreigners in the Letters to the Editor and to proclaim *Nature*'s international importance as a venue for publishing the latest scientific findings.

Chapter 6 discusses two of *Nature*'s less prominent editors: Gregory's former assistants L. J. F. Brimble and A. J. V. Gale. Brimble and Gale led *Nature* through the difficult wartime years and oversaw the publication of some of *Nature*'s most famous papers, including several articles that contributed to the development of plate tectonics in geophysics as well as James Watson and Francis Crick's 1953 article on the structure of DNA. However, under the Brimble and Gale coeditorship, *Nature* was also known for accepting and rejecting papers based on the advice of a handful of prominent scientists and was regarded as a respectable but somewhat dull journal that "might print anything" if the right people recommended it. The Brimble and Gale era gives us a chance to examine *Nature*'s place in the larger postwar publishing landscape. It also gives us a window onto the status of peer review in the mid-twentieth century and shows that as late as the 1960s, a scientific journal could be credible even if it did not send all of its articles out for external refereeing.

Chapter 7 looks at *Nature*'s international status in the 1960s and 1970s under the leadership of two transformative editors: John Maddox and David Davies. In 1966, *Nature* hired Maddox, a physicist and a former science correspondent for the *Guardian*, to replace Brimble after his death (Gale had retired in 1961). The galvanic Maddox changed *Nature*'s format, its submission processes, its communication style, and its news columns almost immediately. In 1973, following a controversial attempt to split *Nature* into three publications, Maddox was pushed out of the editorship. Davies, his replacement, introduced systematic external peer review for all research articles and a wry sense of humor to *Nature*'s editorials. Under Maddox and Davies, *Nature* changed from a British publication to, as one staffer put it, "an international publication with a British accent."[40] By 1980, more than three quarters of the research articles in *Nature* came from outside Britain's borders. Under Maddox and Davies *Nature*'s news pages also expanded their international scope. However, during the Cold War, *Nature*'s internationalism had a significant limitation: it did not extend to countries in the Soviet Bloc. Scientists on the Communist side of the Iron Curtain remained outside of *Nature*'s scientific community, partly by exclusion (*Nature*'s editors and news writers were critical of the USSR) and partly because publishing in *Nature* served little useful function for scientists in the USSR. *Nature*'s example illustrates the consequences these publishing divides had for scientific work on both sides of the Berlin Wall.

The final chapter deals with John Maddox's second editorship (1980–1995) and his unusual actions during two debates over controversial scientific findings. In 1988, Maddox authorized the publication of the French immunologist Jacques Benveniste's study suggesting that there might be laboratory evidence for the efficacy of homeopathy (the practice of treating illnesses with extremely dilute solutions of various substances). Just weeks later, Maddox declared the Benveniste team's results a "delusion" after a highly publicized visit to Benveniste's laboratory in Paris. *Nature*'s readers and contributors, however, were extremely critical of the "circus" atmosphere that they felt had surrounded Maddox's trip to Paris. The next time Maddox faced the opportunity to challenge questionable results, he changed his tactics. Less than a year after the Benveniste controversy, the journal published "Observation of Cold Nuclear Fusion in Condensed Matter," a scientific paper in which Steven E. Jones of Brigham Young University announced that he had produced nuclear fusion at room temperature. Jones had expected to share credit for cold fusion with two other scientists, Martin Fleischmann and Stanley Pons from the University of Utah, but Jones's paper in *Nature* was overshadowed by his competitors' March 23 press conference announcing their discovery and their April 10 paper in the *Journal of Electroanalytical Chemistry*. Despite having printed Jones's paper, *Nature* took a doubtful stance on cold fusion from the outset and quickly became a major publication venue for cold fusion critics. It was the scientific work published in *Nature* that ultimately discredited Pons, Fleischmann, and Jones in the public eye. Maddox's decision to attack results published in his own journal seems unusual at first glance, but it is best understood in light of his vision of *Nature*. Maddox believed a scientific journal should have an opinion on the state of science, and he consciously sought to establish *Nature* as a defender of the scientific journal's importance to the scientific community.

A WEEKLY ILLUSTRATED JOURNAL OF SCIENCE

"*To the solid ground
Of Nature trusts the mind which builds for aye.*"—WORDSWORTH

THURSDAY, NOVEMBER 4, 1869

NATURE: APHORISMS BY GOETHE

NATURE! We are surrounded and embraced by her: powerless to separate ourselves from her, and powerless to penetrate beyond her.

Without asking, or warning, she snatches us up into her circling dance, and whirls us on until we are tired, and drop from her arms.

She is ever shaping new forms: what is, has never yet been; what has been, comes not again. Everything is new, and yet nought but the old.

We live in her midst and know her not. She is incessantly speaking to us, but betrays not her secret. We constantly act upon her, and yet have no power over her.

The one thing she seems to aim at is Individuality; yet she cares nothing for individuals. She is always building up and destroying; but her workshop is inaccessible.

Her life is in her children; but where is the mother? She is the only artist; working-up the most uniform material into utter opposites; arriving, without a trace of effort, at perfection, at the most exact precision, though always veiled under a certain softness.

Each of her works has an essence of its own; each of her phenomena a special characterisation: and yet their diversity is in unity.

She performs a play; we know not whether she sees it herself, and yet she acts for us, the lookers-on.

Incessant life, development, and movement are in her, but she advances not. She changes for ever and ever, and rests not a moment. Quietude is inconceivable to her, and she has laid her curse upon rest. She is firm. Her steps are measured, her exceptions rare, her laws unchangeable.

She has always thought and always thinks; though not as a man, but as Nature. She broods over an all-comprehending idea, which no searching can find out.

Mankind dwell in her and she in them. With all men she plays a game for love, and rejoices the more they win. With many, her moves are so hidden, that the game is over before they know it.

That which is most unnatural is still Nature; the stupidest philistinism has a touch of her genius. Whoso cannot see her everywhere, sees her nowhere rightly.

She loves herself, and her innumerable eyes and affections are fixed upon herself. She has divided herself that she may be her own delight. She causes an endless succession of new capacities for enjoyment to spring up, that her insatiable sympathy may be assuaged.

She rejoices in illusion. Whoso destroys it in himself and others, him she punishes with the sternest tyranny. Whoso follows her in faith, him she takes as a child to her bosom.

Her children are numberless. To none is she altogether miserly; but she has her favourites, on whom she squanders much, and for whom she makes great sacrifices. Over greatness she spreads her shield.

She tosses her creatures out of nothingness, and tells them not whence they came, nor whither they go. It is their business to run, she knows the road. Her mechanism has few springs—but they never wear out, are always active and manifold.

The spectacle of Nature is always new, for she is always renewing the spectators. Life is her most exquisite invention; and death is her expert contrivance to get plenty of life.

She wraps man in darkness, and makes him for ever long for light. She creates him dependent upon the earth, dull and heavy; and yet is always shaking him until he attempts to soar above it.

B

FIGURE 1 *Nature*'s first page, 4 November 1869. Reprinted by permission of the Nature Publishing Group.

CHAPTER ONE

Nature's Shifting Audience, 1869–1875

In the early months of 1869, a thirty-three-year-old British astronomer named Norman Lockyer (1836–1920) began asking his friends and colleagues to write articles he could publish in the first issue of a new weekly scientific publication.[1] The new periodical was not, Lockyer emphasized, a specialized scientific journal. Although Lockyer was soliciting contributions from Britain's most famous men of science and intended to print abstracts of technical papers and reports from foreign scientific societies, the journal was not affiliated with any scientific organization, and the audience was not solely other men of science. Rather, Lockyer hoped that his publication would be read by educated laymen of all trades, and he was publishing the weekly with the commercial London publishing house Macmillan and Company. Most of the people Lockyer consulted about his undertaking had at best modest expectations for the new publication. Lockyer's acquaintance Joseph Hooker, an eminent botanist and the director of Kew Gardens, pessimistically responded to the project by telling Alexander Macmillan, "By all means make public my good will to the Lockyer periodical ... [but] the failure of scientific periodicals patronized by men of mark have been dismal. I do not see how a really scientific man can find time to conduct a periodical scientifically, or brains to go over the mass of trash."[2]

Today, Lockyer's periodical, *Nature*, is arguably the world's most prestigious scientific journal, and most would call the publication an unparalleled success—although not the kind of success its founder had initially envisioned. Early in its life, *Nature* underwent a significant change in form. *Nature* never acquired much of a following among laymen, and the journal quickly abandoned its plan to devote a large portion of its contents to popular science pieces. The first issue of *Nature* was published in November 1869; by 1875, the primary audience for *Nature* had shifted from laymen to men of science. The changes in *Nature*'s content suggest that Lockyer had difficulty balancing the two parts of his initial vision and that the preferences of his contributors drove *Nature*'s transformation into a publication very different from the one its editor had planned.

J. NORMAN LOCKYER: CIVIL SERVANT, ASTRONOMER, WRITER

Joseph Norman Lockyer was born in 1836 in Rugby, England. His father, Joseph Henry Lockyer, was a middle-class physician-apothecary, and his mother, Ann Norman Lockyer, was the daughter of a local squire. Soon after Norman's birth the family moved to Leicester, where he and his younger sister spent their childhood. Following the death of his mother in 1846, Norman was sent to live with relatives in Warwickshire, where he attended private schools and occasionally supported himself with student-teaching responsibilities. At the age of twenty he convinced a local landowner, Lord Leigh, to support his quest for a government position. In 1857 Lockyer began his working life not in science or medicine but as a clerk at the War Office in the London suburb of Wimbledon.

Lockyer's biographer A. J. Meadows dates Lockyer's interest in science to his first years in this job. A significant amount of scientific research in Victorian Britain was carried out by officers in the Army and Navy, and Lockyer probably met colleagues in the War Office who encouraged his interest in science. Furthermore, the overstaffed War Office was not a particularly taxing place to work. Lockyer found himself with plenty of time to pursue other activities, including mountaineering and his growing scientific interests. Several of Lockyer's friends in the Wimbledon village club were astronomical enthusiasts, and the barrister George Pollock appears to have been particularly significant in encouraging Lockyer's interest in astronomy.[3]

In 1861 Lockyer purchased a small (3¾-inch refractor) telescope, which he set up in his garden and used to observe the surface of the moon and Mars. He quickly gained a reputation as a skilled and reliable observer, and in 1862 he was admitted to the Royal Astronomical Society.[4] In the mid-1860s, Lockyer's interests shifted from the observation of planetary bodies to the spectroscopic study of the sun; Lockyer hoped to learn information about the sun's temperature and elemental composition through the

FIGURE 2 Norman Lockyer in middle age. © National Portrait Gallery, London.

analysis of the sun's spectra.⁵ Just five years after the purchase of his first telescope, Lockyer (having upgraded to a much larger telescope with an attached spectrometer) completed an influential study of solar spectra, which indicated that dark sunspots radiated at a much lower temperature than the rest of the sun. The work was the basis for his election as a Fellow of the Royal Society in the summer of 1869.⁶

Astronomy, however, was an expense rather than a source of income, and by the early 1860s Lockyer was married and increasingly in need of money for his growing family. Lockyer supplemented his War Office income by writing articles for a nonscientific audience. By the early 1860s Lockyer had written several articles on astronomy that appeared in lay publications such as the *London Review* and the *Spectator*. In 1862 his Wimbledon neighbor Thomas Hughes, a well-known Christian Socialist, persuaded Lockyer to help him establish a new journal that would discuss British science, religion, and art. Lockyer enthusiastically agreed. The new journal was called the *Reader*, and its first issue appeared in January 1863. Lockyer served as this weekly magazine's science editor. He oversaw a section that mixed popular science articles aimed at laymen with content aimed at researchers, such as reports on meetings of scientific societies and abstracts from specialist journals. Lockyer's correspondence indicates that the *Reader*'s subscribers included many of his fellow scientific researchers and that this readership particularly valued the abstracts and summaries. In 1864, for example, the mathematician Thomas A. Hirst wrote to Lockyer to express concern that the reports of foreign societies had been omitted from the latest issue of the *Reader*:

> Your usual reports of the proceedings of foreign societies are discontinued I see at present, in consequence I apprehend of the press of matter caused by the Bath Meeting. I trust however they will be shortly resumed with arrears for this constitutes an excellent and useful feature in The Reader.⁷

However, like many other Victorian publications, the *Reader* was unable to turn a profit; there were simply never enough subscribers to sustain the costs of publication. In late 1863 Lockyer convinced the other editors of the *Reader* to allow him to increase the size of the science section. A letter to Lockyer from the naturalist George J. Allman suggests that Lockyer wrote to fellow men of science to ask for their aid in expanding the *Reader*'s scientific content:

> I quite approve of your proposal to amplify the scientific section of The Reader and I wish every success to your project. I am personally well-pleased with the idea, for I have recently become a subscriber to the journal.⁸

However, the increased amount of scientific coverage did little to secure more subscribers. The *Reader* continued to spiral into the red. In 1865 the journal was sold to a new owner, Thomas Bendyshe, who eliminated Lockyer's science section.

Although the *Reader* was short lived, Lockyer's work with the magazine brought him into contact with one of his most famous scientific contemporaries: Thomas H. Huxley, "Darwin's bulldog" and advocate of scientific naturalism.[9] Huxley was a member of an informal society of nine British scientists that called themselves the X Club.[10] This influential group included such Victorian scientific luminaries as Huxley, Hooker, John Tyndall, John Lubbock, and Herbert Spencer. Their shared goal was to promote Darwinian evolutionary theory and scientific naturalism, both within the British scientific community and in British society at large. The nine were remarkably successful at winning influential positions in British science. During a visit to England, the American science writer John Fiske described the X Club as "the most powerful and influential scientific coterie in England" and said that the group had "dictated the affairs of the British association for three years past."[11]

Huxley had seen the science section of the *Reader* as a chance to create a forum where the members of the X Club could advance their views before the general public.[12] When the publication began failing, Huxley even arranged for the X Club to assume ownership of the *Reader* in 1864, but the club could not keep the publication afloat and eventually sold it to Bendyshe. Despite the *Reader*'s failure, neither Huxley nor Lockyer gave up the idea of a publication that would allow men of science to promote their work to a lay audience. In fact, the experience seems to have convinced both men of the need for a publication devoted strictly to science.

Meanwhile, Lockyer began work on a book, *Elementary Lessons in Astronomy*, which he published with the London publishing house Macmillan and Company.[13] The book was printed in 1868 and was well received and reasonably profitable. Following the success of *Elementary Lessons in Astronomy*, Macmillan and Company began paying Lockyer to advise them on their scientific publications. Lockyer got along well with Alexander Macmillan, the patriarch of the publishing clan, and quickly became the publishing house's most important scientific consultant. Macmillan once referred to Lockyer as his "consulting physician in regard to scientific books and schemes."[14] Lockyer's financial situation, which had been precarious in the early 1860s, was looking more secure—until 1868, when a complicated series of bureaucratic reorganizations resulted in Lockyer losing a promotion at the War Office and in his salary being reduced by almost half.[15] His career as a civil servant was suddenly looking much less profitable than his science writing. He approached Macmillan about financing a new periodical and hiring him as its editor.

At the time Macmillan and Company was primarily known for publishing books, not periodicals. The publishing house had been printing *Macmillan's Magazine*, a literary monthly, since 1859, but Macmillan's presence in the world of intellectual periodicals was otherwise slim. Fortunately for Lockyer, the 1860s had been an enormously successful decade for Macmillan. The firm had established a new headquarters in London, published an impressive list of profitable novels and series, and opened an American branch office in New York City. Alexander Macmillan was looking to expand his publishing empire even beyond the novels and series that had made his family business one of the leading publishing houses in Britain. Leveraging both his own relationship with Alexander Macmillan and the publishing house's desire to build on their profitable line of scientific books, Lockyer persuaded Macmillan and Company to back his new publication. He immediately began soliciting contributions from the acquaintances he had made during his time at the *Reader*.

LOCKYER'S *NATURE* AND VICTORIAN SCIENCE PUBLISHING

The journal would enter a bustling and highly competitive market in popular science publishing.[16] The market for periodicals had been growing steadily in nineteenth-century Britain thanks to new and cheaper methods of serial production, and scientific periodicals were among the new publications that readers could purchase at London newsstands.[17] In 1815, there were only five commercial science periodicals available in Britain; by 1895, that number stood at eighty.[18] Several shilling monthlies that discussed science, such as the *Cornhill Magazine* and Robert Chambers's *Chambers' Edinburgh Journal*, had successfully established readerships among middle-class Englishmen during the 1850s and 1860s.[19]

Nature was designed to rival such publications on two levels. On the commercial level *Nature* was competing for subscribers, but *Nature*'s editor and contributors were also competing for control of information about science in Britain. Huxley was an early and enthusiastic supporter of *Nature* in part because he and the other members of the X Club were alarmed by the growth of popular science literature written by science journalists.[20] Huxley had become increasingly frustrated by the scientific errors he often spotted in popular science books; even worse, in Huxley's view, many science journalists (especially, he believed, female ones) wove theological overtones into their writings. He came to believe that only scientific researchers could properly educate the public about science.

Lockyer did not share Huxley's low opinion of women, but he did agree that scientific researchers, not journalists or interested dilettantes, should be the ones who told the British public about scientific findings. He envisioned *Nature* as a publication filled with accessible articles by distinguished men of science.[21] This endeavor was intended not only to inform the public but also to place the control of public information about science in the hands of men of science. Furthermore, although the initial price of 4 pence per issue made it a relatively inexpensive weekly, Lockyer and the Macmillans did not intend to market *Nature* to middle-class English families.[22] It would be aimed at an elite audience of highly educated (although not necessarily wealthy) laymen.

Lockyer's inspiration for *Nature*'s format appears to have been *Chemical News*, a publication edited by his good friend William Crookes.[23] Crookes founded *Chemical News* in 1859 with the aim of reaching an audience of chemical researchers, teachers, physicians, and any other readers interested in chemistry or chemical manufacturing. Like *Nature*, *Chemical News* was a weekly with two columns of text printed on each page. When we compare the contents of *Nature* and *Chemical News* in 1869 (table 2), we see that the two journals also contained similar material. Both led with editorials (when available; neither journal published an editorial every week). Both contained book reviews, articles on recent experiments (frequently abstracts of longer papers), reports from scientific societies, correspondence from readers, and a column devoted to miscellaneous pieces of interesting scientific news ("Notes" in *Nature*, "Miscellaneous" in *Chemical News*).

The two were not completely identical, of course; each publication had a few unique features signaling the different missions of the two journals. For example, *Chemical News* included a column specifically devoted to lecture experiments, suggesting that teachers of chemistry formed an important part of Crookes's readership, and also printed a yearly student issue. *Chemical News* also reported on recent chemical patents, a testament to the growing influence of such patents in the chemical industry. *Nature* had a section for reports from different disciplines such as astronomy, botany, geology, or physiology, a feature that would have been unnecessary in the discipline-specific *Chemical News*.

Chemical News was not Lockyer's only model. *Nature* also owed a great

TABLE 2. Contents of *Nature* and *Chemical News* in 1869

Nature	Chemical News
Editorial	Editorial
Lead article(s)	Articles
Book reviews ("Our Book Shelf")	Reports of Societies
Articles	Chemical Notices from Foreign Sources
Letters to the Editor	Notices of Books
Notes	Notes on Lecture Experiments
Scientific Serials	Laboratory Notes
Discipline Reports*	Correspondence
Societies and Academies	Miscellaneous
Diary (Upcoming Meetings)	Patents
Books Received	Answers to Correspondents

*These are brief reports on recent articles and interesting data in various fields, including astronomy, botany, chemistry, geology, physics, and physiology.

deal to Victorian gentlemen's publications, such as the *Reader* and the *Athenaeum* (a periodical associated with the London gentleman's club of the same name). Britain's literary magazines, such as *Fortnightly Review*, *Nineteenth Century*, and *British Quarterly Review*, also provided a source of inspiration and competition. These were the publications that attracted the kinds of subscribers Lockyer wanted—educated men of all trades—and they had been major centers of scientific discussion throughout the nineteenth century. The *Athenaeum*, for example, included reviews of important scientific monographs (such as Charles Darwin's *On the Origin of Species* or Herbert Spencer's *Principles of Biology*) in its Literature section, and like the *Reader*, the *Athenaeum* had a science section that printed reports from scientific societies and schedules of upcoming scientific meetings. In addition, the correspondence section of the *Athenaeum* sometimes contained discussions of scientific questions, such as the proper method of gathering and preserving specimens for the Zoological Society.[24] Literary magazines were even more important as forums of scientific communication in Britain. Although literary magazines generally devoted less than a tenth of their contents to articles on scientific subjects,[25] like the *Athenaeum*, they frequently included reviews and discussions of noteworthy scientific monographs. Furthermore, many famous men of science, including Huxley, Spencer, Crookes, and Alfred Russel Wallace, wrote articles for these publications, and (as we shall see) many scientific controversies began with a man of science criticizing another's theories in a lengthy essay for a literary magazine.

An advertisement in *Athenaeum* described *Nature* to potential readers as both enriching reading for the general public and a useful aid for scientific men.

> The Objects which it is proposed to attain by this Periodical are, first, to place before the general public the results of Scientific Work and Scientific Discovery, and to urge the claims of Science to a more general recognition in Education and Daily Life; and secondly, to aid Scientific Men themselves, by giving early information of all advances made in any branch of natural knowledge throughout the world, and by affording them an opportunity of discussing the various scientific questions which arise from time to time.[26]

Macmillan and Company placed the same advertisement in *Journal of the Society of Arts*, *Cambridge University Gazette*, and a new monthly, the *Academy*.[27] Although the advertisement claimed that *Nature* would be aimed at the "general public," Macmillan's advertising decisions indicate a more elite view of their intended audience. Like Lockyer, Macmillan expected that *Nature* would attract educated readers with a range of intellectual interests.

Nature's narrow intended audience made its chance of success precarious. Joseph Hooker's assessment of its chances was gloomy but realistic, and despite the involvement of Huxley and the X Club, *Nature*'s survival was far from guaranteed. Lockyer's own experience with the quickly indebted *Reader* demonstrated the difficulty of producing a periodical aimed at an educated, elite audience that would attract enough subscribers to be financially viable. Furthermore, by 1869, coverage of science in elite periodicals such as the *Times* had been declining steadily for almost a decade, which indicated to many men of science that interest in their work was also declining among Britain's leaders.[28] It was not clear if *Nature* would be able to find a wide enough audience among its intended readers to ensure its continued existence.

The parallel case of another journal, the *Academy*, provides evidence of the difficulties a new periodical might face.[29] The *Academy*, like *Nature*, was founded in 1869. Its editor, an Oxford graduate named Charles Appleton, wished to summarize the latest scientific research for an educated British audience. Indeed, the *Academy*'s first working title was the *Monthly Journal of Science*, but before publication Appleton decided that a journal devoted to science would not attract the broad readership he desired and added coverage of recent work in literature and philosophy. The new publication went to press under the title *The Academy: A Monthly Record of Literature, Learning, Science and Art*.

The group of scientific reviewers who wrote for the *Academy* included many of the same men who wrote for *Nature*, such as Huxley, Alfred Russel Wallace, and John Tyndall. In fact, a letter to Lockyer from the Cambridge physiologist Michael Foster suggests that Lockyer regarded Appleton's *Academy* as a rival, a fact that concerned Foster:[30]

> I have however been somewhat uneasy in my mind since I saw you about the other journal, Appleton's—I quite understood from Appleton that the two papers would not interfere with each other—but from what fell from you, I fancied that you regarded his effort as in some way a rival to yours. If this were the case I could not work with any great comfort for both of them as at present I am engaged to do. I hope it will not turn out to be so.[31]

At first Appleton did not view *Nature* as a serious competitor, believing that it did not include enough coverage of European research to claim the *Academy*'s readership.[32] But when the *Academy* failed to gain enough readers to become profitable by 1873, Appleton decided that his scientific content did not attract enough subscribers to keep the journal running and began decreasing scientific coverage in favor of more philosophy and literature. Coverage of the physical sciences in particular decreased dramatically over the course of the 1870s. The change in format did little to push the *Academy* into the black, however; by the time of Appleton's death in 1879, the *Academy* was nearly £25,000 in debt to its supporters and subscribers.[33] The journal would survive into the twentieth century, but by 1900 the science section had been entirely eliminated.

NATURE'S CONTENT 1869–1875

After considering the *Reader* and especially the *Academy*, we can easily imagine *Nature* fulfilling Joseph Hooker's pessimistic predictions. Like the *Academy*, *Nature* targeted a limited audience of elite gentlemen; like the *Academy*, it sought to publish popularizing contributions from eminent men of science. Ultimately, neither publication retained this original formula. The *Academy* survived by decreasing its coverage of scientific research. *Nature* developed in the opposite direction: the publication drew away from its nonscientific readers and became a journal by and for men of science.

After receiving Macmillan's backing for the journal, Lockyer began writing to other men of science—some whom he knew personally, others whom he knew by reputation—to ask if they would allow their names to be printed in a list of *Nature*'s future contributors and supporters.[34] Most of Lockyer's correspondents wrote back to say that he was welcome to use their names if he wished but that it was unlikely they could devote time to the publication. The chemist Sir Frederick Abel wrote, "I am afraid there is small chance of my becoming at any rate an <u>early</u> contributor to 'Nature,' but if you consider it worth while to add my name before her certainly—pray do so."[35] Sir George Airy, the Astronomer Royal, said that while he thought the journal would be "extremely useful," it was "totally out of my power to give any assistance in it" because of his busy schedule.[36] A few, including the geologist and archaeologist William Pengelly, expressed concern about the journal's chances of success. Pengelly was specifically worried about whether *Nature* would be able to compete with a less expensive rival, *Scientific Opinion*:

> Thank you for the Prospectus of the new Nature. By all means take my name if it is of any service.... Will not the <u>fourpenny</u> Scientific Opinion interfere with the sale of the <u>sixpenny</u> Nature? Now don't mount your white horse, and say "Nature is to be above Scientific Opinion," but remember that Human Nature prefers to <u>pay</u> a groat rather than a tamer.[37]

Notably, Lockyer and Macmillan decreased *Nature*'s price to four pence per issue before the journal went to print.

Some of Lockyer's friends were more enthusiastic. Charles Pritchard, a fellow astronomer, replied, "You are quite at liberty to include me among the certain allies & the probable contributors to 'Nature.' I wish you good luck."[38] Only a handful, including Michael Foster, promised any kind of concrete support. Foster said he would write for *Nature* and suggested several possible subeditors who might be interested in working for Lockyer's new publication.[39] Despite the concerns he had expressed to Alexander Macmillan, Joseph Hooker also set out to help Lockyer construct his staff; he put Lockyer in contact with a botanist, Alfred Bennett, who could help Lockyer edit contributions that fell outside Lockyer's field of astronomy.[40] The most important contributor was arguably Huxley, who not only wrote several articles himself but convinced his friends in the X Club to write for *Nature* as well. Huxley alone was a significant draw. By the 1860s Huxley's essays were highly sought after by many commercial periodical editors. Well aware of his own popularity, Huxley once told his friend John Morley that when it came to journalism, he was "as spoiled as a maiden with many wooers."[41]

The Macmillans chose Richard Clay and Sons of Suffolk as *Nature*'s printer. Clay's was already responsible for much of Macmillan's scientific printing—not always to the satisfaction of their authors, however. There are several letters in the Macmillan archives complaining about the poor quality of Clay's printing, in particular their handling of scientific illustrations. In 1876, an incensed Alfred Russel Wallace wrote to George Craik, a partner at Macmillan, to complain of the treatment his book was receiving at Clay's hands:

> Your excuse for Messrs Clay is a very bad one, or rather it is <u>no excuse at all</u>. It would be good, if the kind of work had come upon them unawares, but, on the contrary, I shewed the MSS of this very pack (the Birds) to Mr. Clay in Mr. Macmillan's presence early in <u>August</u>, and called his special attention to the headings;—telling him <u>how many of them there would be</u>, and asking whether it would not save time to have stereoforms made instead of setting them up separately. He said he would consider it when they came to it. It was <u>after this full information</u>, that he undertook to print <u>the book through</u>, at the rate of <u>4 sheets a week!</u> ... Now, they come down to one sheet a week, & complain of the difficulty of the work!
>
> Either then, Mr. Clay was, in August last, ignorant of his business and gave an estimate he could not possibly carry out,—or he wilfully [sic] deceived us in order that we might not go elsewhere. They cannot get out of this dilemma now by pleading the difficulty of the work. . . . Had Mr. Clay told me in August that he could not undertake to finish printing till March or April, I should certainly never have begun with him at all.[42]

An 1878 letter from the geologist Archibald Geikie indicated that dissatisfaction with Clay's could extend to dissatisfaction with *Nature*:

> On getting home last evening I found the proof of the first sheet of the Geikie Geology and even after a night's sleep and a good breakfast I haven't recovered from the disappointment not to say disgust with which the sight of it filled me. A more paltry insignificant-looking page I never saw. It is Clay all over and that is saying the worst that can be said for any printing. . . .
>
> Don't think I'm alone in this opinion. The complaints on all sides are loud, both of the carelessness of the typography and the cuts. Look at that chart published in Nature the other day, and compare it with any map printed in the Illustrated London News or any weekly. It is utterly illegible. Or try to make heads or tails of Favre's drawings published a week or two earlier.
>
> I have made up my mind that I shall never have another book printed by Clays. For the honour of Macmillan & Co. I hope something will be done to improve the printing of their books.[43]

Such complaints appear to have had little effect on Macmillan's business relationship with the printer. Clay's would continue to print *Nature* well into the twentieth century.

Nature's first issue was published on November 4, 1869. The issue was twenty pages long and was printed in two columns, with illustrations as needed. It came wrapped in advertisements for scientific instruments, books, and other publications. It is not clear when or how Lockyer settled on his weekly's name, but it was a title that would have resonated strongly with British men of science. Victorian scientific texts often employed the image of "Nature" as the guide to all true scientific knowledge.[44] This image was drawn from Romantic poetry and German *Naturphilosophie* of the early nineteenth century.[45] By choosing *Nature* as the title for his publication, Lockyer was attempting to tap into the British scientific self-image as well as to invoke the Romantic vision of science as a search for all of Nature's truths.[46]

The masthead further illustrated many of Lockyer's hopes for the journal. The half-revealed image of the globe and the Gothic, uneven letters of the title suggested both intense mystery and great potential. *Nature* billed itself as "A Weekly Illustrated Journal of Science" and bore an epigraph from William Wordsworth's 1823 sonnet "A Volant Tribe of Bards on Earth are Found": "To the solid ground / Of Nature trusts the mind which builds for aye." (In Wordsworth's original poem, "Mind" was capitalized and "nature" was not. Lockyer altered the verse to emphasize the preferred word.) *Nature*'s masthead image would remain unchanged until 1958.

In keeping with the Romantic image of science suggested by the title and masthead, *Nature*'s introductory article was a flowery piece by Huxley titled "Nature: Aphorisms by Goethe," in which Huxley expressed the hope that *Nature* would further Goethe's ideal of linking together "all of the phenomena of Nature" and record the best science of the day for the benefit of future generations.[47] The first issue also included an article by Alfred Bennett on plant fertilization, an anonymous piece on the development of science in Australia, and a communication from Lockyer himself about the recent eclipse of the sun.[48] Several book reviews (many of them about books in German), an editorial on science teaching in British schools, an obituary of the Scottish chemist Thomas Graham, a letter to the editor about the Suez Canal, and reports on recent work in astronomy, chemistry, physics, and physiology followed. The journal concluded with reports from scientific societies in London, Manchester, Paris, and Philadelphia, which detailed upcoming meetings and exciting discoveries, such as a report from Manchester that "Dr. Joule, F.R.S." had noticed a band of blue refracted light at the tail end of a sunset "on two or three occasions."[49]

The contrast between material for laymen and material for men of science was evident in the first issue. Huxley's essay and the three articles at the beginning of the journal were written in a journalistic style that assumed little (if any) prior knowledge on the part of the reader. Lockyer and Bennett's

A WEEKLY ILLUSTRATED JOURNAL OF SCIENCE

"*To the solid ground*
Of Nature trusts the mind which builds for aye."—WORDSWORTH

FIGURE 3 *Nature*'s masthead, 1869–1958. Reprinted by permission of the Nature Publishing Group.

articles were written in the first person, reporting personal observations or opinions on recent findings. This was very different from most of the prose in specialist scientific journals at the time. By the 1870s most scientific articles were written as abstract narratives in which Nature, not the researcher, was the actor.[50] The first-person format of these articles would have distinguished them from articles written for scientific researchers in a specialist journal. For example, in the 1869 issue of *Philosophical Transactions of the Royal Society of London* (the flagship periodical of the Royal Society of London), all of the articles, including one by Lockyer, were written in the passive voice with no first-person narrative.[51]

However, several of the book reviews discussed volumes that would only have been of interest to specialists. For example, Michael Foster reviewed the German treatise *Das Hemmungsnervensystem des Herzens* and described it as "a critical and experimental inquiry into the inhibitory action of the pneumogastric nerve;"[52] there was also a review of a new set of German astronomical tables that would enable astronomers to calculate planetary masses.[53] Similarly, the disciplinary reports mixed articles on highly specialized research, such as the synthesis of dichlorinated aldehyde, with less technical reports on the color of wine and possible new treatments for a cholera outbreak in India.

The next handful of issues strongly reflected Lockyer's desire to attract a lay audience. The articles at the front of the journal continued in a journalistic vein; in 1869, *Nature* featured articles on the beauty of science,[54] the agricultural implications of geology,[55] and the importance of Darwinian theory for British national policy (authored, unsurprisingly, by Huxley, under the unsubtle initial "H").[56] Through the end of 1869, the book reviews were less specialized than they had been in the first issue, but the reports from scientific societies and the disciplinary summaries were still aimed at an expert rather than a lay audience. This is particularly evident in the chemistry reports, which often summarized new organic syntheses or gave detailed information on the crystalline structure of newly created salts—information of little use or interest to anyone other than a chemist.

A notice printed on the final page of the 20 January 1870 issue suggests that Lockyer suspected *Nature*'s two sections might be confusing his readers.[57] In an effort to clarify the journal's agenda, Lockyer wrote that *Nature* had two broad aims:

> FIRST, to place before the general public the grand results of Scientific Work and Scientific Discovery, and to urge the claims of Science to a more general recognition in Education and in Daily Life;
>
> And, SECONDLY, to aid Scientific men themselves, by giving early information of all advances made in any branch of Natural knowledge throughout the world, and by affording them an opportunity of discussing the various Scientific questions which arise from time to time.

This part of the notice was nearly identical to the advertisements Lockyer had placed in journals such as the *Athanaeum*. The notice then specified which sections of *Nature* were to be aimed at which audience.

> Those portions of the Paper more especially devoted to the discussion of matters interesting to the public at large contain:
>
> I. Articles written by men eminent in Science on subjects connected with the various points of contact of Natural knowledge with practical affairs, the public health, and material progress; and on the advancement of Science, and its educational and civilizing functions.
>
> II. Full accounts, illustrated when necessary, of Scientific Discoveries of general interest.
>
> III. Records of all efforts made for the encouragement of Natural knowledge in our Colleges and Schools, and notices of aids to Science-teaching.
>
> IV. Full Reviews of Scientific Works, especially directed to the exact Scientific ground gone over, and the contributions to knowledge, whether in the shape of new facts, maps, illustrations, tables, and the like, which they may contain.
>
> In those portions of "NATURE" more especially interesting to Scientific men are given:
>
> V. Abstracts of important Papers communicated to British, American, and Continental Scientific societies and periodicals.

VI. Reports of the Meetings of Scientific bodies at home and abroad.
In addition to the above, there are columns devoted to Correspondence.[58]

This notice would eventually be adopted as *Nature*'s official statement of purpose and would remain unchanged (including the gendered language) until 2000.[59]

Lockyer's stated division of *Nature* into articles, discovery reports, and book reviews for laymen and abstracts and reports of meetings for men of science did not, however, prevent *Nature* from drifting toward its scientific audience over the course of the next five years. Lockyer continued to reserve the front of the journal for more general reports, but between 1870 and 1875 the journalistic articles shrank in number. Instead, more and more space was devoted to articles such as "The Microscopic Fauna of the English Fen District,"[60] "The Source of Solar Energy,"[61] "M. Fizeau's Experiments on 'Newton's Rings,'"[62] and James Clerk Maxwell's two-part piece on "The Dynamical Evidence of the Molecular Constitution of Bodies,"[63] written in the third person and containing a great deal of mathematical data, Latin terminology, and little background information. The article on Fizeau's optical experiments, for example, assumed the reader was already familiar with Anders Ångström's wavelength experiments, and the article on solar energy used extensive mathematical calculations and complex diagrams to advance its argument that the sun's heat output has not been constant over time. Starting in 1870, specialized articles such as the ones mentioned above were printed after the Letters to the Editor; the journalistic articles, by contrast, almost always appeared at the front of the journal.[64] In effect, *Nature* was now split into two sections, separated by the book reviews and Letters to the Editor (which contained a mix of technical and general pieces). The first section contained science articles for laymen and editorials on science education and science policy; the second and much longer section contained reports from scientific societies, summaries of foreign journal contents, and technical articles from respected British researchers that were apparently intended for an audience of their scientific colleagues.

By the mid-1870s *Nature* had assumed a reasonably stable format that would remain largely unchanged until the turn of the century. The journalistic, general-education articles aimed at laymen were now much less prominent. Instead, *Nature* would often (but not always) open with an editorial; most of these essays were anonymous, but on occasion a man of science who was not a member of the *Nature* staff would sign his name to the leading piece. The editorial was followed by book reviews—usually two or three lengthy review essays and a short roundup of other books titled "Our Book Shelf." The Letters to the Editor were printed after the book reviews. These letters were a mix of commentary on articles in *Nature*, questions to be answered, and (as we shall see) combative debates between men of science with opposing views. Like many Victorian gentleman's periodicals, *Nature* allowed editorial writers and correspondents to sign their work with initials or pseudonyms.

Following the Letters to the Editor, *Nature* printed some combination of technical articles, abstracts of recent scientific papers, the Astronomical Column, and summaries of recent lectures at universities or scientific societies. A column titled "Notes" reported recent university promotions, elections to professional societies, and interesting items in other journals and newspapers. In 1871, *Nature* added the "University and Educational Intelligence" column to report upcoming scientific lectures and other news of science in British universities. Each issue concluded with the "Societies and Academies" section, which summarized recent meetings of various scientific academies, and a list of books and pamphlets *Nature* had received for review that week.

As *Nature* transformed into a publication directed largely at men of science, it moved closer in audience to other scientific journals such as the *Philosophical Transactions of the Royal Society*, the *Transactions of the Linnean Society*, or the *Journal of the Chemical Society*. However, *Nature*'s short format and its rapid publication schedule made it distinct from most other specialist journals, which ran hundreds of pages long and were published on a monthly or quarterly schedule. Pieces in *Nature* were necessarily shorter and less comprehensive than articles for other scientific journals, and as a result, the resemblance between *Nature* and these other publications was slight at best. Furthermore, while most scientific journals were affiliated with a scientific society, *Nature* was not. Although by 1875 *Nature* was largely directed at the same audience as the specialized journals, it was clearly an entirely different sort of scientific publication—one that would come to fill a unique function for Victorian men of science.

EDITORIAL CONTROL VERSUS CONTRIBUTORS' INTERESTS: REASONS FOR THE SHIFT

Nature's readers noticed the transition from a layman's periodical to a researcher's publication. One of Lockyer's early readers was his friend Charles Kingsley, a Cambridge clergyman with a strong interest in science.[65] Kingsley was exactly the kind of reader Lockyer had in mind when he began *Nature*: intellectually sophisticated but not a researcher. After reading the first issue, Kingsley wrote a letter of congratulations to his friend, saying, "I am exceedingly desirous that your paper should succeed,"[66] and he even wrote a book review for *Nature*.[67] But just three years later Kingsley found that he was unable to understand most of *Nature*'s content. While Kingsley expressed his continuing admiration for the journal, his remarks to Lockyer make it clear that *Nature* was no longer accessible to laymen:

> I trust that Macmillan did not say that I had a "bad" opinion of Nature. On the contrary, I have the highest respect for it, and I wish I were wise enough to understand more of it. But I fear its circulation must be more limited than you would wish.[68]

These remarks are especially noteworthy because Kingsley was well informed on many scientific subjects, especially evolutionary theory. If even Kingsley could no longer understand much of *Nature*, a London barrister or member of Parliament with no knowledge of science stood little chance of deciphering *Nature*'s content.

Why did *Nature* depart so quickly from Lockyer's goal of a publication for laymen and men of science alike? It does not appear that the editor lost interest in publishing a weekly aimed at a broad cross section of British society or that he decided his initial vision could not be profitable. In contrast to the *Academy*, *Nature*'s transformation seems to have had less to do with the editor's commercial concerns than with his editorial style and with the kinds of contributions he received. Lockyer's vision of *Nature* was divided from the very beginning. He wanted *Nature* to give his fellow Britons a glimpse of the progress and importance of science, but he also wished to include some features such as reports from societies and reprints of abstracts that he and his fellow men of science found useful.

Lockyer initially believed that he could balance his two sets of readers, but *Nature*'s contributors proved more interested in corresponding and debating with each other than they were in writing articles for a lay audience. At this time, some British researchers felt that writing articles about science for laymen was not a high-status undertaking, even science writing for an elite audience of professionals and statesmen.[69] Lockyer and Huxley's interest in science journalism was somewhat unusual for scientific men of their status, and it proved difficult for Lockyer to persuade his colleagues to write journalistic articles for the front section of *Nature*. Furthermore, as we shall see shortly, by 1875 Lockyer had managed to alienate several members of the X Club—exactly the men most likely to write popular science pieces.

Instead, most contributors submitted summaries of their own research (e.g., the aforementioned pieces on "The Microscopic Fauna of the English Fen District" and "The Source of Solar Energy"), reprints of their recent lectures (e.g., Maxwell's "The Dynamical Evidence of the Molecular Constitution of Bodies," originally given as a lecture at the Chemical Institute), or articles critiquing other researchers' work (as G. C. Foster did in "M. Fizeau's Experiments"). Many British men of science found that one of the fastest ways to bring a scientific issue or idea to their fellow researchers' attention was to send a communication to *Nature*. Therefore, while there might have been a drought of the journalistic pieces featured in *Nature*'s first issues, *Nature* soon found itself with a steady supply of reports focusing on specialized research. Lockyer could have rejected some of the specialized pieces in order to maintain the balance between his two intended audiences, but it does not appear that he did so. Perhaps Lockyer was loathe to alienate fellow men of science; in the absence of a healthy supply of articles for laymen, he may also have had trouble filling *Nature*'s pages without printing the specialized pieces. Ultimately, Lockyer's reluctance to turn to authors outside the British scientific community led the journal to focus on the issues of greatest interest to its contributors rather than to the laymen whom Lockyer had seen as his ideal readers.

NATURE AND THE X CLUB

Lockyer's plan to appeal to lay readers was dealt a further blow by a series of fallings-out with members of the X Club. As noted earlier, Huxley and the members of the X Club were some of the most famous early contributors to *Nature*. Huxley in particular hoped the new journal might serve as a spokesman for the X Club's vision of science as a secular, expert discipline. At first glance, one might expect the X Club to have wielded considerable influence over *Nature*. As we shall see, however, their formidable reputation did not shield the X Club from criticism, nor did it ensure the X Club members support in the pages of *Nature*.

Lockyer's trouble with the X Club began less than three years into *Nature*'s existence. In 1872, Joseph Hooker, an X Club member, became embroiled in a dispute with the distinguished paleontologist Richard Owen, a longtime opponent of scientific naturalism in general and the members of the X Club in particular. Hooker was the director of Kew Gardens and had been involved in a battle with First Commissioner of Works Acton Smee Ayrton (a government official charged with managing certain government properties), who wished to convert Kew from a scientific botanical garden into an ordinary public park. Hooker and Ayrton had been arguing for months over Hooker's powers at Kew and the general purpose of the gardens.[70] To further his cause, Ayrton had asked Owen to write a report on Kew. Owen was not a disinterested party to the dispute: he was the director of the natural history departments of the British Museum, including the British Museum herbarium. He and Hooker had often clashed over whose right to collect rare botanical specimens took precedence, and their mutual animosity was well known. Unsurprisingly, Owen's report, published on 16 May 1872, argued that Kew was being grossly mismanaged and suggested that Kew be required to share its government-provided supply of live exotic plants with other organizations, including his British Museum herbarium. In response, Hooker's friend John Tyndall, with the help of fellow X Club members Huxley, William Spottiswoode, and John Lubbock, drafted a letter to Parliament praising Hooker's work at Kew and criticizing Ayrton's attempts to interfere in Kew's business. This letter, along with some of Hooker's correspondence regarding Kew's management, was printed in *Nature* on 11 July 1872.[71] Hooker also wrote a lengthy response to Owen's charges of mismanagement, which *Nature* printed in October.[72] But aside from a brief editorial note accompanying Tyndall's letter to Parliament, *Nature* and Lockyer took no official position on the controversy, although the journal did express "regret" that Owen had become involved.[73]

On 7 November 1872, *Nature* published an article from Owen about the controversy. The famed paleontologist reiterated his charges that Hooker and Kew were hoarding samples and classifying plants incorrectly. Even worse, from Hooker's point of view, Owen implicitly endorsed Ayrton's plan to make Kew a public park when he argued that Kew's mission was to provide a garden of living plants for public enjoyment and that classification and herbarial work was best left to other institutions. Owen did not mention the British Museum by name, but the implication was clear.[74]

Hooker was furious. He was taking the Ayrton controversy quite personally, and he felt betrayed when Lockyer gave Owen the chance to attack him in *Nature*'s pages.[75] Notably, in the ensuing controversy Hooker himself never wrote to Lockyer, nor Lockyer to him; both men used Huxley as an intermediary. Hooker began to suspect that Lockyer was secretly on Owen's side. "My suspicions are strong against Lockyer of whom I have heard much that I do not like,"[76] he wrote to Huxley on November 13. Lockyer, however, told Huxley that he had had no hand in the decision to print Owen's letter. Huxley told Hooker that Lockyer had claimed he had been consumed by other duties besides *Nature* and laid the blame at the feet of his subeditor, Alfred Bennett.[77]

The controversy was somewhat smoothed over after Hooker's reply appeared in the November 21 issue of *Nature* and when Alfred Bennett left the *Nature* staff.[78] However, Hooker and Lockyer never repaired their relationship. In another letter to Huxley, Hooker indicated his frustration with what he saw as poor treatment from the *Nature* editorial staff:

> I do object to Nature's having inserted a scurrilous article, so false & so lowly, that such men as Bentham, Olive, Bab, Dyer, & every other I have seen here should tell me not to answer it in Nature (I ought to underline the 2 last words!). This because they consider it a communication that no respectable paper should have admitted without enquiring as to the truth of its statements, & divestments of personalities—a course doubly due to me as a permanent contributor of Nature & advertised as such from its beginning.[79]

Huxley, wary of the consequences of such debates, warned Lockyer that

> it seems to me that a grave question arises for you as Editor whether 'Nature' ought any longer be made the vehicle of attacks.... It is one thing to give a man fair play and another to afford him the opportunity of publishing a set of scurrilous libels in the hope that some of the mud he throws will stick.[80]

Lockyer did not take Huxley's reproof to heart, as we can see in the very similar case of an 1873 debate between X Club member John Tyndall and Peter Guthrie Tait. Tyndall and Tait had both been regular contributors to *Nature*; Tait tended to focus on scientific epistemology or geology, while Tyndall's articles were usually about heat, light, and spectral research.[81] Tyndall and Huxley had been involved in several skirmishes with Tait and William Thomson (later Lord Kelvin) over glacier theory and thermodynamics. The major stake in these arguments was whether the earth was old enough to allow sufficient time for Darwinian evolution by natural selection. The X Club argued that it was; Tait and Thomson claimed it was not.[82]

In January 1873 Tyndall published a pamphlet titled "Principal Forbes and His Biographers," in which he argued that the late Scottish geologist J. D. Forbes's ideas about glacier formation were both wrong and plagiarized from other geologists. Tait immediately began attacking Tyndall's pamphlet. In September, Tait took his objections to *Nature*. In a letter to the editor, Tait wrote,

> It will probably be considered necessary that Dr. Tyndall's pamphlet, which first appeared as an article in the *Contemporary Review*, be answered at full length.... I hope you will give me space to briefly notice a few of the more obvious inconsistencies of Dr. Tyndall's article.

Only two of Tait's nine listed inconsistencies actually referred to the original pamphlet. Tait's letter was largely devoted to refuting some personal attacks Tyndall had made on Tait in earlier stages of their debate about Forbes's work. In the eighth point, he attacked Tyndall's views on applied research, writing that Tyndall was helping to spread "the monstrous doctrine [that] men who devote themselves to practical applications are men incapable of original research." Tait concluded by praising Tyndall's skills as a popularizer of science (a somewhat backhanded compliment, because Tait followed it by saying that Tyndall had "martyred his scientific authority by deservedly winning distinction in the popular field") and inviting Tyndall to answer him at his leisure.[83]

The combative Tyndall accepted the challenge. He angrily fired back in the next issue of *Nature*.

> He [Tait] asks me to reply to him not according to the letter, but according to the spirit of his attack. If I might use the expression I would say, "God forbid!" for how could I do so without lowering myself to some extent to this level.... It is this man whose blunders and whose injustice have been so often reduced to nakedness, without ever once showing that he possessed the manhood to acknowledge a committed wrong, who now puts himself forward as the corrector of my errors and the definer of my scientific position.

But Tyndall did not save all of his anger for Tait; he also took Lockyer to task for printing Tait's communication in the first place.

> Might I venture, Mr. Editor, to express a doubt as to the wisdom of permitting discussions of this kind to appear in your invaluable journal. Having opened your columns to attack you are, of course, in duty bound to open them to reply, but if I might venture a suggestion, you would wisely use your undoubted editorial rights, and consult the interests of science, by putting a stop to proceedings which dishonour it.[84]

Lockyer made a brief, affronted reply to Tyndall's letter in the September 18 issue, writing,

> If the Editor were to assume the power and responsibility that Prof. Tyndall suggests, NATURE might easily fall from the position of absolute justice and impartiality in all scientific matters which it now occupies and become the mere mouthpiece of a clique.[85]

Lockyer did, however, shut down the correspondence in the Tait-Tyndall case in the September 25 issue. He wrote that while both men had submitted further communications, the journal would not print them because the debate had "assumed somewhat of a personal tone." Lockyer's note included a retraction from Tyndall in which Tyndall apologized for two passages in his previous letter: "The first is that in which I speak of lowering myself to the level of Prof. Tait; the second that in which I reflect upon his manhood."[86] Tait was far from satisfied. In a letter to Lockyer, he castigated the editor for taking sides:

> The fact is that your impartiality as Editor has all along <u>told against me</u>.... Besides, you allow T[yndall], under pretext of withdrawing them, to <u>reprint</u> two of the low things he said. (Enough, however, remains unretracted to make it impossible for me to meet him except with the tip of my toe.) ... There is one little consolation. He doesn't know the difference between manhood & manliness![87]

The contrast between Lockyer's conduct in 1872 and 1873 should not escape our attention. In 1872, Lockyer shifted blame for the Owen affair onto a subeditor and claimed no personal responsibility. A year later Lockyer was far more aggressive in asserting his right as editor to print whatever letters he saw fit. Furthermore, Lockyer's comment about preventing *Nature* from becoming "the mere mouthpiece of a clique" indicates that Lockyer's feelings toward the X Club at this time were not entirely friendly. (The feeling was mutual: in a letter to his friend Rudolf Clausius recounting the controversy, Tyndall described Lockyer as "a man whose conceit has rendered him intolerable to his best friends, and from whom I never disguised my opinion of his conceit."[88]) Lockyer's statement was a public declaration that he would not make editorial decisions based on what the X Club would like to see printed in *Nature*—a much stronger stand than the one he had made during the Hooker-Owen controversy.

Less than six months later, Lockyer upset yet another member of the X Club: the sensitive and volatile Herbert Spencer.[89] The controversy began as an exchange between Spencer and Tait over a recent Spencer pamphlet. Tait himself was not the primary target of Spencer's essay; Spencer was instead responding to a negative review of his book, *First Principles*, in the *British Quarterly Review*. The anonymous reviewer (whom Ruth Barton identifies as the mathematician and judge J. F. Moulton)[90] had taken Spencer to task for his views on the foundation of physics and the laws of motion. In particular, the reviewer argued that Spencer's belief that the laws of motion were a priori truths was patently absurd. Spencer, said the reviewer, clearly had no understanding of the basic principles of physics.[91] Spencer replied in a two-part article for the *Fortnightly Review*, prompting a short response from the Reviewer in the *British Quarterly Review* in which he dismissively said that Spencer had not answered any of his main objections.[92]

Tait became involved when the reviewer, in support of his contention that Spencer's views on the foundations of physics were flawed, cited a passage from Tait's *Thermodynamics*: "Natural philosophy is an experimental, and not an intuitive science. No *à priori* reasoning can conduct us demonstratively to a single physical truth." Consequently, Spencer's pamphlet included a lengthy attack on Tait's view of a priori reasoning.[93] Spencer insisted that physics was littered with a priori truths that could not be proven experimentally, including Newton's three laws of motion. He also attempted to show that Tait's statement in *Thermodynamics* was at odds with previous Tait writings and with the views of William Thomson, the eminent physicist with whom Tait had collaborated.

Tait, irritated by the seemingly unprovoked attack on his work, wrote to *Nature* to express his scorn for Spencer's views on physics—but it appears that Tait may have misunderstood Spencer's argument. Tait's letter suggested that Spencer was skeptical about the reality of the laws of motion rather than their origin.[94] He compared Spencer's apparent uncertainty to the mental agony of a mathematics undergraduate who was not sure whether x was really the unknown quantity in an algebraic equation.[95]

Spencer quickly wrote back to clarify his argument: he was not claiming that the Newton's three laws were not real but that they were not experimentally ascertained, and in fact *could not be* experimentally ascertained. He also argued that the Scottish physicist should have stayed out of his conflict with the anonymous reviewer: "I think it would have been better [for Tait] to keep silence absolutely, rather than to try and dispose of the matter by tearing a sentence from its context, and telling, *à propos* of it, a story not to the point."[96]

This brief exchange between these two famous (and famously prickly) thinkers quickly snowballed into a five-month debate, one that expanded well beyond Spencer and Tait. Tait removed himself from the fray after his initial letter, but several others soon took turns criticizing Spencer's views on the nature of the physical sciences. Most significantly, the anonymous author of the *British Quarterly Review* article on Spencer wrote to *Nature* the week after Tait's letter appeared to accuse Spencer of abusing not just Tait and the reviewer but Sir Isaac Newton himself:

Neither Prof. Tait nor myself are, after all, treated so cruelly as is Newton, who, though his life was spent in maintaining the experimental character of all physical science, is cited as an authority for the *à priori* character of the most important of all physical truths—the well-known Three Laws of Motion.[97]

The English mathematician Robert B. Hayward wrote five letters to *Nature* questioning Spencer's mathematical qualifications and his grasp of the principles of mechanics.[98] Two more anonymous correspondents joined the discussion. One, writing under the nom de plume "A Senior Wrangler," joined his voice with Hayward's and described Spencer as "intensely unmathematical."[99] "Not a Metaphysician" commented that Spencer's arguments were "so purely verbal" that "it is difficult to see how the recognition or non-recognition of [Spencer's point] illustrates the grounds of belief in physical laws."[100] The Scottish chemist Frederick Guthrie also suggested that the mathematics of the laws of motion were beyond Spencer's grasp, although he suggested that the difficulty was due to "the confusion of ideas involved in most mathematical explanation of these laws."[101]

In response to these attacks on Spencer's mathematical ability, Spencer's admirer James Collier wrote in to insist that the debate was really about "inductive logic" and was therefore a psychological issue, not a mathematical one.[102] Collier too soon faced a barrage of criticism from Spencer's opponents. Hayward dismissed Collier's argument by saying it was "almost ridiculous."[103] "A Senior Wrangler" described Collier's letter as

> something like Alice behind the looking-glass; and perhaps behind the looking-glass it may be "a question pertaining to the psychological basis of inductive logic," with which mathematicians, as such, have nothing to do. But in this world, this side the looking-glass, in which forces are measured and effects are measured, Mr. Collier's letter is very perplexing.[104]

One correspondent, Charles Root, suggested that Spencer's view was shared by the great German philosopher Immanuel Kant, but aside from Root's brief letter, the tide seemed most definitely against Collier and Spencer.[105]

Notably, over half of the letters printed about this debate were written in opposition to Spencer, and both sides in the "a priori" debate suspected Lockyer of being more sympathetic to the anti-Spencerites. In May, Spencer complained to Lockyer that his opponents were being allowed to bury him in *Nature*'s pages: "I am the one attacked and in the alternation of attack and defence there is not, up to last week, even an equality of opportunities."[106] On the other hand Tait, who just a year earlier had accused Lockyer of using his editorial power in Tyndall's favor, saw Lockyer as an ally. At the beginning of the "a priori" debate, Tait wrote the following verse in Lockyer's honor:

> Your printers have made but one curious blunder,
> Correct it *instanter*, and then for the thunder!
> We'll see in a jiffy if this Mr. S[pencer]
> Has the ghost of a claim to be thought a good fencer.
> To my vision his merits have still seemed to dwindle.
> Since I found him allied with the great Dr. T[yndall]
> While I have, for my part, grown cockier and cockier,
> Since I found an ally in yourself, Mr. L[ockyer]
> And am always, in consequence, thoroughly willin'
> To perform in the pages of *Nature* (M[acmillan]).[107]

This examination of the X Club's interaction with *Nature* indicates that despite Huxley's intimate involvement with the publication, *Nature* did not owe its success to the support of the X Club members. Far from making *Nature* the primary vehicle for the distribution of their ideas, by the mid-1870s several X Club members had decided that Lockyer could not be trusted and became less eager to contribute to the journal. While Lockyer may have taken some satisfaction in having demonstrated that *Nature* was not "the mouthpiece of a clique," the alienation of such influential contributors did not bode well for the journal's future success.

NATURE IN TROUBLE?

In 1871, Alexander Macmillan indicated to Lockyer that the journal was not yet as successful as they had hoped.

> I shall have much pleasure in discussing with a view to increasing your pay for scientific advice, and of course the advice itself of more value to us. I am sure that it may be put in a satisfactory status. But above all I am very anxious about Nature. I cannot help feeling that a very little more of something would make it a success, and if so of course it would be a permanent benefit to you. I have been thinking of many things. At present we are endeavouring to get it more widely taken at schools.... Wyville Thomson was in this morning, and speaking of its great usefulness.[108]

Macmillan never suggested that the firm might decline further publication of *Nature*, but in the competitive market of Victorian periodicals, Lockyer would have been justified in worrying about its future. By 1875, *Nature*'s situation was arguably even more precarious. *Nature*'s initial target audience could no longer understand most of the material in the journal. Lockyer had managed to alienate some of his most famous and powerful contributors. Most troublingly, the journal had yet to turn a profit (and, in fact, would not clear its debts for another 15 years).

And yet, as we know, *Nature* survived. Although *Nature* was ultimately unsuccessful as a forum where scientists could instruct laymen, it proved to have other uses for its contributors. We see hints of the function *Nature* might serve for its contributors in the X Club debates—the journal's weekly publication schedule made it a faster and more convenient forum for disputes than quarterly journals or dueling monographs. As we shall see, a younger generation of British men of science would soon adopt *Nature* as a primary means of communicating with their fellow men of science, and the journal would come to serve a unique publishing function in the British scientific community.

CHAPTER TWO

Nature's Contributors and the Changing of Britain's Scientific Guard, 1872–1895

In February 1879, Norman Lockyer was nominated for election to the Athenaeum, an exclusive social club whose membership included some of Britain's most eminent men of science. The engineer C. W. Siemens, an Athenaeum member, had put forth Lockyer's name for consideration and was delighted when his friend was elected to the group, but he warned Lockyer in his note of congratulations that the election had been somewhat contentious.

> I have much pleasure to congratulate you upon your election this afternoon into the Athenaeum. This ought to have been a mere matter of course but you have aroused the jealousy or enmity of some persons which made it necessary for your friends to be on the lookout for mischief and it is on this account that I have much pleasure to congratulate you upon your victory![1]

Those familiar with Lockyer's history would not have been surprised that he had aroused "jealousy or enmity" among some Athenaeum members. Lockyer rarely went out of his way to avoid an argument with his fellow men of science. He relished spirited disagreements, even regarded them as essential to spreading his ideas and establishing his reputation within the scientific community.[2] (The possibility that his fellow combatants might not find the same enjoyment in these skirmishes rarely occurred to him.)

We see the same appetite for controversy when we turn our attention to the contents of Lockyer's journal. *Nature* played host to a number of arguments between contributors, and often one or both parties to the argument ended up feeling aggrieved or mistreated by *Nature*'s handling of the discussion. As we have seen, many of these early debates involved the members of the X Club, whose scientific stars were on the rise during the 1870s and 1880s. Between 1869 and 1887, at least one member of the X Club was always on the Council of the Royal Society, and Joseph Hooker, William Spottiswoode, and Thomas Huxley held consecutive presidencies of the Royal Society between 1873 and 1884.[3] The historian Frank M. Turner has grouped Lockyer together with the X Club members as part of "the young guard of science" that sought to bolster the status of men of science by establishing agnostic naturalism as the proper basis for scientific inquiry. Turner argued that *Nature* was the embodiment of the X Club's efforts to "professionalize" science and exclude their opponents from scientific discourse.[4] If Lockyer was indeed part of this "young guard," it might be expected that he would favor the members of the X Club when they clashed with those who did not share their goals. But in reality, while Huxley appears to have remained on good terms with Lockyer, many of the other X Club members did not. Rather than support the X Club over their opponents, Lockyer and his subeditors at *Nature* allowed both sides equal chance to savage one another—a strategy that, as we have seen, led to serious rifts with John Tyndall, Joseph Hooker, and Herbert Spencer.

The key to *Nature*'s success was not the X Club but the following generation of British men of science, a group of scientific practitioners who regarded the X Club generation as their mentors. When we examine patterns of contributions to *Nature*, it becomes evident that the X Club and their contemporaries viewed *Nature* as a place to publish popularizing pieces or participate in debates but not as a desirable forum in which to announce their most important scientific work or to present substantial commentary on scientific theories. In contrast, the younger generation, men born in the 1840s and 1850s, saw *Nature* both as an ideal forum for scientific discussions and, increasingly, as a useful way to spread news of their original work. It was this younger generation who adopted *Nature* as a central organ of scientific communication; their contributions to the journal cemented *Nature*'s status as Britain's most important scientific publication.

NATURE AND A NEW GENERATION

By 1875, several of the X Club members were thoroughly annoyed with Lockyer and had lost much of their earlier enthusiasm for his journal. And yet, losing the X Club's support was not a disastrous blow for Lockyer. Despite their success at obtaining influential positions in scientific societies, the X Club still had many important scientific opponents, and these opponents continued to contribute to *Nature*. Furthermore, the X Club's stay atop Britain's scientific hierarchy would not last forever. By the mid-1880s the members of the X Club were well into their 60s, and many were suffering from poor health. Spencer, Tyndall, Hirst, and Lubbock could no longer travel to the Club's dinners with regularity, and in 1886 Huxley told Lubbock that he feared the club would not endure much longer. In 1889, a disagreement between Huxley and Spencer over land nationalization policy blew up into a serious public argument; the two barely spoke for the next four years. Thomas Hirst's death in 1892 proved the final blow for the declining X Club, and the group ceased to meet after May 1892.[5]

As the members of the X Club aged and grew ill, a new generation, men born after 1840, began to assume positions of leadership in the British scientific community. This new generation included some of *Nature*'s most important and prolific contributors in the nineteenth century. Six individuals in particular stand out: the naturalists E. Ray Lankester, George J. Romanes, and W. T. Thiselton-Dyer; the chemist Raphael Meldola; and the physicists Oliver Lodge and John Perry. Given their importance to the journal, it is worth discussing these men's personal biographies and backgrounds.

The zoologist E. Ray Lankester (1847–1929) was the son of two parents with strong connections to the scientific world: his father was the distinguished natural historian Edwin Lankester, and his mother, Phebe Pope Lankester, was a well-known writer on public health and botany. Lankester obtained his undergraduate degree from Oxford and went on to study physiology in Germany under such distinguished scholars as Ernst Haeckel.[6] In 1872, he was appointed as a fellow and tutor in biology at Exeter College, Oxford, although he was not particularly happy there. Just months after accepting the Exeter position, Lankester wrote a lengthy letter to T. H. Huxley begging the older naturalist to let him know of any opportunities that opened up in London.

> I am afraid that you may as Michael Foster does or did (until I talked to him the other day) regard me as settled and provided for so to speak by my fellowship at Exeter. You may suppose that there is some career open before me—that by exerting myself in teaching I can produce some impression as well as by other work—and that I may thus make my way to a better berth in the place itself. I want to tell you that this is not the case. No one knows who does not live in the place—the inextricable mess of mediaeval folly and corporation-jealousy and effete restrictions which surround all Oxford institutions.

Lankester's letter went on to complain of lazy and indifferent undergraduates, senior fellows unwilling to entertain any suggestion of reform, and the stifling, males-only social life of the university. "Could you have endured the corresponding period of your own life, out of London or some large city, where society is not half-choked by ecclesiastical & aristocratic odours and where there is a normal proportion of the two sexes?" he asked Huxley.[7] In 1875, the same year he was elected as a Fellow of the Royal Society (FRS), Lankester would get his wish to leave Oxford with an appointment to the chair in zoology at University College London.[8]

Lankester's fellow naturalist (and frequent foil) George John Romanes (1848-1894) was a Canadian by birth; he was born in Kingston, Ontario, where his father was a professor of Greek. About the time Romanes was born, his parents inherited a considerable fortune, and they eventually decided to relocate the family to London. Romanes studied at Gonville and Caius College, Cambridge, and after he finished his degree he remained in Cambridge to study under the eminent physiologist Michael Foster. In 1874 he relocated to University College London, where he began conducting experiments on the nervous systems of jellyfish and sea urchins, work that would win him admission to the Royal Society in 1876. Romanes was particularly close to Charles Darwin, and (as we shall see later) after Darwin's death he sought to style himself as the great naturalist's scientific heir.[9]

William Turner Thiselton-Dyer (1843-1928) was the son of a respected Westminster physician. Thiselton-Dyer initially studied mathematics at Christ Church College, Oxford, but he became fascinated with botany and decided to pursue it as a career. After leaving Oxford he taught briefly at the Royal Agricultural College in Circenster and at the Royal College of Science in Dublin. In 1872 he was appointed professor of botany at the Royal Horticultural Society in London. Joseph Hooker (an X Club member) took an interest in the young botanist's career and offered him a second position at Kew Gardens. By 1875 Thiselton-Dyer had been promoted to assistant director at Kew. His connection with Hooker was advantageous in other ways: in 1877, Thiselton-Dyer married Hooker's eldest daughter, Harriet. Thiselton-Dyer was elected FRS in 1880 and succeeded his father-in-law as director of Kew Gardens in 1885.[10]

Raphael Meldola (1849-1915) followed a somewhat different career path. He was the son of a London printer and was named for his paternal grandfather, a renowned British rabbi. He attended neither Cambridge nor Oxford; instead, he was a graduate of the Royal College of Chemistry. After completing his degree, Meldola briefly held a professorship at his alma mater, and he also served as Lockyer's research assistant from 1875 to 1876 before leaving to take a position in the dye industry. He was best known as a highly successful industrial chemist for the firm of Brooke, Simpson, and Spiller, and he was the discoverer of several widely used alkaline blue and green dyes. In 1885, Meldola left Brooke, Simpson, and Spiller to become a professor at Finsbury Technical College. He later expanded his scientific interests to include evolutionary theory. Meldola was elected FRS in 1886.[11]

Oliver Lodge (1851–1940) was the son of a middle-class Staffordshire merchant who sold materials for the manufacture of pottery. Lodge was fascinated with science from an early age and was inspired to pursue further study by a series of lectures John Tyndall delivered at the Royal Institution when Lodge was a teenager. In 1872 he enrolled in an external course at the University of London, and in 1874 he was able to enroll full time at University College London, where he obtained his BS degree. Lodge was well known as one of the "Maxwellians," a group of physicists and engineers who promoted James Clerk Maxwell's complicated theories of the electromagnetic field.[12] In 1881 Lodge would be elected professor of physics at University College, Liverpool, and in 1887 he joined the Fellowship of the Royal Society.[13]

Finally, John Perry (1850–1920), an electrical engineer, was an Irishman raised in Belfast and a graduate of Queen's College. After finishing his degree in 1871, he moved to England to take a professorship at Bristol College, and in 1874 he was appointed as an assistant to the great physicist William Thomson at Glasgow University in Scotland. A year later, Thomson helped Perry win an appointment as professor of engineering at a new university in Tokyo. By 1879 Perry had returned from Japan; he moved to London and took up a position at Finsbury Technical College, where he and his collaborator W. E. Ayrton became well known for producing innovative new instruments for measuring electricity.[14] Perry was elected to the Royal Society in 1885.

As we examine these brief biographies, a few common threads become apparent. All of these men spent at least some time as professors, and five of the six ultimately made their careers in universities (the exception being Thiselton-Dyer). All had university degrees. Three attended Oxford or Cambridge; two took degrees at relatively young London universities. All had at least some connection to London—five of the six chose to pursue their careers there, and four of them did so in London colleges. None of these men had an aristocratic background, although Romanes's family was quite wealthy, and Thiselton-Dyer and Lankester had family connections to science or medicine.

Compare this to a sample of older *Nature* contributors: Peter Guthrie

Tait (1831–1901), Herbert Spencer (1820–1903), Joseph Hooker (1817–1911), Thomas Huxley (1825–1895), Alfred Russel Wallace (1823–1913), and Charles Darwin (1809–1882).[15] In many ways, these men were the older counterparts and mentors of the group we just discussed—Perry and Lodge had strong ties to North British energy physicists such as Tait, while Lankester, Romanes, Thiselton-Dyer, and Meldola frequently corresponded and debated with their fellow evolutionary theorists Spencer, Huxley, Wallace, and Darwin—and there are many similarities between the two generations. Aside from Darwin, who came from a wealthy family, no one from the older group was well-off or aristocratic. The London connection can also be seen in this older group. Hooker, Huxley, Spencer, and Wallace would eventually settle in London, and Darwin lived there for a time as a young man. But there are important differences, most notably in the role universities played in the older generation's scientific work. Of these six, only Hooker, Darwin, and Tait had university degrees (Darwin and Tait from Cambridge, Hooker from Glasgow), and Tait was the only one who made his career in university teaching. Alfred Russel Wallace was a railroad surveyor who indulged his interest in natural history largely in his spare time, and Herbert Spencer also began his career in railway management before deciding to make his living as a writer and theorist in London's literary circles. Three of these men participated in lengthy voyages by sea early in their scientific careers. Darwin famously traveled the world aboard the HMS *Beagle*. Huxley's scientific schooling took place at Charing Cross Hospital, and he conducted much of his early scientific research while serving as a surgeon in the Royal Navy. Similarly, Joseph Hooker served as the assistant surgeon aboard the HMS *Erebus* for four years after finishing his MD degree at Glasgow. With the exception of Darwin, whose family money enabled him to assume the role of gentleman naturalist, and Tait, who stepped into a career in the Scottish university system, these men often had to exercise a great deal of creativity and self-promotional skill in order to find ways to pursue science as a paying vocation. Their younger counterparts seem to have faced fewer difficulties in building their scientific careers.

Turning to *Nature*, we see another contrast between these two groups: the younger group was more attached to *Nature* as a venue for scientific communication than the X Club and their contemporaries. Men such as Huxley, Tait, Wallace, and Spencer were certainly regular contributors to *Nature*. However, their contributions were usually in the form of book reviews, popularizing articles, and responses to other pieces in the journal; they did not use *Nature* to spread news of advances in their scientific work

TABLE 3. Number of pieces in *Nature* by type and generation, 1869-1900

Author	Articles describing own scientific work	Articles creating *new* Nature discussions	Responses to *existing* Nature discussions	Total articles *in* Nature, 1869-1900*
Older generation:				
Huxley (1825-1895)	1	2	7	24
Spencer (1820-1903)	0	0	15	16
Tait (1831-1901)	3	3	21	53
Hooker (1817-1911)	1	0	7	23
Darwin (1809-1882)	6	10	9	35
Wallace (1823-1913)	1	8	65	147
Average no. of articles	2.0	3.83	20.67	49.67
Younger generation:				
Lankester (1847-1929)	12	26	42	135
Lodge (1851-1940)	5	17	51	103
Meldola (1849-1915)	1	9	24	59
Perry (1850-1920)	5	5	10	29
Romanes (1848-1894)	23	14	75	158
Thiselton-Dyer (1843-1928)	8	15	45	92
Average no. of articles	9.0	14.33	41.17	96.0

*This number includes the three types of articles described in this table (articles about one's own work, pieces introducing a new discussion to *Nature*, and articles responding to an existing scientific discussion in *Nature*) as well as other types of contributions not enumerated in this table, such as book reviews, obituaries, popularizing articles, and editorials.

or to print substantial critiques of others' scientific work. Instead, men of the older generation preferred to direct their discussions with fellow men of science to literary monthlies or other general-interest periodicals and saved their own scientific work for society journals, monographs, or a meeting of a scientific society. In contrast, the younger generation adopted *Nature* as their primary forum for debating scientific theories in Great Britain and slowly began the practice of sending a short notice to *Nature* to announce a forthcoming paper in another journal. The members of this younger generation were, in many ways, the scientific heirs of the X Club and their contemporaries, but *Nature* held a different place in their publishing strategies largely because the two generations had very different ideas about the audience for scientific debates.

An examination of the number of articles in *Nature* written by prominent members of the older and younger generations shows that members of the younger generation were, in general, more prolific contributors, averaging nearly twice as many articles between 1869 and 1900 (table 3). However, the number of articles, while illuminating, does not tell the whole story of the contrast between the generations. The most important differences between the older and younger men of science in *Nature* involved not the number but the types of articles submitted to *Nature* and the role *Nature* played in their larger publishing strategies.

Most interestingly, the data strongly reinforce historian Peter Kjærgaard's argument that at the end of the nineteenth century, British men of science began redirecting scientific controversies to *Nature*.[16] The "Responses to existing *Nature* discussions" column in table 3 shows the number of articles and letters to the editor each contributor wrote responding to a debate already taking place in *Nature*; the "Articles creating new *Nature* discussions" column tabulates the number of articles each contributor wrote that were about a scientific issue that had not previously been discussed in *Nature*'s pages.

These figures indicate that while the older generation did engage in discussion of scientific issues in *Nature*, when they did so, they were generally responding to a discussion that was already taking place in that periodical. The older generation tended to prefer literary magazines such as *Nineteenth Century* or *British Quarterly Review* over *Nature* when they sought to initiate a scientific debate. The members of the younger generation, by contrast, were far less active in the literary periodicals than their older counterparts and viewed *Nature* as a more desirable place to carry on scientific debates. They would write their ideas about scientific discussions or issues for *Nature* rather than other periodicals and would redirect discussions taking place in other periodicals to *Nature*.

CASE STUDIES: EVOLUTIONARY CONTROVERSIES AND THE AGE OF THE EARTH

The work of George J. Romanes provides a useful window onto the generational differences in *Nature*. *Nature* played an important role in Romanes's scientific career: it was the publication that called Darwin's attention to the young naturalist in 1873. After reading Romanes's letter to the editor on "Permanent Variation of Colour in Fish," in which Romanes declared his allegiance to the Darwinian theory of natural selection, Darwin wrote a note of congratulations to the young naturalist expressing interest in Romanes's work and future career.[17] Darwin soon became Romanes's scientific mentor and close friend.[18] After Darwin's death in 1882, Romanes took it on himself to defend Darwin's theory of natural selection against critics such as the Duke of Argyll and former Darwinian allies, such as Alfred Russel Wallace, who had come to question whether natural selection applied to human evolution.

Romanes was gifted with a seemingly infinite capacity for correspondence. Between 1881 and his death in 1894, he was almost always engaged in some type of discussion or debate in *Nature*. He and a variety of opponents went back and forth about Darwin's theories, the epistemological implications of natural selection, and Romanes's own ideas about evolution. Romanes's spirited tenacity drew equally impassioned responses from those who disagreed with him and made him a focal point of late Victorian evolutionary controversy in *Nature*.

In the 1880s and 1890s, Romanes advocated three additions to the theory of natural selection: physiological selection, panmixia, and the inheritance of acquired characters. In August 1886, Romanes published a three-part abstract titled "Physiological Selection: An Additional Suggestion on the Origin of Species," a shortened version of a paper he had read before the Linnean Society on May 6.[19] In it, Romanes argued that Darwin had recognized three evolutionary facts that the theory of natural selection could not account for. First, domesticated species (such as different breeds of dogs) bred much more freely with one another than species that had evolved in the wild, which tended to have more selective fertility. Second, the theory of natural selection did not explain how crossbreeding between parents with different characteristics affected the development of species. Finally, the theory of natural selection could not account for the fact that many of the features that distinguished species from one another were useless from a survival standpoint.

Romanes concluded that there must be another evolutionary mechanism operating alongside natural selection. He proposed that this mechanism was something he called "physiological selection." Physiological selection, said Romanes, occurred when a new variety of animal was infertile with its parent form but fertile with other members of its own variety. This limited fertility, said Romanes, would cause a new variety to endure and become a species in its own right.

> When accidental variations of a non-useful kind occur in any of the other systems or parts of organisms, they are, as a rule, immediately extinguished by intercrossing. But whenever they happen to arise in the reproductive system in the way here suggested, they must inevitably tend to be preserved as new natural varieties, or incipient species. At first the difference would only be in respect of the reproductive system; but eventually, on account of independent

variation, other differences would supervene, and the new variety would take rank as a true species.[20]

As we shall see, this new theory did not win many immediate converts, and it provided the basis of an intense ongoing debate about interspecies sterility and whether natural selection was sufficient to explain the origin of species.

A second evolutionary issue Romanes discussed in *Nature* was a theory called "panmixia," a doctrine first proposed by the German naturalist August Weismann.[21] The theory of panmixia stated that when an organ no longer conferred an evolutionary advantage to an animal (e.g., when the horns of a species of sheep ceased to help the animal survive and reproduce), that organ would no longer be the subject of natural selection. Weismann argued that the cessation of selection could result in an organ significantly decreasing in size, or even vanishing altogether. Romanes was a supporter of this theory, but many other British naturalists were skeptical. E. Ray Lankester argued that the cessation of selection would mean that a now-useless organ (e.g., a horn or a tail) was equally inclined to grow and to diminish, and that panmixia could not account for a decrease in the size of an organ unless there was an evolutionary advantage to having smaller horns or a shorter tail.

The debate over panmixia was closely related to a third (and arguably the most significant) point of evolutionary controversy: the inheritance of acquired characters. Like Weismann, Romanes maintained that cessation of selection could account for the dwindling in size of a now-useless organ, but Romanes differed from Weismann in believing that the principle of panmixia alone could not fully explain why a useless organ might vanish altogether. In his first letter to *Nature* on the subject, Romanes wrote,

> While Prof. Weismann believes the cessation of selection to be capable of inducing degeneration down to the almost complete disappearance of a rudimentary organ, I have argued that, *unless assisted by some other principle*, it can at most only reduce the degenerating organ to considerably above one-half its original size—or probably not through so much as one-quarter.[22]

Romanes believed that this "other principle" was the inheritance of acquired characters. If a parent animal did not need its useless organ, suggested Romanes, the effects of this disuse would pass to its offspring, who would be born with an even smaller version of the organ.

Romanes's ideas on physiological selection, panmixia, and the inheritance of acquired characters drew strong opposition from evolutionary theorists of both generations. Wallace was one of the first to criticize Romanes's paper on physiological selection. In an article in the *Fortnightly Review*, Wallace denounced the idea that any principle aside from natural selection was necessary to account for the origin of species.[23] Lankester was deeply skeptical of the principle of panmixia, which led to an eight-week exchange in *Nature* with Romanes in the spring of 1890. Meldola was more favorably disposed than Wallace or Lankester toward Romanes's theory of physiological selection but believed that any physiological selection had to be dependent on Darwinian natural selection:

> But, since Mr. Romanes admits the efficiency of natural selection, the question seems to resolve itself into this: Can physiological selection work independently of natural selection? If not, natural selection must still be regarded as a prime factor, and if physiological selection cannot originate a species independently of the control of natural selection, surely the latter, with its subordinate factors (of which physiological selection *may be* one), is still *the* chief element in the theory of the origin of species.[24]

Wallace, Lankester, and Meldola shared a belief that acquired characters could not be passed down from parents to offspring, but it was Meldola who engaged most extensively with Romanes on this point in *Nature*.[25]

Romanes's opponents often found him a somewhat annoying correspondent, both because of his seemingly insatiable appetite for debate and because of his argumentative tactics. In his exchanges with his fellow scientists, Romanes often attempted to blunt their criticisms by claiming that his opponents actually agreed with him. During the argument with Wallace, Romanes frequently insisted that the older naturalist had acknowledged the importance of fertility and sterility in the evolution of species and that Wallace's opposition to physiological selection was based on a misunderstanding of Romanes's principles.[26] Romanes employed the same strategy during an argument with Meldola in 1891 over whether two apparently unrelated characters, neither of which was advantageous on its own, might combine to provide an advantage and evolve concurrently. Romanes, drawing on the doctrine of use inheritance, believed this could occur; Meldola believed that it was too unlikely for two useless characters to occur in the same animal and combine to produce an advantage. Romanes wrote that Meldola actually agreed with him but did not realize it: "As it appears to me, from his reply, that Prof. Meldola's views on the subject of 'co-adaptation' are really the same as my own, I write once more in order to point out the identity."[27]

This strategy did not endear Romanes to his fellow naturalists. Responding to Romanes's argument that he had adopted the theory of physiological selection without realizing it, Wallace wrote that if Romanes continued to press the claim, "it will show that our respective standards of scientific reasoning and literary consistency are so entirely different as to render any further discussion of the subject on my part unnecessary and useless."[28] Wallace never forgave Romanes for claiming that he had adopted the theory of physiological selection. In an 1893 letter to W. T. Thiselton-Dyer, Wallace declined to write a letter of sympathy to the terminally ill Romanes, explaining,

> he made a very gross misstatement & personal attack on me when he stated, both in English & American periodicals, that, in my Darwinism, I adopted his theory of "Physiological Selection" and claimed it as my own. . . . I told him then that unless he withdrew this accusation as publicly as he had made it I should decline all further correspondence with him, & sh'd avoid referring to him in any of my writings. This is of course very different from any criticism of my theories; that, or even ridicule, would never disturb me—but when a man has done all he can to spread this accusation over the whole civilized world my only answer can be—after showing as I have done (see Nature vol. 43. pp. 79 & 150) that his accusations are wholly untrue—to ignore his existence.[29]

Although Thiselton-Dyer considered Romanes a friend, he well understood Wallace's frustration. Four years earlier, he himself had written to Wallace to complain about a recent conflict with Romanes. "To tell you the truth I was rather cut up about my controversy with Romanes," he admitted. "I will never engage in a discussion with Romanes again. He does not, I am persuaded, grasp his own views, much less those of other people. He is elusive as an eel."[30]

Meldola was equally annoyed by Romanes's persistence. During the 1891 coadaptation debate, Meldola declared that while he had hoped their two-month discussion might come to an end, "I very much regret to find, however, that Dr. Romanes—whose amount of spare time appears to be most enviably inexhaustible—still finds it necessary to prolong the correspondence."[31] At one point, Lankester wrote a letter to Nature about Romanes that was so sharply worded, the controversy-loving Lockyer overruled his subeditors and declined to print it. In a note to Lockyer, Lankester admitted it was probably best the letter was not made public but saved some choice epithets for Romanes:

> You are quite right not to print my letter about Romanes—as it is not argumentative but purely denunciatory. I am glad he has seen it—as he will now know what a humbugging piece of foolery I consider his attempt to say "Darwin-

and-I" and "the Darwin-Romanes theory"—is. It is time that he knew that I consider him a wind-bag.

I think it is perhaps my duty to say so, urbi et orbi—or do you hold that a man is to be allowed to puff himself and falsely pass himself off on the crowd as a second Darwin—without protest.³²

Romanes, a close friend of Lockyer's, also saw Lockyer's hand behind the disappearance of Lankester's letter.³³ He told Lockyer he was unconcerned with the opposition to his theory.

> My own pet theory about Physiological Selection has met, as you will have seen, with a storm of opposition. But this does not affect me in the least; seeing it is obvious that as yet there are no data for an adverse judgment. It can only be made or marred by a long course of verification. Am I right in connecting your return with the non-appearance of Lankester's letter to Nature—proof of which was sent me by letter? I had written such a beautiful reply; but all the while thought it would be a mistake to disfigure Nature with so unseemly a correspondence.³⁴

Men of science of all ages may have been united in their annoyance with Romanes. However, their approaches to contributions in *Nature* indicate that the younger generation was more invested in *Nature* as a vehicle of scientific communication than the X Club and their contemporaries. Notably, although Alfred Russel Wallace was one of *Nature*'s most prolific book reviewers, when he wished to criticize Romanes's theory on physiological selection, he did not write to *Nature*. Instead, his initial reply appeared in *Fortnightly Review*, a prestigious liberal journal that was renowned for its literary and political commentary.³⁵ Only when Romanes attacked him by name in *Nature* did Wallace choose to respond in that journal.³⁶ By contrast, Romanes's younger critics, such as Meldola (born 1849), Lankester (born 1847), and Francis Darwin (Charles Darwin's son, born 1848), sent their thoughts on Romanes's theories to *Nature* first.³⁷ Romanes's response to Wallace's choice is also telling: in his first letter discussing Wallace's criticisms, Romanes strongly implied that the older naturalist had done the *Nature* readership a disservice by moving the discussion to another publication. He expressed surprise that "criticisms on the theory of physiological selection are flowing through channels other than the pages of NATURE."³⁸

This generation gap was not confined to the biological sciences. In 1894 John Perry famously became embroiled in a controversy with his fellow physicists Peter Guthrie Tait and William Thomson (by then Lord Kelvin) over the age of the earth.³⁹ Tait and Thomson had both contributed to

Nature in the past, but like others of their generation (Tait was born in 1831, Thomson in 1824), the two physicists do not appear to have viewed *Nature* as a significant part of their scientific publishing strategies. Tait, as we saw in chapter 1, was involved in several arguments in the Letters to the Editor section of the journal, but he did not communicate news of his own scientific work to *Nature*. Thomson was an even less visible presence in *Nature*. While the journal printed several abstracts of Thomson's scientific lectures, Thomson almost never wrote letters to the editor or involved himself with discussions in *Nature*.[40] Both Thomson and Tait had been debating the age of the earth as early as the 1840s; Thomson in particular was adamant that the earth could not possibly be older than 100 million years.[41] However, before 1895 neither man had published anything in *Nature* on that subject.

It was Perry, born in 1850, who brought the age of the earth discussion to *Nature*. Perry had been Thomson's laboratory assistant at Glasgow in the mid-1870s. Now a professor at Finsbury Technical College, Perry had concluded that Thomson's calculation of the age of the earth was possibly mistaken. Thomson had assumed the earth was a homogenous mass; Perry suggested that an earth with a dense internal mass would have needed more time to form than Thomson's calculations allowed. Perry sent the journal copies of letters between himself, Tait, and Thomson detailing the issues they had been debating. *Nature* printed Perry's submission on 3 January 1895.[42] A month later, Perry wrote again to preempt Thomson's likely response to a recent private communication from the geologist Robert Weber,[43] and he wrote in April to argue that paleontological evidence required physicists to adjust their estimates.[44] Although he continued a vigorous personal correspondence with Perry, Thomson responded to Perry's communications in *Nature* only once, in March 1895; Tait never responded at all.[45] The age of the earth discussion further suggests that in late nineteenth-century Britain, a younger generation of scientists was using *Nature* as a forum for discussion and a platform for promoting their scientific ideas more frequently and enthusiastically than their older counterparts.

Nature was not the only place where there seemed to be a divide between older and younger men of science in Britain. Further evidence of a generation gap in British science can be seen in the *Nature* discussion of what came to be called "the Stokes controversy." Sir George Gabriel Stokes (1819–1903) was a Cambridge-educated Irish physicist who had made important advances in fluid dynamics and optics in the 1840s and 1850s.[46] In 1885, Stokes was elected president of the Royal Society. This was not a surprising development. Stokes had been a Fellow since 1851, had won the Society's

Rumford medal in 1852 for his research, and had served as secretary from 1854 to 1885; in short, he was exactly the kind of man most Fellows considered suitable for the position. But in 1887, Stokes ran for and won the Cambridge University seat in the House of Commons as a representative of the Conservative Party. His previously unremarkable presidency suddenly became quite controversial indeed.

In the lead editorial on 17 November 1887, an anonymous *Nature* contributor stridently questioned whether Stokes had been right to run for Parliament while serving as president of the Royal Society. Of Stokes himself, the editorial said "No man in the scientific world is, or deserves to be, more respected or more popular."

> But, at the present moment, Prof. Stokes is something more than an eminent investigator and teacher: he is President of the Royal Society; and, as such, enjoys all the prestige which is given by the fact that in the eye of the public he has the oldest, the strongest, and the most widely representative body of men of science in the country at his back.... It is therefore important that the freedom of the President's intercourse with Ministers should be in no way trammelled by his political relations.... The occasion is grave and demands action. It is for the President, by the course which he may think fit to adopt, to determine what that action shall be.[47]

A letter Thiselton-Dyer wrote to Huxley about the controversy reveals that Thiselton-Dyer perceived a generation gap in the response to Stokes's dual role as member of Parliament and president of the Royal Society.

> I am greatly relieved too that you approve my letter to Nature. My heart was in my throat when I sent it off. I shd not have ventured into the fray yet for loyalty to you. But Lockyer & Roscoe were good enough to say that they thought well of it. And I really think it has answered its purpose in stiffening up the younger men. Many of the older ones I am sorry to say tell me that they do not look at the matter as I do.[48]

The *Nature* discussion on the Stokes controversy provides further evidence for the generational divide Thiselton-Dyer mentioned. While Thiselton-Dyer spoke out in support of the *Nature* editorial, older correspondents supported Stokes. The physicist Balfour Stewart (1828–1887) argued that Stokes had every right to run for Parliament. "I fail to see what our President has done to incur the reprobation of the writer of this article," Stewart wrote. "He has chosen to be an Englishman first, and a man of science afterwards. Who will blame him for this?"[49] The chemist Alexander William Williamson (1824–1904) argued that Stokes "is as highly esteemed and valued as

President by those among us who may happen privately to differ from him widely in politics as by those who entertain similar political opinions to his own" and that Stokes's "action in political matters concerns us as little as his opinions."[50]

The fact that Thiselton-Dyer wrote to Huxley for advice on the controversy, however, further suggests that his generation viewed themselves as the heirs of the older scientific naturalists.[51] When the members of the X Club began their careers, they saw themselves largely as the opponents of older men of science, such as Richard Owen (1804–1892), who did not share their commitment to scientific naturalism. The generation that followed the X Club, by contrast, cultivated relationships with the older scientific naturalists and saw them as potential allies against men such as Balfour Stewart or Alexander Williamson. Lankester's desperate plea for Huxley's aid in getting out of Oxford; Thiselton-Dyer's close relationships with Wallace, Huxley, and especially Hooker; Perry's professional connection with Thomson; and Romanes's devotion to his mentor Darwin indicate that the scientific ties between these two generations were often quite close. So while Thiselton-Dyer saw a contrast between the "younger men" and the "older ones" in regard to the Stokes debate, he wanted Huxley's advice and approval before entangling himself in the controversy.

WHY *NATURE*? PUBLISHING AND SCIENTIFIC PRESTIGE IN LATE NINETEENTH-CENTURY BRITAIN

Lankester, Romanes, and the rest had close ties to their mentors, but unlike the older generation, they adopted *Nature* as a central organ of scientific communication. One obvious explanation for the gap might be that as a younger journal, *Nature* was more accessible to lesser-known young men than the *British Quarterly Review* or the *Fortnightly Review*, but this argument does not hold up to scrutiny. George J. Romanes wrote prolifically for literary periodicals, suggesting that access to these publications was not the determining factor in his or his colleagues' attachment to *Nature*.[52] Furthermore, even after they became Fellows of the Royal Society and had greater access to more journals, men such as Lankester, Meldola, and Perry continued contributing to *Nature* rather than the literary periodicals.

Instead, *Nature*'s success with these men of science appears to have had a great deal to do with its publication speed. Unlike the literary periodicals, there was almost no delay between the submission of a piece and its appearance in the journal. *Nature* often printed letters and communications the same week they were received. Sir John Maddox, editor of *Nature* from 1966 to 1973 and 1980 to 1995, once suggested that one of *Nature*'s greatest early assets was the speed of the Royal Mail. British men of science knew that a contribution sent to the journal would reach its destination the day after it was posted.[53]

The speed of publication created a sense of immediacy among the contributors to *Nature*—Romanes could write to *Nature* and read responses to his ideas less than two weeks later. *Nature* was the closest print substitute for a meeting of a scientific society, and unlike a discussion at a gentleman's club such as the Athenaeum or a debate at the British Association for the Advancement of Science (BA), a letter to *Nature* would be printed and available to readers outside the membership or who could not attend a particular meeting.[54] (The publication speed may also have contributed to the occasionally combative tone of the publication, as the weekly schedule gave men of science less opportunity to rethink and rewrite harsh words.)[55] Furthermore, an 1873 letter to Lockyer from an American reader, the astronomer Henry Draper, indicates that *Nature* reached audiences outside Britain far more quickly than the older scientific publications:

> I wish that the publications of the great Societies could be made to reach those who are interested more quickly. The Transactions of the Royal Society take an incredible time to make their appearance here and we have really to depend on the abstracts that are published in scientific magazines for fresh information. In this respect "Nature" is invaluable.[56]

Nature was not the only scientific weekly that offered rapid publication, and other weeklies—such as *Chemical News*, *Knowledge*, and *English Mechanic*—all boasted more subscribers than *Nature*'s estimated 5,000.[57] But *Nature* proved more desirable than these other publications because, as Kjærgaard convincingly argues, by the mid-1870s *Nature* had become a specialist periodical with a readership that consisted almost entirely of men of science. Contributors chose *Nature* because it reached a readership that was positioned to evaluate scientific claims. Publishing in *Nature* legitimized one's work or views on a debate as properly scientific.[58] An 1895 letter to Lockyer from William Crookes, the editor of *Chemical News*, strongly reinforces this idea that *Nature*'s readers were considered more scientifically qualified than the readers of other scientific weeklies.

> I have been working night and day to get in type a paper on the spectrum of helium, before my holidays. . . . I should much like to see it in "Nature" if you

can see your way to insert it. It will appear in the Chemical News on Friday, but my circulation is not to the same class of researchers as that of "Nature," and having taken a great deal of trouble about it I want the results to get to the right people.[59]

In order for his results to reach the "right people," Crookes felt that publication in *Chemical News* was not sufficient—the paper had to appear in the pages of *Nature*.

A second reason *Nature*'s speed of publication would have been compelling to men of science is that getting one's work into print quickly had become an increasingly essential part of establishing priority for a scientific finding or theory.[60] Using *Nature* to announce a new finding or a forthcoming paper, which became one of the periodical's primary functions in the twentieth century, was still a developing use of the journal in the nineteenth century. As we can see from table 3, the members of the younger generation were far more likely to submit articles to *Nature* describing their recent scientific work; a few contributors—such as Lankester, Thiselton-Dyer, and especially Romanes—were beginning to use announcements of their work in *Nature* to advance their careers and their scientific reputations. However, some younger contributors (such as Meldola) were still unlikely to do so, suggesting that this was a growing but not yet essential function of *Nature*.

There appear to have been a number of factors that made *Nature* less compelling to the older generation. One explanation for the gap is the disparity in pay between monthly periodicals such as *Fortnightly Review* and *Nature*. Men such as Wallace, Huxley, and Spencer were accustomed to earning their livings from their pens. At the height of their careers these men could earn substantial payments for lengthy scientific essays in monthly periodicals; a letter to the editor in *Nature*, however, was far less lucrative.[61] Furthermore, the members of the older generation were at a vastly different stage in their careers. Men such as Thomson, Tait, and Wallace had already established their reputations and probably did not feel the same need to establish that their work was "scientific." Finally, it is possible that the older generation continued to utilize more established forms of communication because they remembered how Lockyer had initially advertised his journal. They may have viewed *Nature* as a popularizing periodical and preferred to direct their scientific essays to publications they felt were more intellectually prestigious.

It should be noted that men of science who favored *Nature* did not see an article in *Nature* as a substitute for delivering a full paper to a scientific society or writing for one of the scientific societies' journals. An abstract or letter to the editor in *Nature* simply could not convey the same amount of information as a seventy-page article in the *Philosophical Transactions* or a talk at the BA. But although *Nature* was not a replacement for these forums, some of the older institutions began to resent the way in which announcements in *Nature* seemed to steal their thunder. As early as 1880, some within the BA viewed *Nature* as a competitor, complaining that the ease of writing into the weekly journal had stripped the BA meetings of their traditional significance—few saved new or provocative ideas for the annual meeting, instead preferring to initiate discussions immediately by submitting a piece to *Nature*.[62]

While short pieces in *Nature* did not replace the need to publish a longer paper in the journal of a scientific society, contributions to *Nature* do appear to have replaced the literary periodicals for many members of the younger generation. A brief glance at the author listings in the *Wellesley Index to Victorian Periodicals* shows us that between 1870 and 1900, Huxley contributed 70 articles to publications included in the *Wellesley Index*; Wallace wrote 36; Spencer wrote 68.[63] In the younger generation, Lankester wrote only five; Meldola, Perry, and Thiselton-Dyer wrote none. Romanes was a notable and significant exception to this generational trend; in this time period, he wrote 29 articles for the *Wellesley Index* publications.[64] But as Joel Schwartz observes in his article on Romanes's writings for Victorian publications, Romanes used journals such as *Nineteenth Century* to popularize evolutionary theory and to promote his own image as Darwin's heir to a lay audience. Romanes did not view the lay publications as a place to print substantial scientific criticisms of others' theories, as Wallace and Huxley before him had.[65]

The choice of *Nature* versus the literary periodicals as a host for scientific discussions was not a mere aesthetic preference—it represented a choice between two fundamentally different types of scientific debate. When the older generation chose publications such as *British Quarterly Review* or *Nineteenth Century* for an essay questioning a colleague's scientific theory, they were placing their work in a publication read by educated men of all trades and alongside articles on politics, religion, literature, and philosophy. Wallace, Huxley, and the rest would have seen this as a point in the literary periodicals' favor. Publishing in literary monthlies helped them subtly press their belief that science was an intellectual endeavor worthy of equal standing with more "classical" subjects such as history, literature, and politics. Furthermore, the same debate was frequently carried out in several journals at once—when Herbert Spencer wanted to respond to a negative assessment of his work in *British Quarterly Review*, he published his reply in *Fort-*

nightly Review. But younger men of science—such as Romanes, Lankester, or Perry—had reaped the rewards of the older generation's attempts to establish science as a respectable endeavor; they faced far fewer difficulties in constructing their scientific careers than their mentors had. Consequently, they saw less reason to debate scientific questions before the largely nonscientific audience of the literary periodicals and preferred to direct their writings to a publication with a more specialized audience.[66] Romanes's complaint that Wallace had moved the physiological selection debate outside *Nature* seemed valid to Romanes, who assumed scientific issues should be debated for an audience of men of science in their preferred venue, but it would likely have been baffling to Wallace, who was accustomed to scientific debates that spanned several literary publications.

NATURE AND DARWIN'S LEGACY

One important exception to the generational trend was Charles Darwin. The great naturalist occasionally prepared abstracts of his longer scientific papers for *Nature* and saw a short piece in the weekly journal as a useful way to announce a forthcoming study or, more frequently, to disagree with another man of science or call attention to another naturalist's paper that he thought was of particular interest.[67] *Nature* quickly replaced the *Gardener's Chronicle* as Darwin's publication of choice.[68]

After Darwin's death in 1882, his image took on a new life in the pages of *Nature*. The author of *On the Origin of Species* was one of the most revered figures in Victorian science (whatever his reputation may have been in nonscientific circles), and Thiselton-Dyer, Lankester, and especially Romanes were all explicit in their desire to emulate his great scientific career. In the two decades following his death, the correspondents in *Nature* spoke of Darwin with the utmost respect, even reverence. Romanes's physiological selection abstract mentioned Darwin repeatedly, and Romanes attempted to cast his theory as the solution to a problem Darwin himself had identified:

> For he [Darwin] says and he says most truly, "We have conclusive evidence that the sterility of species must be due to some principle quite independent of natural selection." I trust I have now said enough to show that, in all probability, this hitherto undetected principle is the principle of physiological selection.[69]

Many of Romanes's critics also invoked Darwin's name in their discussion of Romanes's theory. Wallace's *Fortnightly Review* critique of physiological selection was titled "Romanes *versus* Darwin" and attempted to show that Romanes's ideas were antithetical to Darwin's work.[70] Meldola wrote that while Darwin had acknowledged that natural selection might not be the only agent at work in evolutionary change, Romanes's contention that physiological selection was equally important was contrary to Darwin's beliefs:

> Darwin to the last considered natural selection as the *chief agency* in the evolution of species, and no one saw more clearly than he did the difficulties which surrounded the formation of incipient species, owing to the obliteration of new characters by intercrossing with the parent form.[71]

Lankester wrote to *Nature* to argue that Romanes's physiological selection paper was an attack on Darwinism: "He [Darwin] considered his theory of natural selection to be a theory of the origin of species. Mr. Romanes says it is not. I say this is an attack on Mr. Darwin's theory, and about as simple and direct an attack as possible."[72] Darwin's son Francis, by then a respected botanist, agreed with Lankester; in a letter to *Nature* he argued that his father had considered and discarded a theory very much like Romanes's.[73] In response to such criticism, Romanes insisted that his theory was an addition to, not a replacement for, natural selection, and that his opponents were the ones who were anti-Darwinian: "My contention from the first has been that upon this point I am in full agreement with Mr. Darwin, and differ only from those Darwinians who differ from their master."[74]

The name of Darwin also came into play during the debates over the inheritance of acquired characters. Once again, both sides attempted to claim the late naturalist as a supporter of their views. Romanes and Spencer, two prominent supporters of the theory of use inheritance, both argued that Darwin's evidence in the *Origin of Species* showed that acquired characters could be passed down from parent to offspring. In a letter to the editor on the subject of acquired characters, Spencer wrote that much of the *Origin of Species* contained evidence in favor of the theory of use inheritance. He said that this clear fact was being conveniently ignored by those who denied the theory: "Clearly the first thing to be done by those who deny the inheritance of acquired characters is to show that the evidence Mr. Darwin has furnished by these numerous instances is all worthless."[75] Romanes pushed the argument even further, suggesting that Darwin had not only collected evidence in favor of use inheritance, but that he himself had adhered to the theory.[76]

Romanes's opponents, however, claimed that Darwin had not regarded the possibility of use inheritance with any great favor. In a March 1890 response to Spencer and Romanes, Lankester wrote,

It is not necessary to show that anything Mr. Darwin wrote was "worthless," but it is necessary to show that certain facts cited by Mr. Darwin admit of another interpretation or explanation than that which he gave to them. Naturally those who have taken up the anti-Lamarckian position have done long ago what Mr. Herbert Spencer says is the first thing for them to do. Of course the cases cited by Darwin were the first thing to be dealt with.

Lankester admitted that Darwin occasionally seemed to endorse the possibility of use inheritance but argued that many of the key passages on the matter had been misinterpreted and that "[Darwin] preferred, where it occurred to him, another interpretation" besides the Lamarckian one. Lankester concluded his letter by writing that contrary to what Romanes argued, Darwin's view on inheritance in the *Origin of Species* "is the essence of the anti-Lamarckian view of the effects of disuse."[77]

This pattern of invoking Darwin as the ultimate authority on evolutionary theory adds a new dimension to Peter Bowler's argument that "Darwinism was besieged on all sides" during the late nineteenth century.[78] It is certainly true, as Bowler argues in his book *The Eclipse of Darwinism*, that many naturalists were advocating changes to evolutionary theory that drew them away from Darwin's original writings. However, the discussions of evolutionary theory in *Nature* suggest that in Britain at least, evolutionary theorists in the immediate post-Darwinian era did not believe they were besieging Darwin (or, at least, had no desire to create that impression). Instead, men such as Romanes, Lankester, Spencer, and Wallace all sought to portray themselves as faithful Darwinians. These naturalists did not see themselves as participants in a "non-Darwinian" revolution; they wished their readers to think they were carrying on Darwin's program of evolutionary work. This observation reinforces the point that the "younger generation" whose contributions were so essential to *Nature* saw themselves as heirs to an established scientific tradition rather than revolutionaries whose ideas represented a break with the previous generation.

NATURE'S PLACE AT THE END OF THE NINETEENTH CENTURY

The mixed reaction to Lockyer's 1879 application for membership in the Athenaeum might suggest that the passionate debates taking place in his journal were winning neither *Nature* nor its editor much affection. Other correspondence about Lockyer seems to reinforce this impression. In 1873, shortly before being elected as president of the BA, Tyndall wrote to Huxley expressing his frustration with *Nature* and "the little man who has hither ruled it at the head of affairs."[79] In 1890, one Oxford mathematician was heard to complain that Lockyer had forgotten "the difference between the Editor of *Nature* and the Author of Nature."[80]

And yet, these contentious discussions seem to have been a crucial part of *Nature*'s success. As a year-round, weekly publication, *Nature* proved to be a convenient host for scientific disputes. Contributors could compose short pieces stating their position and be assured that they would be printed quickly, instead of waiting for the next BA meeting or the next issue of a quarterly journal. Notably, while Tyndall resented the "little man" at *Nature*'s helm, he continued to respond to his critics in the journal. The more scientific discussions were directed to *Nature*, the more the journal became essential reading for a British man of science who wished to remain up to date on the latest issues in the scientific world. *Nature*'s importance had eclipsed that of its predecessors—including its closest inspiration, *Chemical News*, as the 1895 letter from its editor William Crookes to Lockyer made clear.

By the 1880s, *Nature* had managed to cultivate a loyal audience of readers and contributors. But the journal had yet to turn a profit, and after his death, Lockyer's family remembered this decade as the most precarious of *Nature*'s existence.[81] Furthermore, Lockyer was facing a personal crisis—his beloved wife, Winifred James Lockyer, died suddenly in September 1879, leaving Lockyer with seven children between the ages of 6 and 19.[82] Despite Lockyer's worries, there is little indication that the Macmillans were considering withdrawing their support at this time. The Macmillans appreciated the access *Nature* gave them to new writers for their profitable science division.[83] Furthermore, as Frederick Macmillan indicated in a letter to his cousin George, the Macmillans also saw *Nature* as a convenient way to advertise their scientific publications. "I notice in the advertisements of our own books in this week's Nature which has just arrived that there is a certain amount of compression in the arrangement of the subjects.... It would be better to publish lists containing all the important books on any given subject without reference to their size or shape," Frederick wrote, adding, "The plan of having lists of all our scientific books & publishing them in Nature by turns, is, I am sure a very good one."[84]

The Macmillans were not alone in the belief that *Nature* was an excellent means of reaching the British market for scientific books. An 1885 letter to the editor from Sir William Thomson provides another example of *Nature*'s reach among British men of science. As previously noted, Thomson was an occasional contributor to *Nature* but was among the members of his generation who regarded the journal as a lighter periodical unsuitable for publishing significant scientific essays. But when Thomson discovered that a printed copy of his 1884 Baltimore lectures contained several mistakes, he sent out the corrections in *Nature*'s Letters to the Editor column.

> As it is possible that some of your readers may have obtained copies of the Papyrograph Report of my Lectures on "Molecular Dynamics," delivered at Baltimore during October 1884, I should be obliged by your giving publicity to the following corrections.[85]

Thomson's use of *Nature* to correct these errors suggests that Thomson expected that many of the men who had purchased his Baltimore lectures also read *Nature*, and he believed that writing to the journal would be an effective way to communicate his corrections to his readers. This seems to indicate that *Nature* was recognized as a publication with a wide scientific readership, even among scientific workers who did not see *Nature* as a research journal.

The publishing world brought further evidence of *Nature*'s growing influence. In 1880, a New York journalist named John Michels collaborated with the famous inventor Thomas Edison to create the journal *Science*, which, like *Nature*, was a weekly publication intended to draw contributions from distinguished men of science working in a wide range of disciplines. In the introductory issue of *Science*, Michels (who assumed the position of editor) wrote, "It is the desire of the Editor that 'Science' may, in the United States, take the position which '*Nature*' so ably occupies in England, in presenting immediate information of scientific events"—a clear testament to *Nature*'s success in placing itself at the center of the scientific consciousness, as well as to *Nature*'s growing influence among American readers.[86]

Imitation was not always a form of flattery. In 1882, an astronomer named Richard Proctor founded a journal called *Knowledge*. To Proctor, *Nature* represented a sheltered, exclusive body of practitioners who sneered at popularizers such as himself and denied the public any right to participate in scientific discussions. *Knowledge* claimed to welcome contributions to scientific research from anyone, regardless of degrees or connections. The weekly format, short title, price, and even the masthead were designed to position *Knowledge* as a superior alternative to *Nature*.[87] But like his rival Lockyer, Proctor was unable to sustain his original vision for his journal. He soon found himself dealing with submissions from groups like flat-earth proponents, who felt that their contributions ought to be welcome in a jour-

FIGURE 4 *Knowledge*'s masthead, 1882.

nal with such an inclusionary mission. In 1885 Proctor, overwhelmed by the vast amount of correspondence, changed the journal to a monthly publication focusing on the relationship between science and religion.

For some men of science, *Nature* was becoming something more than just a convenient means of reaching their scientific colleagues. *Nature* could also be seen, as an 1893 letter from the physicist Oliver Lodge suggests, as one of the few remaining places where men of science could communicate across increasingly sharp disciplinary boundaries. In October, following the annual British Association for the Advancement of Science (BA) meeting, Lodge wrote to *Nature* to lament that aside from the presidential address, few of those who attended the meeting were seen outside their own sections. Physicists spent the meeting in the company of other physicists; biologists spent the meeting with other biologists. *Nature*, said Lodge, was one of the few remaining places where researchers could communicate across disciplines:

> Whether the British Association can or cannot act as a connecting link between the sciences, there is no doubt but that the pages of NATURE do so act; and long may it be before NATURE (I mean the publication) finds herself also bifurcated or otherwise subdivided, and we on either side cease to hear even an echo of what the other side is talking about.
>
> Perhaps few are able to say that they read NATURE all the way through as Mr. Darwin did, but we all have the chance of doing so.... The fear is lest we drift apart so far that we cease to understand each other's language.[88]

The BA was not the only scientific organization that was beginning to reflect fragmentation in British science. In 1887, the *Philosophical Transactions of the Royal Society,* which had previously been a single unitary publication, split into two journals—series A for mathematics and the physical sciences, and series B for the biological sciences. During an era of growing separation between disciplines, *Nature* was an anomaly, a journal whose readership included astronomers, naturalists, physicists, paleontologists, and statisticians.

Lodge believed that this unitary aspect of *Nature* made it an essential part of British science. Despite the trend toward specialization, Lodge and others like him still felt that there was a wider scientific community that encompassed practitioners from different disciplines, and that it was important for the members of this community to maintain connections with one another. *Nature,* as one of the few remaining publications that welcomed contributions from a variety of disciplines, served as a forum where this could take place. But at the same time that *Nature* was creating a sense of unification among scientific researchers, its contributors were also using the journal to erect the boundaries that defined who would be excluded from their community.

CHAPTER THREE

Defining the "Man of Science" in *Nature*

Norman Lockyer was not the only person who relished scientific controversies in nineteenth-century Britain. George Douglas Campbell, the eighth Duke of Argyll (1823–1900), also enjoyed stirring the proverbial scientific pot. He was a major intellectual voice against Darwin's theory of evolution by natural selection, although this opinion did not prevent him from being friendly with the great naturalist—indeed, Argyll was a pallbearer at Darwin's Westminster Abbey funeral alongside Darwin's close friends and fellow naturalists Joseph Hooker, Thomas Huxley, and Alfred Russel Wallace.[1] Argyll was a Fellow of the Royal Society (FRS), a distinction he had won after publishing his paper "On Tertiary Leaf-Beds in the Isle of Mull" in 1851.[2] He was the chancellor of the University of St. Andrews and the rector of the University of Glasgow, and he had been the president of the BA in 1855 and president of the Royal Society of Edinburgh in 1861. However, Argyll's greatest fame came from his political career; he served as secretary of state for India from 1868 to 1874 and was a member of the cabinet under four different prime ministers.

In 1887, five years after Darwin's death, Argyll read a scientific paper that piqued his interest: "On the Structure and Origin of Coral Reefs and Islands" by his fellow Scotsman John Murray (1841–1914).[3] Murray's paper argued that coral reefs were formed by the accumulation of organic matter—a view that differed from the "subsidence" theory of coral reef formation Darwin had proposed in his 1842 book *The Structure and Distribution of Coral Reefs*.[4] Argyll did what most men of science of his generation would have done in similar circumstances: he wrote an essay for a literary monthly, *Nineteenth Century*, calling attention to the paper. Argyll's article also made the provocative claim that Murray's work had gone unnoticed because the British scientific community had been unwilling to question Darwin.

Argyll expected—and, given his love for intellectual combat, probably hoped—that his essay would be controversial. The controversy he encountered, however, was not quite the one he had anticipated. His essay became the topic of impassioned discussion in *Nature*, but the contributors who wrote in to respond did not argue with his or Murray's scientific claims. Instead, they declared Argyll unfit to comment on coral reef theory at all.

Today it would seem obvious to scientific researchers that a politician who had published a single paper more than thirty years ago should not be considered a valid source of commentary on a recent scientific theory—simply put, Argyll would not be considered a scientist. But in nineteenth-century Britain, the question of whether or not someone like Argyll could be a "man of science" was much murkier. In the 1850s Argyll was welcomed as a man of science, someone whose love of scientific knowledge, past experience investigating Scottish geological formations, and devotion to reading the latest papers marked him as a scientific insider. And yet by the 1880s, *Nature*'s contributors felt confident dismissing his opinions as irrelevant.

The duke's interactions with *Nature* show that something fundamental was changing about science in nineteenth-century Britain: the qualifications for being considered a man of science were becoming more demanding and more specialized. Arguments about who could and could not be a man of science were not petty debates over terminology. They were fundamentally about what science was and who would be allowed to make claims about scientific knowledge. *Nature* was a key site where this battle was fought, where the qualifications for membership in British science were proposed, debated, and established.

This chapter explores instances in *Nature* where the contributors discussed the qualities and background necessary to be a "man of science" (a term they preferred over "scientist"). The essential criterion *Nature*'s participants established for a man of science was that he (or occasionally she) must perform original scientific investigations. Those who simply read about science or who focused on the practical applications of science rather than the creation of new knowledge were not considered the scientific equals of those who devoted themselves to original investigations. *Nature*'s contributors argued that a commitment to investigating scientific truths was both an intellectual and a moral qualification for being a man of science. This narrowing criterion for the man of science would pose a problem when Lockyer set out to choose a successor as *Nature*'s editor, but his assistant Richard Gregory managed to create a niche for himself within *Nature*'s research-centric community by acting as a public spokesman for the contributors' vision of science.

"HE HAS ENTIRELY MISTAKEN HIS VOCATION": MEN OF SCIENCE VERSUS LITERARY MEN AND POLITICIANS

Early in *Nature*'s existence, contributions from laymen were welcomed, even solicited. But by the 1880s the journal's contributors had a much chillier attitude toward laymen interested in contributing their own ideas about scientific theories. In *Nature*'s book reviews, for example, books on scientific subjects written by laymen were almost universally derided as unlearned and irrelevant. Alfred Russel Wallace, the great naturalist who had published his theory of natural selection alongside Darwin's, was a particularly harsh (and quotable) critic of evolutionary tracts by authors with no experience as naturalists. Wallace deemed one book on evolution by the philosopher James Hutchison Stirling "contemptible and worthless."[5] An anonymous 1894 volume, *Nature's Method in the Evolution of Life*, prompted Wallace to write wearily, "Almost every educated man who can write good English, but who cannot understand Darwin's theory of Natural Selection, seems to feel compelled to explain his difficulties and to offer his own preferable theory in the form of a volume on Evolution."[6] Most tellingly, in a review of the anti-Darwinian essay *On the Modification of Organisms* by the literary critic David Syme, Wallace said that Syme had no right to attempt an overthrow of Darwin's work because he was an expert on literature, not science.

> Mr. Syme has a considerable reputation in other departments of literature as a powerful writer and acute critic; but he has entirely mistaken his vocation in this feeble and almost puerile attempt to overthrow the vast edifice of fact and theory raised by the genius and the lifelong labours of Darwin.[7]

The evolutionary theorist Raphael Meldola also had sharp words for authors he felt were unqualified to write about evolution. In 1891 Meldola reviewed *Science or Romance?* by the Reverend John Gerard, a book that questioned Darwin's theory of natural selection. Meldola wrote that while Gerard appeared to have some knowledge of the natural world, he was clearly no man of science—his methods of argumentation were entirely literary.

> But while the purely destructive attacks of the reverend critic may give satisfaction to those who belong to his school, the impartial reader will derive only

amusement, and the man of science will soon perceive that the weapons of attack are not the legitimate implements of scientific warfare, but the tricks of disputation concealed under a somewhat alluring literary cloak.[8]

There was nothing in Gerard's book, said Meldola, that could possibly appeal to or enlighten a man of science—the volume was all cynicism, misrepresented evidence, and literary posturing.

In contrast, when the entomologist Francis P. Pascoe wrote a book disputing Darwin's conclusions in 1891, Meldola's review was much more balanced. Meldola opened the review by acknowledging Pascoe's "special knowledge of certain groups of insects and his general knowledge of other groups of animals," which had enabled him "to collate a large number of difficulties and objections which have occurred to himself and other naturalists." While Meldola clearly disagreed with Pascoe's claim that there was insufficient proof for natural selection, Meldola addressed each of Pascoe's objections to Darwin carefully and concluded by writing, "However much we might differ from the author, it cannot be denied that, as a stimulus to further research, such compilations as that which Mr. Pascoe has produced are distinctly useful."[9] In Meldola's opinion, the naturalist Pascoe, who had obtained expertise through his original investigations of insects, deserved a thoughtful and serious response that the Reverend Gerard had not earned.

Historians of Victorian science have shown that men like the members of the X Club attacked religious authority in part out of a desire to claim the church's cultural authority for scientists.[10] Wallace and Meldola's pointed reviews of books on evolution by literary critics further suggest that some men of science also sought to claim the cultural authority held by literary men. In 1840, Thomas Carlyle described the man of letters as a heroic figure, "our most important modern person,"[11] and literary critics commented on everything from Wordsworth's poetry to political philosophy to scientific books.[12] By the late nineteenth century, however, the intellectual sphere the man of letters occupied was shrinking; a man of letters was identified not as a general intellectual but rather as a literary specialist. *Nature*'s book reviews indicate that men like Wallace and Meldola sought to claim the right to comment on scientific theories solely for men of science. Significantly, in his review of Syme's book, Wallace wrote, "This little book is one of a class that was more common twenty years ago, when any acute literary critic thought he could demolish Darwin"[13]—indicating that in Wallace's eyes, a literary critic who thought he could write about scientific theories was an irrelevant relic of a bygone age.

Meldola's more favorable review of Pascoe's book illustrates one of the emerging criteria for scientific expertise. In order to be qualified to make scientific claims, it was not enough to be well read on scientific topics or to be passionate about scientific knowledge; a qualified commenter had to have performed original scientific investigations and made contributions to scientific knowledge. The importance *Nature*'s contributors placed on original scientific investigations can also be seen in *Nature*'s discussions about female scientific practitioners. It was, of course, no accident that most (male) nineteenth-century scientific workers preferred the term *man of science* as a descriptor—the term, like its parallel *man of letters*, was quite deliberately gendered. Defining science as a masculine endeavor implied greater social and intellectual respectability; furthermore, some prominent members of the scientific community, including Huxley, had very low opinions of women. However, women were beginning to gain ground in science at the end of the nineteenth century and made significant contributions to scientific knowledge. The fact that women were engaged in original scientific work quickly became cited as a reason to admit them as full members of the scientific community.

Female researchers found an ally in Lockyer, whose second wife, Thomazine Mary Broadhurst Lockyer, was a noted suffragette.[14] When the physicist Hertha Ayrton was considered for fellowship in the Royal Society, Lockyer was a strong supporter of her candidacy.[15] Under Lockyer, *Nature* was similarly supportive of female investigators. When female chemists petitioned the Chemical Society to admit female fellows in 1908, for instance, *Nature* ran a lead editorial in support of their petition. Notably, the editorial's anonymous author (possibly Lockyer himself) wrote that women should be admitted because they had written original papers:

> It cannot be denied that women have contributed their fair share of original communications. Indeed, in proportion to their numbers they have shown themselves to be among the most active and successful of investigators. The society consents to publish their work, which redounds to its credit.

The author then contrasted these admirable contributors to knowledge with the uselessness of Chemical Society members who had not performed original investigations but could be fellows because they were men:

> Why, then, should the drones who never have done, and never will do, a stroke of original work in their lives be preferred to them simply because they wear a distinctive dress and are privileged to grow a moustache?[16]

When the council decided to admit women as "subscribers" instead of fellows, *Nature* published another editorial criticizing the council's decision as "wholly irregular and unconstitutional." The author accused "a self-constituted oligarchy" of ignoring the will of the majority of the Chemical Society in order to "gratify its personal prejudices." This second editorial praised the women chemists who had petitioned the society for rejecting the council's offer and holding out for full fellowship.[17] A (possibly female) *Nature* reader, writing under the initial "T.," submitted the following poem to express support for the women chemists' quest:

> Daughters of Eve! So zealous to pursue
> The work in Life by which you seek to live!
> When F.C.S. you claim, as is your rightful due—
> The S alone is what they, grudging, give!
>
> Be patient! Time is on your side.
> Reason and justice will our cause defend.
> Ignoble spite and arrogance of pride
> Shall meet their retribution in the end![18]

Nature and its contributors were similarly supportive of women scientists' quests to join the Geological Society, the Paris Academy of Sciences, and the Royal Society. Most of the *Nature* material on these efforts emphasized that women had produced enough original scientific work to qualify for membership alongside their male colleagues.[19] The fact that supporters pointed to the female investigators' original scientific work as a reason to grant them full membership illustrates how central research had become to scientific identity. However, it would be many more years before women were admitted to any of these societies, suggesting that for many, being male was still considered an essential qualification for being a man of science.[20]

The discussions about men of letters and female chemists show that many *Nature* contributors believed that original scientific investigations were a key prerequisite for being qualified to comment on science. The Duke of Argyll's interactions with *Nature* also reflect this increasing emphasis on original contributions to scientific knowledge, but they reveal another criterion for the man of science as well: he (or she) had to be devoted to scientific truth above all other goals. Despite his impressive list of scientific achievements and titles, in the 1880s many of the duke's fellow *Nature* readers and correspondents argued that he was an outsider in their community—not just because he was no longer an active investigator, but because he was a politician.

The debate began as it often did in the late nineteenth century: with Charles Darwin. This time, the theory under discussion was not natural selection but the subsidence theory of coral reef formation. In brief, Darwin's theory was that coral reefs first formed in the shallow waters around volcanic islands. As those islands sank, the reefs around them subsided into the ocean, providing a chance for further coral growth on both the old coral reef and the newly underwater portions of the island. Darwin had developed his theory of coral reef formation during his legendary voyage on the HMS *Beagle* in the 1830s, and his book *The Structure and Distribution of Coral Reefs* was the first scientific monograph of Darwin's storied career.

Nearly forty years later, the geologist John Murray would also join a sea voyage, this one aboard the HMS *Challenger*. Murray spent much of the voyage examining reefs in the deep ocean, and he came to the conclusion that coral reefs were formed by the accumulation of organic sentiment, including the remains of various deep-sea creatures. Darwin's subsidence theory, he argued, was unnecessary to explain the formation of coral reefs. Murray published his theory in 1880 in the *Proceedings of the Royal Society of Edinburgh*, and by the mid-1880s he found himself embroiled in a debate with the American geologist James Dwight Dana over whether his theory or Darwin's best explained the existing observations about coral reefs.[21]

Argyll, whose son George Granville Campbell had also been on board the HMS *Challenger*, decided to champion his countryman's theories.[22] In September 1887, Argyll published an article in *Nineteenth Century* in which he declared Murray's work a triumph and its reception an object lesson in scientific "idolatry." He argued that the "slow and sulky" reaction to Murray's paper showed that men of science were unwilling to acknowledge any work that might contradict Darwin's. He accused men of science of "reluctance to admit such an error in the great Idol of the scientific world."[23]

Darwin had been challenged, and Darwin's bulldog entered the fray. Thomas Huxley published a response to Argyll in the November issue of *Nineteenth Century* in which he argued that Dana's work had thoroughly refuted Murray's theories. Huxley described Dana as "the most competent person now living to act as umpire" in the coral reef debates and reprimanded Argyll for his apparent unfamiliarity with Dana's work. He demurred from taking a position on the coral reef debate himself, explaining that coral reefs were a very difficult area of study and that "until I had two or three months to give to the renewed study of the subject in all its bearings, I must be content to remain in a condition of suspended judgment." Huxley was, however, quite willing to take a position on Argyll: he accused

Argyll of lying about Murray's reception in order to cast "aspersions of the honour of scientific men."[24]

T. G. Bonney (1833–1923), the outgoing president of the Geological Society, carried the protest against Argyll's article to *Nature*. In an editorial for *Nature*, Bonney wrote that Argyll's charges were an attack on "the honour and good faith" of all men of science.[25] Bonney argued that Murray's theory had won some converts but that coral reef theory was a subject of many "differences of opinion among those best qualified to judge"—the implication, of course, being that Argyll was not among those best qualified to judge. Much of Bonney's response to the duke focused not on Argyll's scientific claims, however, but on his political activities. Even though the duke had performed impressive scientific work in the past, said Bonney, his work as a politician effectively canceled out his scientific qualifications.

> The Duke of Argyll is eminent as a statesman, and has won distinction as a man of science. The mental qualities, however, which lead to success in these capacities are widely different; nay, in the opinion of some, almost oppugnant. To the man of science, truth is a "pearl of great price," to buy which he is ready to part with everything previously obtained; to the statesman, success is the one thing needful, for the sake of which hardly any sacrifice appears too great. . . . The Duke of Argyll has recently afforded a remarkable instance of the extreme difficulty of combining in one person these apparently opposite characters.

Argyll wrote to *Nature* to justify his language, saying that he had not intended any slight on men of science as a group. Indeed, initially he attempted to align himself with the *Nature* community's views on the inferiority of politicians, writing that scientific men "are, I admit, immensely superior to politicians, especially just now." But, Argyll insisted, even men of science could make errors: "everyone who knows the history of science must be able to call to mind not one instance only, but many instances, in which the progress of knowledge has been delayed for long periods of time by the powerful and repressive influences of authority."[26]

Argyll's opponents were not mollified by this explanation. They continued to flood *Nature*'s Letters to the Editor column with correspondence insisting that Argyll had insulted all men of science when he claimed that Murray's theories had been ignored to protect Darwin.[27] Many of these letters criticized the duke's "political" style of argument. The geologist T. Mellard Reade, for example, wrote that the duke was attempting to bring political tactics to the realm of science: "I was pleased to see Prof. Bonney's article on the Duke of Argyll's strictures on scientific men. It is to be hoped that the rhetoric and methods of Parliamentary debate will not become common in scientific controversy."[28]

Eventually Argyll grew frustrated with the repeated attacks on his claim of a "conspiracy" and attempted to shift the discussion back to scientific theory. In a later letter, Argyll protested against the focus on "a few words" and challenged his opponents to deal seriously with his science.

> May I ask your correspondents who have been good enough to read my article on "Darwin's Theory of Coral Islands" . . . to begin addressing themselves to the merits of the scientific question there dealt with, and to cease wasting their own time and your space upon scolding me for a few words[?][29]

Notably, few of the men who wrote to *Nature* to censure the duke engaged with his scientific claims. Like Bonney, most correspondents seemed to agree that even though Argyll was an FRS, he could not be a true man of science and a politician at the same time. Argyll's scientific arguments deserved no serious answer not only because he was not an expert on coral reef theory but also because they had come from the mind of a politician.

Argyll met with a similar reaction in December 1889 when he expressed his support for the theory of the inheritance of acquired characters. Argyll claimed that biologists had failed to give sufficient weight to the inheritance of acquired characters in their evolutionary theories and that this was another instance of Darwin's admirers prematurely dismissing non-Darwinian ideas.[30] W. T. Thiselton-Dyer wrote a lengthy response to the duke for *Nature*. Once again, Argyll's opponent argued that the duke was a politician and therefore by definition not a true man of science.

> It has a curious and not uninstructive effect to see the pages of this journal invaded by the methods of discussion which are characteristic of political warfare. . . . In politics, the personal rivalry which is bound up inextricably with the solution of great problems may make it a necessary part of the game to endeavour to belittle one's opponents. But in science it is not so.[31]

Unlike Argyll's opponents in the coral reef discussion, Thiselton-Dyer dealt at length with Argyll's scientific claims about the evidence for the inheritance of acquired characters, but the main thrust of his criticism was that Argyll was poorly informed and generally unqualified to participate in the debate. In response to Argyll's claim that he had waited 30 years for Darwinians to deal with the obvious problems with Darwinian theory, Thiselton-Dyer dismissively wrote, "One can only wonder what Darwinian literature has been the subject of his studies during that time."[32]

The Duke of Argyll's participation in *Nature* provides one of the journal's most striking examples of scientific boundary drawing. Although Argyll was an FRS, a man with a substantial scientific paper to his name, and a past president of two of Britain's most prestigious scientific organizations, it had been many years since he had performed any scientific work. Furthermore, he had not undertaken any original investigations into coral reefs or acquired characters—and, indeed, he seemed to be unfamiliar with some of the most recent literature on these subjects. *Nature*'s community of researchers therefore expressed indignation when Argyll attempted to insert himself into debates on these topics.

But when Argyll's opponents denounced him, they did not merely dismiss him as unqualified or ignorant, as they had done with the men of letters. The duke's opponents also used the opportunity to underline what they saw as the differences between politicians and scientific workers and to argue that a politician's tactics were unsuitable in a scientific setting. According to *Nature*'s contributors, politicians simply did not play by the same set of rules as men of science. Politicians used rhetoric and insults to achieve victory over their opponents; men of science presented facts in support of their arguments and did not seek to cast doubt on their opponents' motives in order to strengthen their case. Being a politician did not just signal a lack of intellectual qualifications but a lack of moral qualifications necessary to be a man of science. It is also instructive to note the similarities and differences between *Nature*'s letters about Argyll and the letters about George Gabriel Stokes, the physicist who ran for Parliament while serving as the president of the Royal Society (see chap. 2). The *Nature* contributors who were uncomfortable with Stokes's election to Parliament, like many of Argyll's opponents, believed the role of member of Parliament was fundamentally incompatible with the role of a man of science—especially a man of science as significant as the president of the Royal Society. But interestingly, several of *Nature*'s readers submitted pieces defending Stokes's decision. Argyll received no such support. The essential difference seems to have been that Stokes devoted his life to science, entering politics only at a very late date, while Argyll had once dabbled in scientific research but had ultimately chosen to devote himself to politics.

WHAT COUNTS AS LEGITIMATE INQUIRY? SPIRITUALISM AND ENGINEERING

More complicated cases of boundary drawing in *Nature* involved subjects that some contributors felt were not legitimate areas of scientific inquiry. One interesting example from the journal's early decades is that of spiritualism. In the second half of the nineteenth century, many Britons became interested in the possibility of communicating with the dead and demonstrating the reality of the afterlife. Several respected men of science became involved in the spiritualist movement and adopted the study of psychic phenomena as a major part of their research agenda. The Society for Psychical Research (SPR), for example, counted several eminent physicists among its membership.[33]

One of the SPR's most prominent members was also an important early contributor to *Nature*. William Crookes was a dedicated psychical researcher and a close friend of Lockyer's; he was also the editor of *Chemical News*, the weekly publication to which *Nature* owed much of its inspiration. Crookes was a significant presence in the *Nature* discussions on spiritualism. He authored letters in defense of psychical research,[34] and he also chose *Nature* as the venue for a two-part article, "On Radiant Matter," in which he explicitly stated that his research into a "fourth state of matter" might lead to better understanding of unknown "Ultimate Realities."[35]

Crookes was neither unusual nor alone in his interests. Other major *Nature* contributors, including Alfred Russel Wallace, Edward Dixon, and Oliver Lodge, also wrote articles on spiritualist studies for *Nature*. Many of these pieces made forceful arguments for the discipline's acceptance as a science. In 1877, Crookes argued that his "scientific honour" required him to investigate the unexplained phenomena associated with spiritualism, saying that "every uninvestigated phenomenon is a probable mine of discovery."[36] Alfred Russel Wallace, as we saw in chapter 2, preferred to publish major scientific tracts in periodicals other than *Nature*, but he was not willing to let challenges in *Nature* go unanswered. Wallace frequently wrote to *Nature* to strike back against articles criticizing spiritualism, as he did in 1877 when *Nature* printed a letter from the physiologist William Carpenter that criticized Wallace and Crookes's work:

> I beg to refer your readers to a reply to Dr. Carpenter's attack, and a full exposure of his false accusations against Mr. Crookes and myself, which will appear in the next issue of [*Fraser*]. They will then see *who* has been led by "prepossession" to adopt "methods which are thoroughly *un*-scientific," and *whose* are "the statements which ought to be rejected as completely untrustworthy."[37]

In the 1890s, Dixon and Lodge took on the task of trying to convince the *Nature* readership that psychical investigations were a legitimate field of scientific inquiry. In an 1895 letter to the editor, Lodge wrote that he welcomed criticism of the SPR's work so long as it took the society's methods and data seriously: "One of our main difficulties is that our critics will not take the trouble to study or even read our evidence."[38] Dixon's contribution to the debate, however, suggested that he, Lodge, and the SPR members were on the defensive within the scientific community: "Most people, I am afraid, fight shy of psychical research, either because they are afraid that *if* there is anything in it it is the devil, or because they have a scientific reputation which they are afraid of losing."[39] Dixon's words implied that by 1894, one's scientific reputation might well be damaged by declaring an interest in psychical research.

Although Crookes, Wallace, Dixon, and Lodge were frequent *Nature* contributors and respected researchers, other contributors to *Nature* were skeptical about attempts to use scientific methodology to investigate psychic phenomena. The anonymous reviewer of an 1879 volume on spiritualism, for example, claimed that the subject matter was clearly unscientific:

> This is an essay of 150 pages by a thoroughgoing "spiritualist," according to the most "modern" signification of the term. As such it is not a book very easy to review in the pages of a periodical devoted to the consideration of modern science.... We feel that our function as reviewer ends, when we say that in all his statements of and references to the facts of physical science the essayist is accurate.[40]

In the 1890s, the mathematician Karl Pearson became one of the leading critics of psychical research in *Nature*.[41] Pearson expressed extreme disdain for many of the methods used by psychical researchers, complaining that their results were often statistically insignificant and betrayed a lack of mathematical understanding. In response to the SPR's account of a card-guessing experiment in which several alleged psychics showed the ability to correctly guess which card an SPR member had drawn from a deck, for example, Pearson wrote that the SPR had failed to take precautions to ensure that the testers were not duped. He claimed that this omission demonstrated a lack of "scientific acumen."[42] The anonymous 1879 reviewer had said outright that psychical research was not science. Pearson's point was slightly different: while the experiments the SPR was conducting were not obviously unscientific, the SPR's members lacked the ability to carry them out properly. Pearson's skeptical analysis of the SPR's statistical methods strongly implied that competent scientists would have reached a different conclusion.

It would be inaccurate to portray *Nature* as a major publication venue for psychical research. SPR members generally preferred to publish their results in their own journal, *Proceedings*, and the amount of space *Nature* devoted to psychical research was quite limited compared with the pages on evolutionary theory (see chap. 2), x-rays, or radioactivity (see chap. 4). Nonetheless, psychical research provides us with an interesting case in which *Nature*'s editors and contributors dealt with a field that was struggling to make a claim to scientific status. Faced with this dilemma, *Nature*'s editorial staff seems to have taken a fairly neutral editorial stance, neither encouraging nor condemning psychical contributions to its pages. Although the *Nature* book reviews generally supported the view that psychical research was not a scientific discipline, overall the journal appears to have given equal space to both sides of the spiritualism debate in the Letters to the Editor and in the abstracts and articles. Furthermore, while opponents of psychical research frequently argued that its practitioners did not employ proper scientific methods, they did not take *Nature* itself to task for printing such pieces. Unlike the Duke of Argyll or the men of letters who had attempted to weigh in on evolutionary theory, Crookes, Wallace, Dixon, and Lodge were respected scientific researchers. Their status within *Nature*'s community appears to have gained them a hearing even on a subject as controversial as psychic phenomena.

Discussions in *Nature* about engineering and its relationship to the physical sciences further underline the importance *Nature*'s contributors placed on research. They also reveal another necessary qualification for a man of science: he had to pursue and value theoretical knowledge. When *Nature* was founded in 1869, British engineers had managed to carve out a reasonably secure place in Britain's social hierarchy; they were praised for their work constructing railroads and factories, and several had been knighted or had become wealthy from their work.[43] But in the 1840s and 1850s, some British engineers, most notably the railroad engineer I. K. Brunel, began pushing to create a more standardized curriculum for training engineers and to enhance the theoretical components of an engineer's education. By the 1870s engineers interested in theory, such as William J. M. Rankine and Fleeming Jenkin, had successfully convinced several universities to award academic degrees in engineering. Academic engineers began trying to convince "practical" engineers of the value of "scientific" engineering.

The discussion about practical and scientific engineering found its way to the pages of *Nature*, and *Nature*'s contributors were not shy about choosing a side.[44] In the 1880s and 1890s, *Nature* ran several articles that expressed skepticism, even disdain, for engineers who relied on practical experience rather than theoretical knowledge. A particularly striking example is an editorial from April 1889, which began,

At the last meeting of the British Association an energetic attempt was made to prove that the progress of the human race has been chiefly due to the "practical man," and this teaching was quickly caught up and explained to mean that the triumphs of industry have been achieved without the help of workers in the field of pure science. We have before us a periodical which is instructive reading when viewed in the light of the discussion on this subject. It is a recently issued number of the Transactions of an Institute connected with one of the most important of our national industries.[45]

The article went on to point out several mistakes in the mathematics and unit conversions used in the unnamed *Transactions*. This would not be a matter of concern if the practical men stuck to practical matters, said the anonymous author, but when these practical men "print their opinions on 'Sir William Thompson's' (sic) address to the Institute of Electrical Engineers, the matter becomes serious." According to the author, "practical" men were too ignorant of basic physics to have any right to comment on the work of men like Sir William Thomson (whose name they could not even spell correctly). Furthermore, the claim that "practical men" were the ones responsible for recent advances in industry was borderline ridiculous. Some engineers had contributed to such advances, "but these ranked among them not because they were practical men who did not 'want to know what electricity is,' but because they had risen above such wretched cant, and become not only 'practical' but scientific."[46]

The clash over whether theoretical or practical knowledge was superior frequently manifested itself in the form of arguments over physics terminology and units. In 1887, for example, P. G. Tait found himself in conflict with another periodical, the *Engineer*, over a negative review Tait had given a volume on engineering in a previous issue of *Nature*. Tait had criticized the book for dividing foot-pounds per minute by foot-pounds and giving an answer in horsepower (i.e., foot-pounds per minute). When the *Engineer* attacked his criticisms, Tait insisted that his remarks were not "the pedantry of the 'professor'" but a serious objection that cast doubt on the author's competence. Tait presented the following challenge to the writers in the *Engineer*: "I wonder what the *Engineer* would assign as the result of dividing 10 eggs per minute by 2 eggs. Would it, or would it not, be 5 eggs per minute?"[47]

Another argument over units and terminology began with an 1889 letter from the engineering professor A. M. Worthington, who wrote to *Nature* to suggest a new way of using the terms *mass* and *inertia*. Worthington argued that physicists and physical textbooks often used the word *mass* in two different senses—one to mean "a lump of matter" and the other to mean "inertia." Worthington regarded this as unacceptably confusing for his engineering students and blamed "scientific men" for the confusion.[48] The debate in *Nature* over Worthington's proposals lasted a full two months and included several swipes at engineers and "practical men." Oliver Lodge wrote that Worthington had no one to blame but himself if the terms *mass* and *inertia* were unclear to his students; he should, said Lodge, employ the more precise term *coefficient of inertia* instead of *inertia* when teaching introductory courses.[49] Andrew Gray, a professor of physics in the University College of North Wales (and a former assistant to William Thomson), also did not care for Worthington's proposals and said that students only needed to be "properly taught" in order to understand scientific units.[50]

Worthington was not without his supporters. The mathematician and engineer A. G. Greenhill, a professor at the Royal Artillery Officers academy, agreed that men of science, not engineers, had been the ones to muddle the definitions of various physical terms.[51] But notably, although there was some defense of "practical men" in *Nature*, the majority of the contributors who wrote in on this subject defended theoretical knowledge, indicating that most of *Nature*'s contributors did not consider "practical" knowledge sufficient to be considered an expert on physics.

Overall, the discussions in *Nature* about men of letters, women, politicians, psychical research, and engineers suggest that it was not necessarily the subject on which a contributor wrote that determined whether other contributors would respect his or her work. A former BA president such as the Duke of Argyll could write to *Nature* on a perfectly respectable topic such as evolutionary theory, but his identity as a politician and the fact that he had not performed original investigations recently meant that other contributors rejected his attempts to participate in the scientific discussion. Engineers who had a thorough knowledge of the practical applications of physics were not considered qualified to write about William Thomson's work or offer suggestions on how to teach physics because they were not interested in theoretical knowledge. In contrast, William Crookes, Oliver Lodge, and other psychical researchers contributed pieces to *Nature* on a topic that many of their fellow *Nature* contributors considered obviously unscientific, but their psychical interests did not prevent them from being considered eminent physicists. At a moment when the identity of the man of science was still being negotiated in Great Britain, *Nature*'s contributors used the journal as a forum to discuss the appropriate qualifications for a British man of science. Most of them agreed that the essential requirement for being a competent participant in a scientific discussion was having conducted original scientific investigations.

Arguments about who was entitled to call himself a man of science in Great Britain have traditionally been linked to the "professionalization" of science in the nineteenth century.[52] Thus far, I have deliberately avoided using the word *professionalization* because recent scholarship has suggested that the term is not entirely appropriate in the context of nineteenth-century Britain. In his recent biography of Joseph Hooker, Jim Endersby argued that *professionalization* is a problematic term because nineteenth-century men of science in Britain did not see themselves as "professionals." Endersby observes that even late in the nineteenth century, it is difficult to discuss who was a scientific "professional" and who was not because *professional* was not a term men of science used to describe themselves. In fact, *professional* carried negative connotations in a society where gentility and social respectability were often considered incompatible with working for a living.[53] Similarly, Ruth Barton has shown that the essential distinction between competent and incompetent participants in scientific discourse through the 1870s was not whether they were "professional" (i.e., whether they were able to make a living from their scientific work); rather, men of science fashioned their identity based on moral and intellectual qualities. In order to be a man of science, it was not necessary to earn money for scientific work, or even to perform scientific work full time—the essential qualification was the desire to pursue scientific truth. British men of science sought to justify their claims to moral and intellectual authority by fashioning an image as servants of the public good who cared only for the improvement of knowledge.[54]

Professionalization seems to be the wrong term to describe what was happening to science in nineteenth-century Britain, but as Argyll's example shows, the nineteenth century saw major changes in the way British scientific practitioners thought about their community and who was qualified for membership in it. *Nature*'s content gives us a window onto the changes that were taking place and how contemporaries viewed them. Men of science in Britain did not wish to become or be seen as "professionals," but they used other criteria to define who was and was not a true man of science. The material from *Nature* strongly supports Barton's point about the importance of the man of science's moral identity; the discussion of men of science versus politicians is particularly illuminating on this point.

The examination of *Nature* also indicates that the intellectual qualifications necessary to be considered a man of science were growing more stringent in the late nineteenth century. The litmus test for whether or not someone was treated as a man of science in *Nature* was not merely whether he sought to know scientific truths (which, as Barton shows, was the case in the 1850s) but whether or not he devoted his time to original scientific investigations and made contributions to scientific knowledge. If a contributor had acquired his scientific knowledge solely through reading others' scientific writings—as Argyll had in the coral reef debate—*Nature* might still print his letters and essays, but the researchers who read *Nature* would soon write in to question his qualifications. An original investigator, on the other hand, was clearly a man of science even if he occasionally pursued projects that some considered outside the bounds of science. *Nature*'s contributors sought to establish science as an exclusive and demanding pursuit in which only a limited number of devoted individuals attained expertise, not as a casual intellectual endeavor that welcomed experts and laymen alike.

RICHARD GREGORY, SPOKESMAN OF SCIENCE

At the very moment when *Nature*'s contributors were using the journal to sharpen the boundaries of their scientific community, their publication was increasingly being managed by someone who met almost none of their criteria for a man of science: Richard Arman Gregory (1864–1959).[55] Gregory was the son of a Bristol poet and literary man named John Gregory. The Gregory family was respectable but not affluent, and at the age of fifteen Richard was apprenticed to the Bristol shoemaker James Parson. The young man did not much like making shoes, but he was a voracious reader, and during his apprenticeship he began attending classes at the Bristol Trade School. At nineteen, Gregory paid Parson to be released from the final year of his apprenticeship and took a position as a laboratory assistant. Two years later, Gregory won a scholarship through the Bristol Trade School to attend the Royal College of Science at South Kensington as a teacher-in-training. The affable, charming Gregory made friends easily in London, among them a fellow student named Herbert George Wells, who would remain one of Gregory's closest friends throughout his life.[56] After two years at South Kensington, Gregory married Kate Florence, a widow with two young children. In 1889, Gregory learned that Sir Norman Lockyer was looking for an astronomical assistant to work in his London observatory. Gregory obtained the position and moved his family to London.

Like Lockyer thirty years earlier, the young Gregory was perpetually juggling several positions in order to support his family. In the four years he worked at Lockyer's observatory, Gregory was also a lecturer for the Oxford University Extension Delegacy and marked examinations for the University Correspondence College. His real passion, one he discovered while working for Lockyer, was for science writing. Gregory was a regular contributor to numerous literary periodicals, most notably the *Leisure Hour,* for which he wrote a signed column called "Science and Discovery." He also contributed to the *Fortnightly Review,* the *Academy* and even *Nature*'s rival *Knowledge.* In the early 1890s Gregory was able to turn his lectures for the Oxford extension program into a series of textbooks for the London publisher Joseph Hughes. Between 1891 and 1893, Gregory wrote or cowrote five textbooks on science (one of which, *Honours Physiography,* was coauthored with H. G. Wells).[57] Gregory also wrote several articles for *Nature* on behalf of Lockyer's observatory.[58]

In 1893, Lockyer asked Gregory to leave his assistantship at the observatory and join the staff at *Nature.* Gregory welcomed the relatively well-paying position at *Nature,* which enabled him to give up marking exams and focus on lecturing, editing, and science writing. He also welcomed the closer contact with Macmillan and Company. His relationship with the publishing house would be one of Gregory's most important assets over the course of his unusual career. Shortly after Gregory began working at *Nature,* Macmillan took over publication of Gregory's textbooks. He succeeded Lockyer as the publishing house's science editor in 1905.[59]

Gregory seems to have kept a relatively low editorial profile while Lockyer was still at the journal. In the years when he was technically Lockyer's editorial assistant, Gregory—perhaps with Lockyer's encouragement—was careful not to let his editorial fingerprints show too clearly. Evidence of Gregory's low-profile editorial strategy can be seen in the journal's treatment of female scientific workers in Britain. As noted earlier, Lockyer was known for his support of women's rights. During Lockyer's editorial tenure, *Nature*'s columns and editorials often spoke in favor of increasing the status of women within the scientific community, as we saw in the debate about fellowship in the Chemical Society.[60]

Gregory, in contrast, was no friend to women in science. He appears to have shared Thomas Huxley's belief that women were too emotional and superstitious to comprehend higher scientific truths. For example, in two *Nature* reviews of books by Agnes M. Clerke, a popularizer of astronomy, Gregory argued that Clerke's sex prevented her from truly understanding her subject. "A cynic has said that it is a characteristic of women to make rash assertions, and in the absence of contradiction to accept them as true. Miss Clerke is apparently not free from this weakness of her sex," he wrote in 1903.[61] A 1906 review of Clerke's book *The System of the Stars* made Gregory's views about the female intellect even clearer:

The intuitive instinct of a woman is a safer guide to follow than her reasoning faculties; and although in these days it is considered ungracious to make this suggestion, evidence of its truth is not difficult to discover in most literary products of the feminine mind. It is no disparagement to Miss Clerke to say that even she shares this characteristic of her sex, so that sometimes she lets her sympathies limit her range of vision in the field of stellar research.[62]

Gregory was also skeptical of female researchers. In 1923, four years after he assumed the official editorship, Gregory approved the publication of an obituary of the physicist Hertha Ayrton. The obituary's author, Gregory's friend Henry Armstrong, suggested that Mrs. Ayrton owed her scientific success entirely to her indulgent husband, the physicist William Edward Ayrton. "I never saw reason to believe that she was original in any special degree," wrote Armstrong; "indeed, I always thought that she was far more subject to her husband's lead than he or she imagined."[63] Ayrton's daughter, Barbara Ayrton Gould, wrote a lengthy letter to the editor castigating Armstrong for numerous factual errors and for casting doubt on her mother's accomplishments. Gregory dismissively replied that "everyone who knows Professor Armstrong will realise that it was characteristic of him and that, therefore, much of what was said in the article was of the nature of mild chaff, and should not be taken too seriously." He declined to print Mrs. Gould's letter in the journal.[64]

But in spite of Gregory's own disdain for women in science, *Nature* in the early twentieth century mostly continued to be a source of support for women seeking advancement in the sciences. This support suggests that Gregory kept his own beliefs out of such discussions, instead deferring to Lockyer's well-known sympathies and to *Nature*'s history of supporting female scientific workers. Notably, Gregory only signed his name to negative statements about a female science *popularizer*. Criticizing a fellow science writer was one thing. Criticizing a scientific worker, when Gregory possessed scant claim to any scientific authority of his own, would have been quite another.

Even for his closest friend, H. G. Wells, Gregory could not—or perhaps would not—override Lockyer's established editorial practices. Gregory admired Wells's novels and essays, and he frequently secured Wells's books prominent reviews in *Nature*. (One famous review of Wells's 1926 novel *The World of William Clissold* took up 11 columns, an unprecedented length for any *Nature* review.)[65] But in 1904 Wells published a set of essays titled *Mankind in the Making*, in which he discussed educational theories and influences, and a book called *Aspirations*, which outlined a vision of a utopian society in which Darwinian evolution had perfected mankind. The *Nature* contributor who reviewed these books, F. W. H., was unimpressed; he argued that Wells's imagination was "apt to run away with him" and that Wells did not fully grasp Darwin's theories.[66]

Wells's irate reply to F. W. H.'s review did not reach *Nature*'s pages as quickly as most contributions. When Wells wrote to Gregory to complain that his letter had not yet been printed, Gregory said his hands were tied:

> My dear H.G.,
> It is the invariable rule of "Nature" to send letters referring to reviews to the reviewers, so that the letter and the reply may appear together. Unfortunately, your reviewer is in Algeria now & a copy of your letter has had to be sent to him there. I have asked the Editor whether in these circumstances he will print your letter now, but he has just replied that we must await the reply of the reviewer as usual. I am very sorry, but I am quite helpless in the matter.[67]

In the absence of an editorial archive, it is difficult to determine how much control Lockyer ceded to Gregory or when any official transfers of responsibility occurred. But it appears that during his tenure as assistant editor, Gregory deferred to Lockyer's established practices. Moreover, when Gregory's influence was visible, it was restricted to areas where he could securely claim authority. While he could and did use *Nature* to criticize the British government for inadequate support of science, or to call out a popularizer he felt was not qualified to write about science, Gregory never interceded in scientific discussions or criticized men (or women) of science for their research. Scientific workers could work out their own theoretical arguments on the pages of *Nature*; Gregory's job was to take the message of science's importance to the British public.

COULD A POPULARIZER EDIT *NATURE*?

By 1918, Lockyer was eighty-two, and his health was wavering. He and Frederick Macmillan both knew it would soon be time to choose a successor, and Gregory was the obvious man—after all, he had already succeeded Lockyer elsewhere at Macmillan. But while Lockyer valued Gregory's work, an October 1918 letter to Macmillan made it clear that he had misgivings about the prospect of Gregory as the editor of *Nature*.

> My dear Fred,
> Since I received your letter Mr. Gregory, who has been spending some part of his holiday in the South, has been to see me & I have had a general talk with him about Nature affairs. . . .

> In our talk, to which I have already referred, we discussed the possibility of my successor as Editor being preferably a Fellow of the Royal Society. But bearing the recent experiences in mind I do not think now we could do better than entrust the work to him although he is not a Fellow.
>
> Fellows of the Royal Society are very difficult to lead & very few of them have the slightest idea of administrative work or possess business capacity.[68]

But a few weeks later, Lockyer told Macmillan that he thought it best to postpone a firm decision about his successor.

> In your absence I could not say anything very definite to Gregory about changes of condition, but I gathered from him that he is quite content with things as they are & does not at present desire any change. So I think that we might perhaps agree to postpone action for a few months when the war will be over, a condition which we contemplated in our talk, & I shall have been Editor for half a century.[69]

The problem was obvious: Gregory was not an FRS and therefore lacked the scientific prestige Lockyer would have liked to see in his successor. And Gregory was extremely unlikely to win the FRS distinction because he had never performed original scientific work. Lockyer seems to have worried that Gregory's lack of research credentials might lead *Nature* contributors to question whether he was the right man to choose what was printed in Britain's most important scientific weekly.

Gregory, however, was able to construct a role for himself within this exclusive community: he cast himself as science's spokesman, an articulate and charming emissary to the general public. He was skilled at cultivating scientific allies to aid his efforts. For instance, Gregory had a longstanding interest in educational reform, and in the late 1890s he set his sights on bringing educational theory to the British Association for the Advancement of Science. He recruited the chemist and educational reformer Henry Armstrong (who would later author the Ayrton obituary) to support his efforts to create a separate section for the discussion of scientific education before university.[70] In 1901, in the midst of a heated discussion about educational reform in Great Britain, Gregory and Armstrong convinced the BA council to create a section for Educational Science. Section L met for the first time in 1902 with Gregory as its secretary. Gregory soon became a gregarious fixture of BA meetings; he would not miss another until the mid-1950s.

Gregory found further opportunity to interact with prominent men of science through an organization called the British Science Guild. In 1905, Lockyer founded the British Science Guild with the aim of educating Britons on the importance and usefulness of science (the goal he had once had for *Nature*). Lockyer and his allies, among them Raphael Meldola, Sir William Ramsay, W. E. Ayrton, and Sir John Lubbock, sought to convince the British government to apply scientific methods to government and industry.[71] Gregory was an extremely active member of the British Science Guild, especially after Lockyer stepped down from its management in 1914. Gregory, often speaking on behalf of the British Science Guild, was a highly visible participant in the famous "neglect of science" debate during the First World War, in which politicians, scientists, and other social commentators discussed whether Britain's ill fortunes in the war were a result of neglecting scientific research. Unsurprisingly, Gregory answered the question in the affirmative.[72]

Gregory's greatest public fame, however, came from his 1916 book *Discovery; or, the Spirit and Service of Science*. The volume was a tremendous success for Macmillan and Company and went through 12 printings in 10 years. *Discovery* embodied everything Gregory wished to tell the British public about science and the researchers who devoted their lives to its pursuit. It combined quotations from famous researchers such as Michael Faraday, Lord Kelvin, and Huxley with stories of scientific greatness, praise of technological advances, and lengthy passages describing the extraordinary intellectual and moral qualities necessary to pursue scientific study. Gregory's frequent use of religious imagery in the volume underlined his argument that science was a noble calling that required special moral qualifications:

> Scientific truth is not won by prayer and fasting, but by patient observation and persistent inquiry. Nature, like the rich man of the parable, requires importunate pleading before she will bestow any of her riches upon a suppliant at her temple. . . . It is necessary to believe in the holiness of scientific work in order to persevere to the end; for without the encouragement which such belief gives, many investigators would fall by the wayside. But no man of science who has put his hand to the plough of research ever turns back.[73]

The man of science's distinguishing characteristic, wrote Gregory, was a "love of truth" that led him to devote his life to science. Furthermore, men of science had provided the world with many of the modern advances it now enjoyed. And yet, said Gregory, this "holy" undertaking was "deplorably neglected" among modern men. He criticized "literary men" for consistent errors when discussing scientific subjects and said that most journalists had "no acquaintance with the most elementary vocabulary of science." This problem, Gregory said, would only be solved "when it is realized that an educated man must know something of science as well as of literature."[74]

In short, *Discovery* sought to tell the public what many contributors to *Nature* had been saying for years: that scientific workers were a moral and noble group privy to a special body of knowledge; that scientific advances led to greater quality of life; and that scientific researchers ought to be accorded a more prominent place in British society. The book launched Gregory's career as a public spokesman for science, and he would soon become one of Britain's most widely quoted figures on scientific matters.

SIR RICHARD GREGORY FRS, EDITOR OF *NATURE*

Toward the end of the First World War, Macmillan and Company was coping with rapidly rising paper prices and decided to compensate by increasing *Nature*'s subscription price. Notably, one of the first people Maurice Macmillan notified about the change was Gregory, who replied by saying that he was "not at all surprised.... I am a little anxious about the effect of the increase upon the sales but I hope subscribers & others will recognise that the action was inevitable."[75] Gregory was clearly a trusted member of *Nature*'s editorial team. His public profile was increasing as well. In 1919, Gregory was knighted in recognition of his work organizing British industrial exhibitions. He was also elected to the Athenaeum that same year, and his books (in particular, *Discovery*) continued to sell extremely well.

In spite of Gregory's established usefulness at *Nature*, his strong relationship with Macmillan and Company, and his growing public fame, the correspondence between Lockyer and Frederick Macmillan shows that there was some doubt as to whether Gregory would be an appropriate successor for the ailing Lockyer. The standards *Nature* contributors had set for a man of science were ones Gregory could not meet. Lockyer was not entirely comfortable with the idea of leaving a non-FRS in charge of his journal, however qualified Gregory might be as an administrator.

Interestingly, Lockyer had some concerns about appointing an FRS as well. Lockyer wrote that an FRS would be "very difficult to lead," suggesting that the Macmillans might have trouble working with a new editor who was also an FRS. Furthermore, Lockyer indicated that an FRS was unlikely to have "the slightest idea of administrative work or possess business capacity." In many ways, Lockyer's comments about the potential difficulties of choosing an FRS were an outgrowth of the narrowing definition of a *man of science*. According to the criteria expressed in *Nature*, a man of science, the kind of man who would have been elected as an FRS, had to be a researcher.

FIGURE 5 Sir Richard Gregory. Image held by the University of Sussex Library. © National Portrait Gallery.

A time-consuming editorship of a weekly journal was almost certainly incompatible with conducting significant research, and it seems unlikely that the Macmillans could have persuaded someone at the height of his scientific power to give up his laboratory or his theoretical work in order to select each week's Letters to the Editor. The Macmillans could have chosen an older FRS at the end of his research career, but there would have been little point in having the elderly Lockyer step down in favor of a man who was only slightly younger. Furthermore, because science was a much more specialized pursuit in 1919 than it had been in 1869, *Nature* was unlikely to find a new editor who was both an FRS and a man with knowledge of business and administration.

Business considerations ultimately trumped the question of scientific prestige. When Lockyer retired in November 1919, fifty years after the publication of *Nature*'s first issue, Gregory assumed the editorship. Notably, while *Nature*'s jubilee issue included an essay by Gregory on "The Promotion of Research," the journal never specified the identity of Lockyer's successor.[76] Once he had inconspicuously assumed the editorship, Gregory continued to maintain his image as impartial advocate for science, rarely venturing into any editorial decisions the contributors might have considered questionable (with the notable exception of the Ayrton obituary).

This is not to say that *Nature* underwent no changes after Gregory officially assumed the editorship. Most changes were minor. For example, Gregory switched *Nature*'s publication day from Thursday to Saturday on 1 April 1922. He also shifted the journal's table of contents from the last page to the first page of each issue, changed some of the typefaces, moved all obituaries to a designated Obituary column (rather than printing long articles on prominent deaths and shorter notices in the Notes column about less famous men and women), and renamed the journal's Notes column "Current Topics and Events." He later changed the column's title again, to "News and Views."

Other changes, however, reflected Gregory's ambitions for both *Nature* and for his own career. The journal began addressing itself much more directly to British political leaders. Lead editorials, which had been published every four or five issues under Lockyer, were printed in nearly every issue once Gregory assumed editorial control.[77] These editorials were usually on topics such as improving connections between science and industry, the best way for scientists and the government to organize British scientific work, or ways in which science could improve the administration of British colonies. All of these topics invoked one of Gregory's favorite themes: science was a boon to British society and should have more support from British citizens, especially the British government. Gregory also increased the number of letters to the editor, in part to accommodate increased correspondence about the more politically charged editorials, in part to accommodate the column's new popularity for announcing preliminary research results (see chap. 4). In 1913, there were an average of four letters to the editor per issue; by 1920, the average had jumped to eight.[78]

Gregory also saw his editorship of *Nature* as a way to enhance his credibility as a spokesman for science. During the early years of his editorship, he kept a slightly lower public profile, but by the late 1920s, Gregory began asserting himself more as a public authority on science. He was quoted frequently in articles for a wide range of publications, including the *Times* and the *Guardian*. Early in his editorship, articles quoting Gregory rarely identified him as the editor of *Nature*, instead using one of his other titles—such as president-elect of the South-Eastern Union of Scientific Societies—or simply describing him (somewhat inaccurately) as "an eminent scientist."[79] By the 1930s, however, Gregory was consistently identified as the editor of *Nature* when newspapers sought his commentary on scientific issues.[80] Prime Minister James Ramsay MacDonald even made Gregory a baronet in 1931, perhaps hoping to indicate the Labour Party's support of Britain's scientific research. Gregory's personal coat of arms included, at his direction, the image of *Nature*'s masthead.[81]

Gregory's carefully crafted image and his tireless advocacy for increased social support of science eventually won him a unique honor: in 1933, Gregory was elected an FRS under a special regulation. The Royal Society charter contained 11 statutes governing the election of those who contributed to the advancement of scientific knowledge. Statute 12 created an additional category of members: persons who "have rendered conspicuous service to the cause of science."[82] Gregory became only the eleventh man elected to the Royal Society under Statute 12. Naturally, congratulatory letters followed, including one from a friend at King's College London, the botanist and eugenicist Reginald Ruggles Gates.

> Dear Gregory:
> I was delighted to hear the announcement at the Royal Society yesterday afternoon that you had been elected a Fellow under the special rule. Your editorship of <u>Nature</u> alone is quite sufficient for this. It is not only the leading scientific journal in the world, but I feel that its policy in recent years of emphasizing the need for scientific men to take a leading part in solving the social and international development problems arising from the development of science is of great value for the future.[83]

In Gates's estimation, *Nature* was by now of such significance that the act of running the journal was in itself a service to the cause of science in Great Britain. But could *Nature* make a claim to be the world's leading scientific journal, not just the most prominent British one?

CHAPTER FOUR

Scientific Internationalism and Scientific Nationalism

In the summer of 1910, the great physicist Ernest Rutherford was preparing for an international scientific congress in Brussels that would establish a standard unit of radiation. Rutherford's peers had chosen him as the chairman of the Radium Standards Committee, and he already had the size and name of the unit in mind: he wanted the unit to be the number of decays per second in 10^{-8} grams of radium and was pushing to call it the "curie" in honor of Pierre and Marie Curie. Rutherford hoped that his committee would accomplish more if he brought his most prominent colleagues to an agreement on the unit before the congress began. In his quest to establish the curie, Rutherford exchanged letters with researchers from across the Western world, including Stefan Meyer in Vienna, Otto Hahn in Berlin, Arthur S. Eve in Montreal, and Marie Curie in Paris, all of whom he saw in Brussels that September.[1] He also discussed the matter frequently with his friend Bertram Borden Boltwood, an American from Yale University who was spending a year with Rutherford at Manchester and who was also a delegate to the Brussels conference.[2]

Rutherford's multinational correspondence about the curie was not a mere show put on to demonstrate an ideal of scientific internationalism. It was an accurate reflection of the history and status of his field. Discoveries in the early history of radioactivity research came from every major scientific center in Europe, and in 1910 important work on radioactivity was being performed in Paris, Berlin, Vienna, London, Cambridge, Montreal, and New Haven, Connecticut. Radioactivity was, unquestionably, an international undertaking.[3]

In contrast, the *Nature* we have seen thus far was unquestionably not an international undertaking. Although a handful of foreign scientists did write articles or letters for *Nature*, and while the journal could prove useful to men such as the American astronomer Henry Draper as a source of abstracts, *Nature* in the nineteenth century was a publication by and for British men of science.[4] Its contents were dictated by the concerns of the British scientific community. And yet one of *Nature*'s most prolific early twentieth-century contributors was Rutherford, who wrote dozens of letters to the editor detailing his latest work. Why would Rutherford, who corresponded regularly with colleagues from across the globe and was on the board of two foreign radioactivity journals (*Le Radium* and *Jahrbuch der Radioaktivität und Elektronik*), choose to publish in the heavily British *Nature*?

In fact, *Nature* played a major role in spreading news of the latest research in the international science of radioactivity. Much of this was due to Rutherford, who chose *Nature* as a venue for some of his most important early work both because of its rapid publication schedule and in order to further his goal of obtaining a position in the United Kingdom. Rutherford's talent for attracting promising foreign physicists to his laboratory led scientists such as Otto Hahn and Bertram Boltwood to follow his example and publish their latest work in *Nature* as well. However, in another similarly international field, Mendelian genetics, *Nature* did not attract nearly as many international contributions; the scientists who published in *Nature* on the subject of heredity were almost entirely British. Evaluating *Nature*'s role in these two fields shows that even during an era of increasing scientific internationalism, and even in fields as international as Mendelian genetics and radioactivity, a scientific worker's national origin still shaped his or her publication strategies.

RÖNTGEN RAYS, URANIUM RAYS, AND RADIOACTIVITY

Several important developments in the physical sciences give us the opportunity to examine *Nature*'s place in British scientific publishing at the end of the nineteenth century. In 1895, the German physicist Wilhelm Conrad Röntgen (1845–1923) noticed an interesting phenomenon while experimenting with a vacuum discharge tube (an evacuated glass vessel in which metal electrodes have been sealed): when he placed his hand between the tube and a screen coated with barium platinocyanide, the darkened image of the bones in his hand appeared on the screen. It quickly became apparent that Röntgen had discovered a new kind of wave, and "Röntgen rays" became a scientific and popular sensation.[5] (Most Anglophone scientists eventually adopted Röntgen's preferred name for his discovery, x-rays.)

One of the many scientists inspired to study Röntgen's new phenomenon was Henri Becquerel (1852-1908), a member of the French Académie des sciences and a professor at the prestigious École Polytechnique in Paris. Becquerel was interested in whether naturally phosphorescent minerals might also produce X-rays or other unknown rays. In March 1896 he reported an unusual finding to the Académie: one night, he had placed uranyl-potassium sulfate (a salt of uranium that phosphoresces when exposed to sunlight) in a drawer with wrapped photographic plates, and the next morning, a silhouetted image of the salts had developed on the plates. Subsequent experiments had revealed that the salts also developed photographic plates even when they had not been exposed to sunlight—the production of "uranium rays" (or, as some scientists called them, "Becquerel rays") was not linked to the salt's phosphorescence at all.

Marie Skłodowska Curie (1867-1934), working in her husband Pierre's laboratory at the École Municipale de Physique in Paris, took up the study of Becquerel's uranium rays. Curie was interested in finding other materials that might emit similar rays. She soon discovered that there were several materials—most famously, pitchblende—that not only emitted Becquerel's "uranium rays" but emitted them much more strongly than uranium salts. Curie adopted the term *radioactivity* instead of "uranium rays" to describe the phenomenon that she was studying. In 1898 the Curies and the chemist Gustave Bémont announced the discovery of two new elements, polonium (named for Marie Curie's native Poland) and radium, both of which were hundreds of times more radioactive than uranium.[6]

Meanwhile, across the English Channel research into x-rays and radioactivity was also taking place in Cambridge's Cavendish Laboratory. J. J. Thomson (1856-1940), the laboratory's head, was personally more interested in electricity, but he encouraged Ernest Rutherford (1871-1937), who had recently come to the Cavendish from New Zealand, to study Becquerel's uranium rays. In 1898, Rutherford reported that there were two distinct varieties of uranium rays, which he called "alpha" and "beta." Alpha rays were positively charged and readily absorbed by most substances, but beta rays were negatively charged and could pass through metal without being hindered. Negatively charged beta rays were quickly identified as electrons; the nature of alpha particles was less clear. In 1900, a colleague of the Curies' named Paul Villard (1860-1934) discovered a third type of radiation, "gamma," which was even more penetrating than beta radiation but carried no charge.

In 1898, Rutherford accepted the Macdonald Chair of Physics at McGill

University in Montreal, Canada. Two years later, the Oxford-educated chemist Frederick Soddy (1877-1956) accepted a position as a demonstrator in Rutherford's department, and the two undertook an extremely fruitful collaboration on the study of thorium radioactivity.[7] In 1902-1903, Rutherford and Soddy embarked on study of alpha radiation and discovered that alpha particles had a 1:1 mass-to-charge ratio. The two concluded that they were either hydrogen ions with a +1 charge or helium ions with a +2 charge.

Their theory that alpha particles had an elemental identity of their own led Rutherford and Soddy to suggest that radioactivity was the result of atomic disintegration and that radioactive atoms released matter as alpha, beta, and gamma rays. As a result, radioactive elements changed their elemental identity.[8] The idea met with some resistance; many physicists, including the eminent William Thomson (now Lord Kelvin) and Dmitri Mendeleev, the creator of the periodic table, dismissively equated Rutherford and Soddy's theory with the old alchemical idea of elemental transmutation.[9] But J. J. Thomson and Marie Curie both came to agree that radiation was an emission of matter accompanied by a loss of weight in the radioactive substance. Following a 1903 experiment by Pierre Curie and Albert Laborde, which found that one gram of radium could heat 1.33 grams of water from the melting point to the boiling point in one hour, J. J. Thomson, the Curies, and others argued that the emission of matter must also be accompanied by an emission of energy.[10]

Rutherford and Soddy's ideas about radioactive change received a boost later that year when Soddy, who had moved to University College London and was collaborating with Sir William Ramsay (1852-1916), performed a spectroscopic study of the emanations from radium salts. He detected helium, lending experimental weight to the theory that radiation was an emission of matter. In 1907-1908, Rutherford and his colleagues in Montreal undertook a spectroscopic study of alpha rays and were able to show that alpha rays were composed of helium particles.

Nature played a significant role in this remarkable thirteen years (1895-1908) of physics. Röntgen's discovery made the news in many English-language newspapers and journals, and *Nature* was no exception. The Notes column of 16 January 1896 mentioned Röntgen's findings and announced that Röntgen had used his new waves to obtain pictures "showing only the bones of living persons." The anonymous *Nature* staffer predicted that "The scientific world will look forward with interest to the publication of the details of Prof. Röntgen's work."[11] *Nature*'s very next issue, published on January 23, prominently featured the new discovery. The physicist Arthur Schuster wrote to the editor to urge his fellow physicists not to discard the idea that Röntgen rays might be an unusual manifestation of light waves, a conclusion Röntgen's paper had rejected. J. T. Bottomley's letter noted that Röntgen's paper had concluded with the speculation that his rays might be longitudinal vibrations in the luminiferous ether and called attention to a passage in Lord Kelvin's Baltimore lectures that seemed to anticipate the discovery of such a wave. Most famously, this issue of *Nature* printed the first English-language version of Röntgen's paper, translated by the Manchester physicist Arthur Stanton.[12] The electrical engineer A. A. C. Swinton supplemented Stanton's translation with an article in which he reported that he had "repeated many of Prof. Röntgen's experiments with entire success." Swinton's article also included the first x-ray photograph taken in England.[13]

These letters and articles were the first in a rapid flood of pieces on the new phenomenon. In 1896, there were 139 articles in *Nature* that mentioned Röntgen rays, an average of nearly 3 per week. In contrast, the *Philosophical Magazine* contained 8 articles about Röntgen rays that year; *Proceedings of the Royal Society of London* ran 12 articles on the rays in 1896–1897; *Philosophical Transactions of the Royal Society* did not run any articles on Röntgen rays until 1897. Of course, the comparison is not entirely fair. *Philosophical Magazine*, *Proceedings*, and *Philosophical Transactions* ran far fewer articles than *Nature* and had a much longer delay between submission and publication. A more direct comparison can be drawn between *Nature* and two other British scientific weeklies, *Chemical News* and the *Electrician*. In 1896, *Chemical News* mentioned Röntgen rays 28 times (an average of once every other week), the *Electrician* 86 (an average of 1.65 mentions per week).

Nature thus stands out for the sheer amount of its Röntgen ray coverage, but *Nature*'s material about Röntgen rays was not necessarily unique among scientific weeklies. The pieces in *Nature* included abstracts from other journals (usually foreign ones), reports of lectures about the rays, theoretical speculations like Bottomley's letter, summary articles written by *Nature*'s staff, and original reports of experimental results like Swinton's. *Chemical News* and the *Electrician* both published material on Röntgen rays similar to *Nature*'s. *Chemical News*'s coverage relied heavily on reprints of articles published elsewhere or summaries of foreign research, but a handful of researchers did send preliminary experimental results to *Chemical News* or submit letters about the rays to *Chemical News*'s occasional correspondence columns.[14] Similarly, the *Electrician* included a large number of abstracts of foreign articles on Röntgen rays (most frequently from the *Comptes Rendus*) as well as summary articles about the current state of Röntgen ray research and reports on lectures about the rays. The *Electrician* also published correspondence discussing the nature of the rays and a few original research articles about Röntgen ray experiments.[15]

The data from both the weekly and the monthly or quarterly publications underscore two important points: first, in 1896 *Nature* was not the only British scientific weekly where a researcher could or would submit interesting new research results, and second, scientific weeklies played a unique role in researchers' publishing strategies at the end of the nineteenth century by offering researchers a forum where short articles could be printed quickly. *Nature*'s closest monthly competitor for these pieces was the *Philosophical Magazine*, whose Intelligence and Miscellaneous Articles section contained short pieces similar to the articles in weeklies. But the *Philosophical Magazine*'s monthly publication schedule usually meant a longer wait time between submission and publication than the wait for the weeklies. A few researchers sent the *Philosophical Magazine* pieces on Röntgen rays for Intelligence and Miscellaneous Articles in 1896, for example, but none of these pieces appeared before April.[16] This suggests that specialist weeklies were able to capitalize on the intense interest in Röntgen's discovery by offering researchers a forum where preliminary observations and theories about the nature of the rays could reach an audience of scientific specialists within a week of submission.

When compared with the avalanche of research that followed the discovery of Röntgen rays, British periodicals' response to Becquerel's uranium rays was quite mild. Becquerel's discovery was mentioned several times in *Nature*'s Notes and in the Societies and Academies column in 1896, and J. J. Thomson devoted an article to a discussion of Becquerel's experiment and how it might cast light on the nature of Röntgen rays, but Becquerel rays do not appear to have excited a Röntgen-like frenzy among *Nature*'s readers or contributors.[17] Similarly, the *Electrician* mentioned Becquerel's discovery only once in 1896, in a short note about recent experiments related to Röntgen rays.[18] *Chemical News* was more enthusiastic; Crookes and his team reprinted the full version of Becquerel's original paper on "uranium rays" and also abstracted his subsequent articles on the rays.[19]

It might be tempting to accuse British publications of ignoring Becquerel (with the exception of *Chemical News*), but in fact the limited coverage of "Becquerel rays" was typical for the physics community at the time. Initially, Becquerel's rays appeared to be an odd phenomenon confined to uranium salts with no useful application.[20] In 1896 the Académie des sciences heard approximately 100 papers about Röntgen rays but only three about Becquerel rays.[21] Becquerel himself followed his discovery of Becquerel rays with 11 months of unrelated research on optics. Most physicists did not begin to think of these rays as a phenomenon worthy of intense investigation until the Curies discovered that materials besides uranium salts were capable of emitting Becquerel rays in 1898.

Interestingly, although the Curies' discovery sparked the imagination of the British and American popular press, specialist periodicals in Britain contained only limited coverage of the Curies' work or their 1898 discovery.[22] The *Electrician*, which had so thoroughly summarized the *Comptes Rendus* articles on Röntgen rays, contained almost no material on the Curies' radioactivity work before 1903 (the year the Curies won the Nobel Prize). Short descriptions of the Curies' papers appeared in the Societies and Academies and Notes sections of *Nature*, but in contrast to the material on Röntgen rays, no one translated their papers on radium or polonium for *Nature*, and few researchers in England contributed their own radioactivity work to *Nature* before 1900.[23] In fact, the discovery of polonium and radium occasioned scarcely more coverage in *Nature* than Marie Curie's work on the magnetic properties of steel in the mid-1890s.[24] The *Philosophical Magazine* ran only one article from the Curies, a summary of a recent paper in the *Comptes Rendus de l'Académie des Sciences*, in February 1900.[25] As with Becquerel, *Chemical News* led the British coverage of the Curies' radioactivity work, reprinting both a shortened English version of the Curies' November 1899 *Comptes Rendus* paper on radioactivity and a full translation of Marie Curie's doctoral thesis in 1903.[26]

The coverage of radioactivity in British publications was limited compared with the coverage of Röntgen rays largely because few researchers in England contributed their own radioactivity work to British weeklies or scientific society journals before 1903. Before 1903, there were only four articles in the *Philosophical Magazine* about radioactivity, and articles on radioactivity in the *Proceedings* or the *Philosophical Transactions* did not become a regular occurrence until after 1903. In March 1902, the staff of the *Electrician* even remarked on the limited British coverage of Becquerel rays and the Curies' research in their "Notes" column, writing,

> WHETHER from lack of interest or merely through ignorance of the interesting experiments which had been carried out by M. BECQUEREL, M. and Mdme. CURIE, and a few others, surprisingly little attention has been devoted in this country to the subject of Becquerel rays and the theory underlying these remarkable "radio-active" bodies.[27]

The Curies' absence in *Nature* and other British journals was not simply a matter of British publications ignoring French scientists. Unlike Henri Becquerel, who occasionally wrote articles for *Nature*, the Curies chose not to publish in *Nature* or, indeed, in any English-language publications. Their publications of choice were the *Comptes Rendus* and, after 1904, the new publication *Le Radium*.[28] The Curies' focus on publishing in France rather than overseas is not surprising when we consider their career trajectory in the late 1890s. Biographies of the Curies have consistently pointed out that Marie and Pierre considered themselves outsiders in the French academic community. Pierre was the son of a Communard with no family ties to the academic establishment, and Marie was Polish by birth (and a woman besides). Once again, the well-connected Becquerel provides us with a useful contrast to the Curies: Becquerel was the son and grandson of men who held the chair in physics at the Sorbonne, and in time his son would follow in the family business and hold that chair as well. In the insular and hierarchical French academic community, personal connections and professional status were often closely tied, and neither of the Curies had the sort of social connections that might have smoothed their path to academic success.[29] In February 1898 (five months before they submitted their paper on the discovery of polonium to the Académie), Pierre's application for a professorship at the Sorbonne was denied, and the couple was having some difficulty making ends meet financially. When the couple's radioactivity research first began attracting attention in the scientific community, they were unquestionably more focused on obtaining recognition in France than from the rest of the world.

Only after they were awarded the Nobel Prize in 1903 did Pierre and Marie Curie achieve significant recognition for their work in France. (The Curies shared the Prize with Becquerel; the three were acknowledged jointly for their work on radioactivity.) But their scientific recognition came with a great deal of public attention. The Curies found much of the newspaper coverage of their discovery and the subsequent public interest in their domestic life disruptive and irritating. It is possible that they were simply not interested in courting wider international fame by reaching out to English-language journals. Moreover, by 1903, there was arguably a new international leader in radioactivity research who had overtaken both Becquerel and the Curies: Ernest Rutherford at McGill University in Montreal, Canada.

RADIOACTIVITY, ERNEST RUTHERFORD, AND *NATURE*

Ernest Rutherford was a New Zealander who had come to Cambridge's Cavendish Laboratory with the support of an 1851 Exhibition Scholarship in 1894—the first year the scholarships were open to students born in the colonies.³⁰ J. J. Thomson, the Cavendish's celebrated leader, would remain Rutherford's most important mentor throughout his career. The young New Zealander worked at Cambridge until 1898 when he was named to a vacant professorship of physics at McGill.³¹

Rutherford, who was only twenty-seven years old, had not expected to receive the McGill position despite Thomson's enthusiastic recommendation. When he wrote to his fiancée Mary Newton about the post, he warned her that "There would probably be big competition for it, all over England. . . . I think it is extremely doubtful that I will compete for it."³² McGill was a highly desirable appointment for a research physicist; the university was well funded, and its Macdonald Physics Laboratory was one of the best-equipped research laboratories in the world.³³

The major disadvantage of Rutherford's job at McGill was its location. As we shall see from Rutherford's correspondence, despite the presence of colleagues such as Frederick Soddy, the young physicist felt intellectually isolated in Montreal. The Macdonald Physics Laboratory, however well equipped with instruments, could not replace the sense of intellectual community Rutherford had experienced at the Cavendish Laboratory. But intellectual isolation was not the only perceived consequence of Rutherford's remote "colonial" appointment. In November 1899, Rutherford sent a paper to the *Philosophical Magazine* titled "Radioactivity produced in Substances by the Action of Thorium Compounds." Rutherford believed that his results, which suggested that radioactive thorium could induce radioactivity in other substances, were some of his most interesting and important findings to date. But just two weeks after he submitted his paper, he discovered something most unwelcome: his chief competitors in the field of radioactivity research, Becquerel and the Curies, had just published a new article in the *Comptes Rendus* in which they argued that radium and polonium could induce "excited radioactivity" in other substances.³⁴ By the time Rutherford's paper appeared in the February issue of the *Philosophical Magazine*, his findings were no longer as groundbreaking as he had initially believed. His paper ended in a morose footnote acknowledging that the Curies had been the first to publish about excited radioactivity.³⁵

Being scooped by his French rivals was a blow to the ambitious Rutherford, who was extremely concerned about establishing priority for his work. Rutherford's biographer David Wilson writes that losing the race to be first on excited radioactivity taught Rutherford "the hard lesson of the sheer distance of Canada from the centers of scientific activity in Europe where researchers could get their results printed, and claim priority of discovery, within a few days of submitting their work."³⁶ A letter Rutherford wrote to his mother in 1900 underlines his competitive spirit as well as his desire to publish his work quickly:

> I have to keep going as there are always people on my track. I have to publish my present work as rapidly as possible in order to keep in the race. The best sprinters in this road of investigation are Becquerel and the Curies in Paris who have done a great deal of very important work in the subject of radioactive bodies during the last few years.³⁷

But Rutherford's misfortune was *Nature*'s gain. There were three major publications where Rutherford sent his work: *Proceedings of the Royal Society of London*, *Philosophical Magazine*, and *Nature*. Before the 1899 excited radioactivity scoop, Rutherford had not contributed to Macmillan's weekly. That would soon change. Both the *Proceedings* and the *Philosophical Magazine* had significant lag times between submission and publication (the *Proceedings* averaged six months, the *Philosophical Magazine* three), which made *Nature* and its weekly turnaround uniquely valuable to the priority-conscious Rutherford.³⁸ *Nature*, which still published its letters to the editor the same week they were submitted, fulfilled both Rutherford's desire to publish in Europe and his need to minimize the delay caused by sending submissions across the Atlantic. Rutherford contributed over a dozen letters to the editor between 1901 and 1908 (when he moved to the position at Manchester).³⁹

Interestingly, Rutherford's desire to publish quickly did not lead him to submit to other weeklies besides *Nature*. Although *Chemical News* regularly covered Rutherford's papers and lectures, and although he and Soddy published a multipart article on thorium emanations in *Chemical News* in 1902, after Soddy left McGill Rutherford ceased to contribute articles to *Chemical News*.⁴⁰ Similarly, although Thomson had sent an abstract of one of Rutherford's forthcoming articles to the *Electrician* while Rutherford was at Cambridge, Rutherford did not send his own experimental findings to that publication either.⁴¹ Why *Nature* and not *Chemical News* or the *Electrician*—especially given that *Chemical News* had been far more interested in radioactivity than *Nature*? It was partly a question of discipline. *Chemical News* catered to Britain's chemists. Soddy identified as a chemist, but Rutherford viewed himself as a physicist (and was, indeed, rather bemused when he won the Nobel Prize in Chemistry), which explains why Rutherford coauthored with Soddy but did not write pieces for *Chemical News* on his own. Similarly, the *Electrician* aimed itself at an audience of engineers and industrial scientists; Rutherford considered himself neither. Rutherford's choice of *Nature* over other weeklies was probably also about the relative prestige of these publications. As early as the 1870s and 1880s, scientific researchers in Great Britain were choosing *Nature* over other weeklies because it reached "the right people" (as *Chemical News* editor William Crookes put it). Publishing in *Nature* legitimized one's work in a way that publishing in other weeklies did not.[42]

Interestingly, Rutherford also did not make use of another weekly publication, *Science*, which also had a correspondence column (called Discussion and Correspondence) and whose New York editorial offices were much closer to Montreal than *Nature*'s London offices. Nor did Rutherford publish much work in Canadian journals, even though his submissions could have reached those journals faster than they could reach journals in Great Britain. Rutherford's British-focused publishing strategy suggests that in addition to concerns about priority and prestige, Rutherford also sought to reach the right national audience. Rutherford's personal correspondence reveals a strong feeling that Montreal was distant both geographically and intellectually from the centers of the physics world. In March 1901, Rutherford wrote to his mentor Thomson to seek advice on whether he should apply for the chair of physics at the University of Edinburgh, which had recently been vacated by the retirement of Peter Guthrie Tait. The letter acknowledged the excellent facilities at McGill but expressed some frustration at Montreal's distance from other scientific centers.

> After the years in the Cavendish I feel myself rather out of things scientific, and greatly miss the opportunities of meeting men interested in physics. Outside the small circle of the laboratory it is seldom I meet anyone to hear what is being done elsewhere. I think that this feeling of isolation is the great drawback to colonial appointments, for unless one is prepared to stagnate, one feels badly the want of scientific intercourse.[43]

Rutherford ultimately decided not to apply for the Edinburgh chair, reasoning that the field of candidates was likely to be quite large and would include some of Tait's former protégés. But it appears that he continued to feel isolated in his "colonial appointment" over the next few years. In 1906, Rutherford again brought up his feeling of isolation when telling his mother of the offer at Manchester:

> I have received the offer of the Physics Chair at Manchester. I think it quite likely I shall accept. I think it is a wise move for a variety of reasons. I shall receive a better salary and be director of the laboratory and what is most important to me, will be nearer the centre of things scientifically.[44]

Rutherford's career goals probably explain his reluctance to direct his work to North American periodicals even though they might have been able to get his work into print more quickly than British periodicals. Although Canada was a self-governing British colony, the political connection did not guarantee that British professors back in the "home country" would be aware of papers published in Canadian journals. Correspondence in Lockyer's personal papers—including a letter from Rutherford about *Nature*'s coverage of a recent appointment to the Canadian Geological Survey—suggests that there was a strong feeling in the Canadian scientific community that Canadian accomplishments did not always receive their due in the United Kingdom.[45] Rutherford may have been concerned that publishing in North American journals would cause his work to be overlooked in Britain and in Europe. Publishing in British journals increased the likelihood that radioactivity physicists in Europe's scientific centers would read Rutherford's papers. It also helped ensure that Rutherford's name was familiar to British universities who might be looking for a new professor of physics. Notably, before moving to Manchester, Rutherford turned down offers of physics professorships from Victoria University College in New Zealand, the University of Western Australia, and Columbia University in New York, suggesting that his true goal was to return to the United Kingdom.[46]

RADIOACTIVITY AND CHANGES IN *NATURE*

Rutherford's use of *Nature* affected not only his own career but the journal as well. Lawrence Badash has argued that Rutherford "transformed the letters-to-the-editor section of *Nature* from one of genteel comments on scientific activity and reports of the first robin of spring, to announcements of the greatest fundamental importance and hard-hitting scientific controversy."[47] This observation is not entirely accurate—as we have seen, *Nature* was a forum for "hard-hitting scientific controversy" well before Rutherford started contributing, and many of those discussions were far from "genteel." However, Rutherford did make a significant mark on *Nature* by establishing the Letters to the Editor column as a major venue for priority claims, a development that helped set *Nature* even farther apart from other scientific weeklies in Great Britain.

When George J. Romanes, E. Ray Lankester, John Perry, and the rest of *Nature*'s prolific nineteenth-century contributors composed letters to the editor of *Nature*, their letters were generally counterparts to longer forthcoming papers or focused commentaries on someone else's work. But Rutherford, like the researchers who had written to *Nature* about Röntgen rays, used letters to the editor as an end in themselves. Rutherford did not wait until he had a full paper in press for the *Philosophical Magazine* or the *Proceedings* to send an abstract to *Nature*; he sent his most interesting experimental results immediately, both as a way of keeping his colleagues updated on his work and as insurance against being scooped as he had in 1899. During his years at McGill, Rutherford wrote frequent letters to *Nature* filled with data and experimental observations about the heating effects of radioactivity, the amount of helium emanating from radium, the dependence of radioactivity on the concentration of radioactive materials, and the electrical charge on the alpha rays emitted from radium. These updates from one of the world's preeminent radioactivity physicists made *Nature* indispensable to anyone working on radioactivity—not just in Britain, but in Canada, the United States, Paris, Berlin, and Vienna. Rutherford would continue submitting letters even after returning to the United Kingdom, albeit less frequently than he had before, suggesting that *Nature* was still part of his publishing strategy but no longer as vital a part as it had been when he worked at McGill.[48]

Rutherford's personal correspondence also shows that *Nature* played a significant role in spreading news of the latest radioactivity research, especially among researchers living outside Europe. In the correspondence between Rutherford and the American physicist Bertram Borden Boltwood, for example, both men frequently mentioned *Nature* as a place to print their own articles and an important source of information about others' research.[49] Boltwood, arguably the most important radioactivity physicist in the United States, had obtained his PhD from Yale in 1897 and began his career as a consulting chemist.[50] He conducted research out of his own private laboratory in New Haven until his appointment as an assistant professor of physics at Yale in 1906.

Like Rutherford, Boltwood struggled with the disadvantages of being at a distance from major radioactivity research centers such as Paris and Cambridge. *Nature* proved invaluable as a source of pertinent abstracts and as a place to publish his work. In an April 1905 letter to Rutherford about his work on the radioactive decay series, Boltwood wrote, "I have held up this letter somewhat, hoping to find some details of the R.S. paper on the new (?) element 'which gives off thorium emanation' in the Nature which came last night. Now that I know that it comes from thorianite, I am also willing to bet that it is Th-X."[51] A 1906 letter from Boltwood illustrates the American's practice of sending preliminary results both to American journals and to *Nature*. Boltwood wrote to share some new results on the radioactive decay series and told Rutherford, "I have sent off a brief communication to the Editor of Nature and a note for the December number of the Am. Jour."[52] Similarly, Rutherford frequently referred to recent pieces in *Nature* when discussing scientific matters with Boltwood. In June 1904, for example, he assured Boltwood that Soddy's recent findings in *Nature* did not disprove Boltwood's own theories, writing, "I would not lay any especial stress on negative results of attempts to grow radium as in Soddy's letter to Nature."[53] Rutherford's letters also refer to sending early results to *Nature*; in October 1906 he wrote, "I have done a few expts. recently which show that the emanations are completely absorbed in cocoanut charcoal at ordinary temperatures.... You will see an account in Nature of the same in a week or so."[54] Even after Rutherford's return to England, *Nature* continued to feature heavily in the letters between Rutherford and Boltwood.[55]

The Rutherford-Boltwood correspondence further reveals that *Nature* remained a host for controversial discussions in the twentieth century. After Soddy left McGill in 1902, Rutherford and Boltwood found themselves butting heads with both Soddy and William Ramsay over the radioactive decay series.[56] Rutherford retained a fondness for the eccentric Soddy, but his opinion of Ramsay was quite negative, even hostile; Boltwood appears to have been largely indifferent to Ramsay but found himself the target of Soddy's spirited attacks more often than he would have liked. In July 1907, Rutherford wrote to his friend to tell him that the *Nature* editorial staff had consulted him about a recent letter Boltwood submitted to the journal:

> By the way, *between ourselves,* your letter to Nature re Soddy's attack was forwarded to me to report *whether it was not too personal for publication*. I replied that Soddy's letter was extremely personal & provocative and that your letter was far more restrained!! than Soddy's; otherwise your letter would have been returned with thanks.[57]

Boltwood wrote back to thank Rutherford for encouraging publication of his letter:

> I spent over a week trying to compose a decently polite reply to his communication.... I suspected that it had been held up somewhere because of the delay in its publication. By your approval of it you not only did me a good turn but you saved for me my high opinion of the fairness of Nature which would certainly have gone by the board if they had been unwilling to let me defend myself.[58]

The Boltwood-Soddy exchanges nicely highlight some of the continuities in *Nature* between the nineteenth and twentieth centuries. *Nature* remained a center for controversial scientific discussions, and some of those discussions could become personal. However, it is worth noting that unlike the members of the X Club, who had accused Lockyer of editorial bias, Boltwood praised *Nature*'s "fairness." It is also worth noting that Lockyer and Gregory consulted another contributor, Rutherford, when they questioned whether to publish Boltwood's letter—a contrast to the days when Lockyer simply made the decisions about which letters might be too personal.

While the Curies remained at most an ephemeral presence in *Nature*, other international radioactivity scientists followed Rutherford, Soddy, and Boltwood onto the pages of *Nature*. The most notable among these was Otto Hahn, a future Nobel Prize winner (for the discovery of uranium fission) who worked at McGill with Rutherford in 1905-1906. Like Rutherford and other Anglophone colleagues, Hahn soon adopted the practice of writing to *Nature* about interesting preliminary results.[59] This was probably due to Hahn's desire to keep his English-speaking colleagues updated on his work but also speaks to a perceived underdevelopment of radioactivity research in his native Germany. In June 1907, for example, Hahn wrote to Rutherford to share the good news that he had passed his examinations to become a *privatdozent* (an instructor at the university level), which he attributed to the fact that his examiners "were so terror stricken that they asked only some simple radioactive questions about the matter and did not want to hear anything else." In the same letter, Hahn also spoke of feeling "lonely among all these chemists who don't really believe in radioactivity" now that he had returned to Germany.[60]

NATIONAL AND INTERNATIONAL CONTRIBUTORS: MENDELIAN HEREDITY AS A CONTRASTING EXAMPLE

In the nineteenth century, *Nature* was a publication that was strongly focused on scientific issues in the United Kingdom. The journal did have subscribers abroad, but such readers were not its target audience, and *Nature* was not regarded as a publication with any special international prominence. In 1879, for example, Lockyer's Parisian friend Maurice Berthelot encouragingly told Lockyer that "[*Nature*] is, I think, the most informative scientific journal in England," but Berthelot himself did not contribute to it.[61]

It might be tempting to watch men such as Boltwood and Hahn follow Rutherford onto *Nature*'s pages and conclude that *Nature* was becoming more international. But a closer examination of the journal reveals that *Nature* was still a very British publication. Even a passing glance through

Nature's articles and editorials reveals a strongly British focus. *Nature* editorials frequently addressed themselves to the British scientific community or the British government specifically. Examples include a lead editorial in 1902 urging the British crown to charter an organization for humanistic studies similar to the Royal Society, or a 1904 editorial by Sir William Abney that criticized Britain for not giving its sciences sufficient financial support.[62] The journal's reaction to the deaths of Queen Victoria in 1901 and King Edward in 1910 was perhaps the clearest indication of *Nature*'s British roots. On both occasions, *Nature* devoted its lead editorials to the praise and mourning of both monarchs, and the journal's pages bore black outlines.[63]

Notably, *Nature* contributors did not write articles bemoaning the status of science in non-British countries or calling on foreign governments for more support of science. Such calls and complaints in *Nature* were aimed exclusively at Great Britain. Articles about science in other countries were not entirely absent, of course. But these articles were written largely by British contributors and presented as either human-interest stories for the edification of British readers or as excuses to compare British support of science unfavorably with other countries.[64] A good example of both aims is a 1906 article on Rutherford's laboratory at McGill. Arthur S. Eve, the piece's author, was an English scientist who had recently moved to Canada to work with Rutherford.[65] The article opened with effusive praise—not for Rutherford, but for William Macdonald, the wealthy Canadian industrialist whose generosity had paid for the construction of the McGill physics building. Eve went on to chide wealthy fellow Englishmen for failing to follow Macdonald's example.[66]

Otto Hahn's reminiscences about his days in Rutherford's McGill laboratory include a story about this very article:

> Early in the year 1906, a photographer came to the Macdonald Physics Building to take a photograph of Rutherford working in his laboratory, for publication in the columns of *Nature*. . . . In the opinion of the photographer, however, the already famous professor was not dressed elegantly enough for the readers of *Nature*. Not even cuffs were to be seen peeping from the sleeves of his coat! But the photographer found a way out; I was to lend Rutherford my loose cuffs. They were so arranged that they protruded well beyond the end of the sleeves. The photographer expressed satisfaction with the new photograph. As a result . . . we see not only Professor Rutherford seated alongside the apparatus with which he carried out his epoch-making experiments on the alpha-rays, but also one of the cuffs of a young research student.[67]

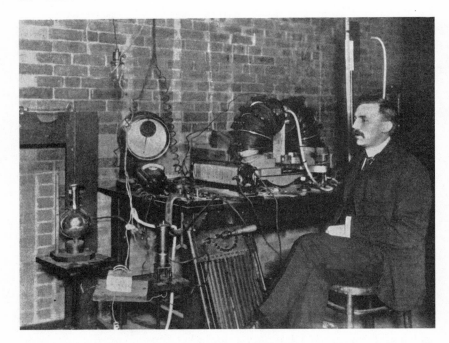

FIGURE 6 *Nature* photograph of Ernest Rutherford, complete with Otto Hahn's cuffs. From Arthur S. Eve, "Some Scientific Centres: VIII. The Macdonald Physics Building, McGill University, Montreal," *Nature* 74 (19 July 1906): 273. Reprinted by permission of the Nature Publishing Group.

The photographer's insistence that Rutherford be "elegantly" attired for the sake of the *Nature* readership also reflects the journal's British orientation. Whether or not Rutherford wore cuffs in the photograph seems to have been of little concern to Rutherford himself, but for *Nature*, he had to abide by the standards of dress British readers would expect from a famous professor of physics.

We find further evidence of *Nature*'s continuing Britishness in the journal's discussions of Mendelian inheritance. Like radioactivity, genetics was an international field from its very beginning. Between April and July 1900, three European scientists—Hugo de Vries in Amsterdam, Carl Correns in Munich, and Erich Tschermak in Vienna—each published a paper in a German botanical journal, the *Berichte der deutschen botanischen Gesellschaft*, detailing experimental work on heredity and citing the work of a then obscure Austrian monk, Gregor Mendel.[68] Many scientists across Europe were fascinated by the rediscovered Mendelian theories of heredity, including the English botanist William Bateson (1861-1926). Bateson, a good friend of de Vries's, read the work with great interest and began incorporating Mendelian ideas into his own research and lectures.[69]

Bateson, much like the radioactivists, made frequent use of *Nature*'s Letters to the Editor. The Letters to the Editor often played host to debates between Bateson and British biometricians, most notably W. F. R. Weldon (1860-1906), who argued that the statistical evidence in favor of Mendel's theories was inadequate.[70] The biometricians preferred Francis Galton's theory of ancestral heredity, which stated that an individual's genetic makeup incorporated hereditary material from all of the individual's ancestors, not just the parents. The biometricians also argued that an individual's characters (height, hair color, skin color, etc.) were the result of the blending of the characters from his progenitors. They rejected the Mendelian implication that inheritance of a particular character (such as wrinkly or smooth skin on peas, which had been one of Mendel's test cases) was a yes-or-no, all-or-nothing proposition. In response, Bateson argued that the biometricians did not attribute enough evolutionary significance to "sports" (animals displaying sudden, discontinuous variation from their parents) and that Mendel's laws might hold the key to a complete understanding of evolution.

Bateson's conflict with Weldon had started well before Bateson took up the mantle of Mendelism. The two biologists had attended Cambridge University at the same time and were close friends in the 1880s. However, they fell out in the early 1890s when Bateson began arguing more forcefully, both in print and in personal correspondence, for the importance of discontinuous variation. The friendship was finally severed following a debate—partially conducted in *Nature*—about the ancestry of cineraria, an unusual cultivated plant. At an 1895 meeting of the Royal Society, Bateson criticized W. T. Thiselton-Dyer's suggestion that the wild *Cineraria cruenta* might give rise to cultivated cineraria. In early May both men wrote to *Nature* to tell their version of the debate. Weldon wrote in to *Nature* in support of Thiselton-Dyer, a move that Bateson took very personally.[71]

When Bateson began championing Mendelian theory in Britain, he once again found himself in conflict with his old friend. The two debated bitterly in *Biometrika* (a publication Weldon helped found in 1900) and in *Nature* until Weldon's premature death from pneumonia in 1906. After Weldon's death, his colleague Karl Pearson assumed the leadership of the British biometricians as well as the responsibility of arguing with Bateson.[72] The prickly Pearson took Bateson's criticisms as personally as Weldon had. In fact, Pearson's feelings about Bateson grew so strong that in 1910, when two members of the *Biometrika* editorial board (the American biologists Raymond Pearl and Charles Davenport) published comments favoring Mendelian theory, Pearson responded by abolishing *Biometrika*'s editorial board altogether.[73] The intensity of the conflict was fueled both by personal pride and the desire for professional survival. Biometry and genetics were new fields attempting to establish a foothold in Great Britain's limited number of scientific institutions; their supporters may have felt that the success of their field depended on discrediting the competition.[74]

Nature was one of Bateson's favorite publication venues. But when we compare Bateson's use of the Letters to the Editor with Rutherford's, we see that Bateson did not use the column to announce new results as Rutherford did. Instead, Bateson mainly wrote letters to the editor to criticize anti-Mendelian opponents, a pattern of contribution more consistent with nineteenth-century contributors such as George J. Romanes or E. Ray Lankester. In 1892, for example, Bateson wrote two letters to the editor arguing with a recent book review by Edward B. Poulton.[75] Similarly, in 1903 Bateson wrote to *Nature* with an analysis of Weldon's latest article in *Biometrika*. He sharply criticized Weldon's conclusions and argued that Weldon's study of the inheritance of eye color and coat color in mice produced results that were perfectly in accord with Mendel's predictions.[76] Weldon wrote back to cite another *Biometrika* article by a Mr. Darbishire, who had described breeding albino and brown mice and obtaining hybrids with a "lilac" coat. Weldon presented the "lilac" mice as evidence of the blending theory of heredity.[77] The two exchanged another pair of letters before ending the correspondence in April.[78]

The controversy between Bateson and the biometricians has received a great deal of attention from scholars and has tended to overshadow other participants in the debate over Mendelian inheritance in Great Britain. A number of other scientists also contributed to *Nature*'s Mendelian discussions, including Bateson's student R. H. Lock, the biologist G. Archdall Reid, a lecturer at London Medical College named George P. Mudge, E. Ray Lankester's former pupil J. T. Cunningham, and W. T. Thiselton-Dyer. These men engaged in debates as lengthy and as passionate (though not as personally tinged) as those of Bateson, Weldon, and Pearson. In October 1907, for instance, Reid began a ten-week correspondence on Mendelian inheritance with a letter arguing that Mendel's laws did not apply to parthenogenic reproduction and therefore could not be the key to unlocking the puzzle of heredity.[79] Lock, Mudge, and Thiselton-Dyer all joined the discussion in

Nature about Mendelism, sex, and the evidence for Mendel's laws. Interestingly, only Reid wrote in opposition to Mendelian doctrine; the other biologists all expressed their belief in the accuracy of Mendel's laws.[80]

A quick examination of the parties involved in these discussions reveals something important: despite the wide international reach of genetics, the scientists writing to *Nature* about Mendelian inheritance were all living and working in the United Kingdom. Geneticists from outside the United Kingdom were not represented in these discussions. Why was genetics so different from radioactivity in its ability to draw international contributors to *Nature*?

In fact, the gap between the two fields may not be as pronounced as it first appears. If we return our attention to radioactivity, we see that the most frequent international contributors to *Nature* were researchers who had a personal connection to Rutherford—in particular, scientists such as Hahn who had studied for a time in his laboratory. Scientists who spent their careers in their native countries—for example, Marie Curie or Stefan Meyer—seldom contributed to *Nature*. In contrast to radioactivity, foreigners rarely came to work in British genetics laboratories in the early twentieth century. At first glance this seems quite odd. Genetics has a long and rich history of international congresses; it was a 1906 congress in London that formally accepted the name "genetics" for the discipline that had grown out of Mendelian theories of heredity. We might expect that a field so quick to embrace international congresses and international communication would be more likely to send scientists abroad to study with foreign colleagues. But in Britain, Germany, France, and the United States, geneticists were struggling for national recognition of their new discipline, which may have dampened the enthusiasm for sending their students and papers abroad.[81] Furthermore, because few of them had studied in Britain, foreign geneticists would have had less exposure to *Nature*'s submissions process than scientists who had watched Rutherford dispatch a new missive to *Nature* every other month.

JOURNALS, NATIONALISM, AND INTERNATIONALISM

Many historians have written that the twenty years preceding the outbreak of World War I were an era of increasing international ties between scientific workers. The number of international scientific congresses increased dramatically between 1870 and 1914, fueled in part by a desire to standardize terminology and units and in part by the enormous boom in railway networks across Europe.[82] The rhetoric of "scientific internationalism" was also growing—more and more scientists were speaking of science as a fundamentally international endeavor, one that transcended political boundaries even when governments were at odds or at war. (As Debra Everett-Lane points out, this "internationalism" was heavily Eurocentric.)[83] Furthermore, scholars have observed that many of the new fields that emerged in the late nineteenth and early twentieth centuries, such as radioactivity and genetics, were international sciences from the very beginning, with researchers from multiple countries making significant contributions to these fields almost simultaneously.

With scientific internationalism on the rise, it might be expected that scientific journals would take on a more international character, but in the early twentieth century *Nature* was far from alone in its national orientation. In fact, many of the most influential scientific journals in Europe were similarly focused on serving a particular national scientific community. One good example is the *Annalen der Physik und Chemie*, Germany's most important journal of the physical sciences. The *Annalen*, founded in 1790, rose to prominence in the early nineteenth century under the leadership of Johann Christian Poggendorff, who assumed the editorship in 1824.[84] Fifty years later Poggendorff declared that his *Annalen* was the "only organ of physics for Germany"—a bold statement, but one few would have challenged.[85] The journal's important place in German physics endured even after Poggendorff's death in 1877. In 1925 one German scientist commented that the *Annalen* "unite[s] in itself the entire physical life in Germany."[86]

Throughout the nineteenth century, the *Annalen* regularly published translations of important papers by foreign physicists. But German and Austrian scientists such as Hermann von Helmholtz, Georg Ohm, Gustav Kirchoff, Heinrich Hertz, and Max Planck were responsible for the vast majority of the *Annalen*'s content, especially toward the end of Poggendorff's editorial tenure. Poggendorff and his successor Gustav Wiedemann believed that the main purpose of their journal was to publish the latest work by Germanophone physicists in order to strengthen physical research in the German-speaking lands. While the editors and contributors would not have been displeased that foreign scientists read their journal, building an international readership was not their primary goal. The *Annalen* was meant to serve the physical sciences in Germany first and foremost.

A similar case is that of *Le Radium*, a Parisian publication devoted to radioactivity research. *Le Radium* was founded in 1904 under the editorship of Jacques Danne and published by Masson et Compagnie. It was a relatively short monthly magazine (an average issue was about 35 pages long) that published both original articles on radioactivity and shorter summaries of articles in other journals. *Le Radium* published steadily through June 1914, but the outbreak of World War I resulted in a five-year gap in publication. The journal resumed publication of its final volume in May 1919 and ceased to publish after December of that year. *Le Radium*'s demise was probably linked to the death of its editor, Danne, who died shortly before the journal resumed publication in 1919.[87]

Like the *Annalen*, *Le Radium* regularly published pieces by foreign scientists such as Rutherford, Boltwood, Soddy, and Hahn as well as articles by lesser-known British, German, and American scientists.[88] Foreign scientists prepared and submitted some of these articles doubtless with the intent to publicize their work in the French radioactivity community. But *Le Radium*'s editorial staff translated most of the articles by foreign scientists themselves, and French scientists were responsible for the majority of *Le Radium*'s original articles.[89] Furthermore, *Le Radium*'s board of directors included only three foreigners: Rutherford, the German physicist Heinrich Rubens, and the Danish physicist Niels Finsen. The other eleven directors were all French (one, Jean Danysz, was Polish-French).[90] Even though radioactivity was an international science, *Le Radium*, like *Nature* and the *Annalen*, was aimed at a national readership.

The continuing national focus of prominent journals has important implications for the history of internationalism in early twentieth-century science. Although scientific internationalism was gaining strength in the late nineteenth and early twentieth centuries, many historians have observed that internationalism often had to contend with the other major political movement of the late nineteenth century: nationalism.[91] While international scientific congresses and correspondence networks bore witness to increasing scientific internationalism, journals continued to reflect national scientific concerns, and a scientist's choice of where to publish his or her work was often dictated by national background and career ambitions. Rutherford and the Curies are excellent examples of scientists who had a wide network of international colleagues but remained focused on publishing within their own national context. Despite the growing rhetoric of scientific internationalism and the increasing importance of international scientific colleagues, both Rutherford and the Curies placed great emphasis on asserting their scientific talents within their national scientific communities.

It is important to recognize that nationalism and internationalism were not always antagonists. National pride might motivate a country to host an international scientific congress, for example, or prompt scientists in a previously isolated nation to make their work available to international colleagues. Even the Nobel Institute, the embodiment of early twentieth-century scientific internationalism, acknowledged and made use of scientific nationalism. Historian Elisabeth Crawford writes that the Nobel Institution was "built not only on the coexistence of nationalism and internationalism but also on the essential tension between the two." She notes that the scientists who nominated their colleagues for the Nobel Prize were expected to act as representatives of their own nations—and that this appeal to nationalism was not seen as being contrary to the Nobel Institute's international mission.[92] Similarly, while early twentieth-century journals were largely national rather than international in orientation, the publications still contributed to the exchange of information across international borders. A heavily British publication such as *Nature* could still have international significance, as Rutherford's correspondence with Boltwood reveals.

The increasingly international makeup of *Nature*'s radioactivity contributors shows that the journal's community of contributors and readers was beginning to include scientists outside Britain's borders in the early twentieth century. The outbreak of the Great War in August 1914, however, proved to be a serious challenge to ideals of scientific internationalism. Despite the claim that scientific brotherhood transcended political conflicts, the war shattered many of the ties between scientists who now found themselves on opposite political sides. As we shall see in the next chapter, there was no question about where *Nature*'s editors and contributors felt their political— and scientific—loyalties lay.

CHAPTER FIVE

Nature, Interwar Politics, and Intellectual Freedom

In January 1938, the *Nature* offices at Macmillan received unexpected news of a recent development in Germany: their journal was no longer acceptable reading in Germany's libraries. On 12 November 1937, Bernhard Rust, the German Minister of Science and Education, had issued an executive order removing *Nature* from "general usage" at all German universities and research libraries. According to Rust, the new restrictions on *Nature* were due to the magazine's "unprecedented and low attacks against German science and the National Socialist state."[1] Richard Gregory and his staff at *Nature* were almost certainly not surprised at Rust's feelings about their journal—*Nature* had published a number of articles strongly criticizing the Nazi government's academic policies, some of which accused Rust, by name, of being a destructive influence on German science. But why did Rust consider the British weekly worth restricting at all?

In the early twentieth century, *Nature* was largely a journal written by and for British scientific workers. By the 1920s and 1930s, however, *Nature*'s Letters to the Editor column attracted a much larger number of international contributions from a much wider range of fields than it had before World War I. This was not the only change in *Nature*. Before the war, *Nature*'s editors and contributors had often used the journal to complain that British science lagged behind science in other countries—particularly Germany. After the war, in contrast, *Nature*'s contributors shifted from criticizing their own society and government to criticizing foreign nations. *Nature*'s contributors and editors were particularly vocal when they felt foreign governments hindered scientific research or attacked intellectual freedom. International scientific issues such as the Scopes trial in the United States, the treatment of geneticists in Soviet Russia, and especially the academic policies of the National Socialist government all found a prominent place in *Nature*'s pages. Emboldened by both the increased international reach of the Letters to the Editor and post–World War I patriotism in Britain, *Nature*'s editor and contributors used *Nature* to portray Britain as a nation that exemplified respect for scientific truth and to criticize other countries for political interference with science.

NATURE AND GERMANY

One of *Nature*'s recurring themes under both Lockyer and Gregory was the alleged inferiority of British governmental support of science. *Nature*'s editorials and news articles often claimed that Britain was "falling behind" its Continental competitors, and Germany was the most frequent target of the contributors' envy. In the very first volume of *Nature*, Henry E. Roscoe, a university administrator who had trained in Heidelberg with the great German chemist Robert Bunsen, wrote a two-part article on "Science Education in Germany." Roscoe's article said that while England had done much to encourage the teaching of science in primary school, its secondary schools and institutions of higher education were woefully inadequate when compared to Germany's.[2] In the second volume, a contributor named S. Stricker published a three-part article titled "The Medical Schools of England and Germany" that outlined a number of deficiencies in the British system of medical education. Stricker praised the superiority of German medical schools, which he said had better financial support and better facilities.[3]

The comparisons with Germany abated somewhat during the 1880s as *Nature*'s contributors focused instead on France and the great Louis Pasteur as their point of comparison, but by the 1890s the British scientific community had returned to comparing Britain unfavorably with Germany.[4] For example, in August 1896 the physical chemist William Ramsay sent the London *Times* a letter from his good friend Wilhelm Ostwald, a professor at the University of Leipzig. Ostwald's letter confidently stated that German education was superior to British education and outlined a number of reasons for the gap.[5] Far from taking exception to Ostwald's suggestion that English scientific education was not up to snuff, Lockyer approved a lead editorial endorsing Ostwald's observations and reprinted his *Times* letter.[6] After the volatile Kaiser Wilhelm II assumed the German throne in 1888, political relations between Germany and Britain became increasingly strained.[7] Despite this tension, *Nature*'s attitude toward German science remained enthusiastic and admiring. In April 1914, just months before the Great War broke out, Richard Gregory himself wrote an article on primary education that held Germany forth as an example for Britain to emulate.[8]

This admiration for German science—and the ideals of scientific internationalism that we saw in chapter 4—faced a severe test after the outbreak of war in Europe in August 1914. In Lockyer's last editorial for *Nature*, printed on 10 September 1914, he stridently denounced Germany for causing the war, but could not resist a dig at the British government for failing to heed the British Science Guild's warnings about the need to catch up to Germany in technical education and chemical industry.[9] Notably, German scientists were at first largely exempt from Lockyer's criticism; he said that he "did not doubt" that German researchers had sought to advance knowledge and civilization, rather than German militarism.

But as the war progressed and news of German atrocities in neutral Belgium reached the British press, the British attitude toward Germany and its citizens quickly deteriorated. On 4 October 1914, ninety-two German intellectuals—including twenty-three eminent scientists—signed the infamous "Appeal to the Civilized World," claiming that Germany had been unfairly blamed for starting the war. The letter declared that German "militarism" was in fact a national spirit that had made Germany a great nation. The petition added fuel to the growing Germanophobia in Britain and convinced some British scientists that German science was as corrupt as the rest of Germany.[10]

On October 8, *Nature* published a blistering editorial about the war by Ramsay, the chemist who had endorsed Ostwald's letter about German educational superiority. Ramsay had a number of professional and personal links to Germany. A vocal advocate for the improvement of British science education, Ramsay frequently argued that the German model was one Britain ought to admire and emulate. As a young student he had studied briefly in Heidelberg with Bunsen. He was a foreign member of the Berlin Academy of Sciences, and in 1911 he had proudly accepted the Prussian Order of Merit from the German emperor. But Ramsay's 1914 editorial contained no hint of sympathy for Germany. He opened the editorial by painting the contrast, as he saw it, between England and Germany: the Anglo-Saxons were devoted to "fair-dealing" and had "never been a race of oppressors." Germany's dedication to the principle of "Deutschland über Alles in der Welt" (Germany over all else in the world), said Ramsey, had given the fair-minded Britons no choice but to declare war.

So far this was a fairly mild justification for Great Britain's conflict with Germany. When Ramsay turned to the subject of German science, his editorial became much more barbed. He argued that Germans, "in spite of certain brilliant exceptions," had never been original—their scientific achievements had all come through the "exploitation of the inventions and discoveries of others." Ramsay declared that the German nation needed to be "bled white" of its militarism and despotic tendencies. This would mean no loss for science: "The greatest advances in scientific thought have not been made by members of the German race; nor have the earlier applications of science had Germany for their origin. So far as we can see at present, the restriction of the Teutons will relieve the world from a deluge of mediocrity."[11] In Ramsay's view, Germans had placed their national loyalties above their scientific ones, thus demonstrating that their science was hollow.[12]

Hostility toward Germany and German scientists was also evident outside *Nature*'s pages. German-born scientists who had lived and worked in the United Kingdom for years were suddenly subjected to suspicion and even persecution. The prominent Frankfurt-born physicist Arthur Schuster (1851–1934), whose family moved to Manchester in 1870, was serving as the secretary of the Royal Society in 1914 when the war began and had been elected as president of the British Association for the Advancement of Science for the 1915 annual meeting. Such prestigious offices clearly indicate Schuster's prewar status within the British scientific community. But once the war was underway, Schuster and his family came under attack for their "foreignness." Police confiscated the wireless apparatus Schuster had installed in his home to receive time signals for meteorological observations on the grounds that it might be used to receive messages from Berlin.[13] Several members of the British scientific community protested Schuster's presidency of the BA because of his German name and birth; only the direct plea of the BA council convinced Schuster not to resign. (Ironically, given the furor over his allegedly suspect loyalties, just before delivering his presidential address, Schuster received the devastating news that his son, a soldier in the British army, had been wounded in action.[14]) The BA controversy did not end Schuster's troubles. Throughout the course of the war, three prominent Fellows—Henry Armstrong, E. Ray Lankester, and A. B. Bassett—collaborated on a highly public (though unsuccessful) campaign to have Schuster removed from his post as secretary of the Royal Society.[15]

Personal loss inevitably intensified the ill feelings toward Germany. In June 1915, Ernest Rutherford told the Swedish chemist Svante Arrhenius that "it seems to me ... that all the social and scientific intercourse with Germany will be practically stopped for this generation."[16] Rutherford's own feelings of antipathy toward Germany would soon deepen. Communication between Rutherford and his German colleagues, including former close friends such as Otto Hahn, all but ceased after the August 1915 death of Rutherford's protégé Henry Moseley (1887-1915).[17] Moseley and Rutherford had worked together in Manchester after Moseley's graduation from Oxford in 1910. In 1914 Moseley had demonstrated that each element in the periodic table had a characteristic atomic number, and he was able to show that several recently discovered "new elements" were actually compounds of known elements. Mosely also identified seven gaps in the new atomic number periodic table, predicting that new elements with the missing atomic numbers would soon be discovered.[18] After this discovery Moseley was widely regarded as the most promising physicist of his generation. But when Britain declared war on Germany, Moseley felt it was his duty to enlist, and he died less than a year later at Gallipoli. Rutherford blamed "German aggression" for the loss of his young friend's life and promising scientific career.[19] Moseley's death even strained Rutherford's friendship with the American Bertram Boltwood. Unlike Rutherford, Boltwood continued to correspond with German colleagues such as Hahn and persisted in telling Rutherford that Germany and Britain were both to blame for the escalation of hostilities.[20]

As the war progressed, *Nature*'s contributors became more strident in their condemnation of Germany. The journal's lead article for 14 January 1915 opened with an angry denunciation of the recent German shelling of Yorkshire, an attack on British seaports that had resulted in seventeen deaths, including that of a 14-month-old baby. The article also criticized a German professor of Celtic studies, Kuno Meyer, who had spent most of his career at the University of Liverpool but was now writing to Irishmen and Americans of Irish descent urging their sympathy with the German cause. "Savages have a code that, after breaking bread in a man's house, it is treacherous to war against him; not so Prof. Kuno Meyer. This is evidently another instance of 'Kultur.'" The editorial closed by arguing that Germans had been overstating their contributions to the scientific world for years. Although the Germans had improved many technical applications of science, wrote the author, German contributions to "scientific truth" were minimal—a sentiment that strongly echoed Ramsay's earlier editorial.[21]

By the early fall of 1918, it was clear that Britain and the Allies would emerge victorious in the Great War (the Germans officially surrendered on November 11, the date that became known as Armistice Day), and the conversation changed from condemnation of Germany's crimes to a discussion of how to deal with German science and scientists once the war was over.

In September 1918, Thomas de Gray, the sixth Baron of Walsingham and a noted entomologist, wrote a letter to the editor suggesting that more naturalists ought to follow the example set by the zoologist Sir George Hampson, who had refused to use German nomenclature or cite German authors since 1914. "None but a German would use the German language by preference for scientific descriptions of species or genera," declared Walsingham. Furthermore, Walsingham argued that Germans should be excluded from future scientific congresses.

> Are American, English, French, or Italian naturalists to be expected to meet Germans and to join them in friendly discussion on the various questions that may arise? . . . Let us trust that for the next twenty years at least all Germans will be relegated to the category of persons with whom honest men will decline to have any dealings.[22]

An entry in the Notes column for October 17 made it clear that most naturalists disagreed with Walsingham's proposal to eliminate German nomenclature, largely because changing well-known terminology would be confusing.[23] But the geologist H. H. Godwin-Austen vigorously supported Walsingham's wish to exclude Germans from scientific congresses. In his own letter to the editor, Godwin-Austen praised Walsingham's "excellent letter," adding that the Allied nations could not simply meet the Germans "just as if nothing had happened since 1914." Although he had known many honorable German men of science when he was young, Godwin-Austen wrote, "In those days they were quite different men in every way from those of today, so complete a change has come over the whole German population. It is sincerely to be hoped they will never be employed again in any capacity."[24]

Unsurprisingly, relations between *Nature* and German science did not improve overnight following Armistice Day.[25] The attitude toward German scientists in *Nature* softened somewhat over the course of the 1920s; many *Nature* contributors began suggesting that ties with Germany should be renewed, and Gregory himself wrote an editorial urging the International Research Council to admit former Central Powers to their organization.[26] But *Nature*'s contributors would never again use their most prominent scientific weekly to complain that Britain's science was inferior to Germany's.

INTERWAR LETTERS TO THE EDITOR

Hard feelings during and after World War I led *Nature* contributors to cease comparing Britain unfavorably with Germany. Another trend also contributed to a shift in *Nature*'s tone: the increasingly international origins of *Nature*'s Letters to the Editor column. In the early twentieth century, *Nature*'s Letters to the Editor continued to follow the pattern we saw in chapter 4. The column printed a few letters from international scientists, but British contributors were still responsible for an overwhelming majority of the letters. *Nature*'s discussion of relativity provides a helpful example of this trend. Although little was published in *Nature* about relativity before the First World War, the journal was a major host of discussions about relativity following Arthur Eddington's eclipse expedition in 1919. The relativity discussions in *Nature*, however, were almost entirely confined to British participants such as Eddington, Oliver Lodge, Norman R. Campbell, and Herbert Dixon.[27] Even the special issue of *Nature* devoted to relativity contained contributions from only three foreign sources—Albert Einstein, the German mathematician Hermann Weyl, and the Dutch physicist Hendrik Lorentz.[28]

In the field of quantum mechanics and atomic physics, however, the material in *Nature* continued building on the international trends we saw in radioactivity research in chapter 4. Rutherford still contributed, as did his colleagues at the Cavendish Laboratory (where Rutherford moved in 1919), but international contributors took on a new prominence in the years following World War I. The continued growth in international physics contributions was closely linked to the career of yet another Rutherford student: the Danish physicist Niels Bohr (1885–1962).

Bohr had come to Rutherford's lab in Manchester as a postdoctoral researcher in 1912 following a successful stint at the Cavendish Laboratory under J. J. Thomson. He returned to Denmark in 1913 and published his famous model of atomic structure later that year.[29] At the beginning of World War I, Bohr returned to Manchester, and while in Rutherford's laboratory he adopted the practice of sending preliminary results to *Nature* as letters to the editor.[30] In 1916 Bohr accepted a professorship in theoretical physics at the University of Copenhagen; four years later, the university approved the foundation of an Institute for Theoretical Physics, which would become one of the most influential scientific centers of the mid-twentieth century. Bohr also won the 1922 Nobel Prize in Physics for his work on atomic structures. He and Rutherford remained close friends and frequent correspondents until Rutherford's sudden death in 1937.[31]

Bohr was one of the most prolific scientific mentors of the twentieth century, and Bohr's pupils followed him onto the pages of *Nature* as Rutherford's pupils had followed him. Contributions from Bohr's center at Copenhagen were extremely frequent in the interwar years.[32] Most of these contributions were letters to the editor and followed the pattern of using *Nature*'s correspondence columns to establish priority for exciting results. An especially noteworthy piece from the institute was a 1923 letter coauthored by the Danish physicist Dirk Coster and the Hungarian physicist George Hevesy announcing the discovery of one of the missing elements Henry Moseley had predicted. Coster and Hevesy proposed naming the element with atomic number 72 "hafnium" in honor of Copenhagen.[33]

Bohr was not the only Rutherford student to make his mark on *Nature* during the interwar years. Another Rutherford protégé, the British physicist James Chadwick (1891–1974), wrote one of the most influential *Nature* papers of the 1930s. Chadwick had been a student of Rutherford's at Manchester in the early 1910s. He was in Berlin on a scholarship when the war broke out and spent the war interned in a German camp for enemy aliens; the poor food and conditions in the camp would cause Chadwick lifelong health problems. After Chadwick's release in 1919, Rutherford hired his former pupil at Manchester, and shortly afterward Chadwick followed his mentor to the Cavendish Laboratory at Cambridge.

In 1932, Chadwick read a paper by Irène Curie (Marie and Pierre Curie's daughter) and Frederic Joliot suggesting that gamma radiation from beryllium disintegration had the power to force the ejection of alpha and beta particles from lighter elements.[34] Chadwick felt the Joliot-Curie results were not consistent with lighter elements bombarded by gamma rays. He hypothesized that the emission from beryllium was in fact an as yet undiscovered atomic particle with a mass of one and a neutral charge. After three weeks of intense work with the beryllium emissions, Chadwick felt his suspicion had been confirmed and submitted a letter to *Nature* on the "Possible Existence of a Neutron." The letter was printed on 27 February 1932.[35] Physicists and chemists quickly embraced Chadwick's discovery. A large amount of material on neutrons would be published in *Nature* through the end of the 1930s, including letters from such diverse locations as Paris, Vienna, Chicago, Warsaw, Calcutta, Kiev, Leningrad, Moscow, and Osaka.[36]

The tactic of using *Nature*'s Letters to the Editor to announce results also spread to those who were not Rutherford or Bohr students. One example is the Italian physicist Enrico Fermi, who published several letters in *Nature* during the 1930s, including the announcement of his discovery of radioactivity induced by neutron bombardment.[37] (*Nature* also famously rejected Fermi's 1933 submission on his theory of beta decay, now regarded as one of the most important papers in the history of particle physics, on the grounds that it was too speculative.)[38] Scientists working in new physics research centers in Japan began publishing letters in *Nature* and contributed several pieces on atomic theory and artificial disintegration.[39] Even in the field of genetics, where the contributors before World War I had been almost entirely British, scientists from diverse locations such as Prague, Helsinki, Berkeley, Princeton, Stockholm, Hamburg, and Moscow also began sending their latest findings to *Nature* during the interwar period.[40]

By the 1930s, the volume of contributions to the Letters to the Editor column had become so large that Gregory and the *Nature* staff felt the subject was worth discussion in the journal's News and Views column. A notice on Letters to the Editor on 10 February 1934 examined the expansion in the number of letters and the new purpose the column was serving for *Nature*'s contributors.

> During the year 1933, no less than four hundred communications appeared in NATURE under the heading of 'Letters to the Editor,' the big majority of which were the first announcements to be published of new work—news from the actual contributors to advances in science. Of this total, 201 were from scientific workers in universities and similar research centres in Great Britain and Ireland, and the remainder, 199, were from workers abroad distributed by continents as follows: Europe 78, America 57, Asia 37, Australia 14, Africa 13. . . . Science truly is not confined by national boundaries. We think it is a high compliment that scientific workers all over the world should regard our columns as the appropriate place to announce the progress of their labours and to discuss scientific matters and topics in which science and its methods are involved.
>
> Already this year we have printed 88 columns of 'correspondence', including the 20 columns appearing in this issue. Yet the waiting list is still large. The amount of space which can be given to 'letters' in a normal issue of NATURE must of necessity be limited if the journal is to discharge the remaining part of its function as a general journal of science, and we may even be obliged in the future to ask correspondents to limit their 'letters' to about five hundred words, or one column of space. For the present, we would urge them most strongly to be concise and precise in their communications.[41]

This short News and Views notice is worth further analysis for several reasons. First, the data on the Letters to the Editor for 1933 indicate that half of the published letters to the editor now came from countries other than Great Britain. Nearly 70 percent of the non-British letters were of European or American origin, but 19 percent had come from Asia, indicating that *Nature*'s influence had spread to new centers of scientific research in China and especially Japan.[42] Second, the notice indicates the expansion of the trend we saw in chapter 4. The Letters to the Editor column was now largely devoted to announcements of new work and was so popular for this purpose that even though the column had expanded greatly under Gregory's editorship, the journal had a substantial waiting list of letters that had not yet been printed.

The expanding size of the Letters to the Editor column supplied a recurring topic of discussion in News and Views over the next couple of years as the journal tried to clear a backlog by devoting expanded columns and even supplemental sections to correspondence.[43] Most of these short News and Views notices proudly mentioned the international origins of the Letters to the Editor. The global reach of the Letters to the Editor was particularly noteworthy in light of the political tension of the 1930s, as a February 1936 note made clear. "In these days of political upheaval," said the notice, "it is an encouraging thought that among men of science there is still a strong bond of common interest in original investigations and results, and we are gratified that they should select NATURE as the vehicle of their communications."[44]

Even as a second European war loomed, *Nature* maintained its influence as a center for the publication of recent findings. In January 1939, the journal received a submission from two Austrian-born physicists, Otto Frisch (1904–1979) and Lise Meitner (1878–1968). Their Letter to the Editor proposed a startling explanation for some recent experimental findings. Meitner's former colleagues in Berlin, Otto Hahn and Fritz Strassmann, had bombarded uranium nitrate with neutrons and discovered that their sample subsequently contained barium.[45] "Hahn and Strassmann," wrote Frisch and Meitner, "were forced to conclude that *isotopes of barium ($Z = 56$) are formed as a consequence of the bombardment of uranium ($Z = 92$) with neutrons.*"[46] Frisch and Meitner offered a hypothesis on what had occurred. They suggested that the uranium nucleus had, in fact, split in two—resulting in one barium nucleus and one krypton nucleus. They also proposed a mechanism for how the nucleus could have split: "On account of their close packing and strong energy exchange, the particles in a heavy nucleus would be expected to move in a collective way which has some resemblance to the movement of a liquid drop. If the movement is made sufficiently violent by adding energy, such a drop may divide itself into two drops."[47] Meitner and Frisch's hypothesis was a remarkable one; it had long been taken for granted that the dense atomic nucleus simply could not split. The Meitner-Frisch letter proposing the "liquid drop" model of nuclear fission has become one of the most famous papers in the history of physics.

For Meitner, *Nature* was not a natural choice. The Austrian physicist had spent most of her career at the Kaiser-Wilhelm-Institut in Berlin, where she worked closely with Rutherford's former student Otto Hahn. She had remained in Germany after Adolf Hitler's rise to power in spite of her Jewish ancestry, but when Germany annexed Austria, Meitner fled Germany and accepted a job at the Nobel Institute in Stockholm.[48] Although Meitner was known and respected in Britain—*Nature* published a congratulatory note on her sixtieth birthday along with a glowing account of her scientific career[49]—she had never published an article in *Nature*, preferring instead to direct her work to German scientific journals. Such journals, of course, were now less accessible to her after her flight to Sweden.

Frisch, both a fellow physicist and Meitner's nephew, probably drove the decision to publish in *Nature*. Frisch was visiting Meitner in Stockholm in the winter of 1938 when Hahn sent Meitner a copy of his new results and asked for her thoughts on how to interpret the findings.[50] Frisch and Meitner examined Hahn's results and worked out a model for how the nucleus might split. It seems likely that Frisch suggested *Nature* as the best journal for their theory. He had been at Copenhagen since 1933, when Hitler's rise to power prompted him to leave his post at the University of Hamburg for one at Bohr's Institute for Theoretical Physics, and he had coauthored many letters to the editor in *Nature* with his colleagues in Denmark.[51]

Following the publication of their letter in *Nature*, Meitner and Frisch were frequently credited as the "discoverers" of nuclear fission. This dismayed Hahn, who (like his former mentor Rutherford) was conscious of establishing scientific priority, especially for this discovery.[52] Hahn was particularly annoyed by a *Nature* article by the physicist Norman Feather that seemed to suggest that the French physicists Irène Curie and Paul Savitch had produced nuclear fission in the lab before he and Strassman had.[53] Hahn was agitated enough to ask Feather to pass on his own letter to *Nature* clarifying that Curie and Savitch's results were not the same as those he and Strassman had obtained.[54] Although Hahn had largely ceased to publish his own results in *Nature* following the First World War, his interest in the *Nature* coverage of the fission discovery indicates that he worried *Nature* might be influential enough to strip him of credit for the discovery. Hahn's concern over *Nature*'s material on fission was especially remarkable given that *Nature* had been functionally banned in Germany for over a year.

NATURE AND INTELLECTUAL FREEDOM IN THE USSR, THE UNITED STATES, AND GERMANY

In the nineteenth century, *Nature* had frequently engaged with political and scientific issues in Great Britain, but before the First World War, the editors and contributors limited commentary on scientific issues outside the British Isles to chatty human-interest pieces on science in other countries and complaints about how Great Britain lagged behind other nations in its financial support of science. The international crisis of World War I and British outrage over German actions during the war opened the door to criticism of other countries. Furthermore, the expanding influence of Letters to the Editor made *Nature* more visible internationally and may have emboldened the newly promoted Richard Gregory to pursue more critical discussions of scientific issues outside Great Britain.

Nature remained grounded in its British roots—the 4 May 1935 issue was devoted to the progress of science during the reign of King George V, and the King's death the following January occasioned black borders on *Nature*'s pages, as had the deaths of Queen Victoria and King Edward VII before him.[55] But in the years following the end of World War I, *Nature* expanded the scope of its commentary beyond Britain's borders. Much of this international commentary was grounded in *Nature* contributors' conviction that science should not be subordinated to a religious or political agenda, an ideal that came to be articulated as a concern for preserving intellectual freedom. Restricting science, the contributors argued, was wrong because scientific truths were more valuable morally than either political or religious beliefs—an argument very much in line with the vision of science Richard Gregory had promoted in *Discovery*.[56]

After the end of World War I, the newly formed USSR was the first country to receive *Nature*'s scrutiny for restricting science. Much of *Nature*'s commentary on what was happening in the USSR came from citizens of the affected countries, especially scientific workers who had left their homes after the Russian Revolution. In the early 1920s Soviet émigrés wrote several pieces drawing *Nature* readers' attention to the situation of science under the Bolshevik regime. Boris Sokoloff, a noted socialist writer and a former professor of biology at Petrograd University who immigrated to the United States in 1920, contributed a 1921 article describing a grim situation for science under Lenin's leadership. "Science in Russia is now passing through difficult times," he wrote. The Soviet government was campaigning against "bourgeois science" and "Science [was] struggling with politics for its freedom."[57] Several more Russian and Eastern European correspondents wrote in to *Nature* to express their agreement with Sokoloff. The English professor J. W. Mellor sent in a letter from an anonymous Russian professor of chemistry who said that the situation of science under the Bolsheviks was extremely dire and that a vast number of Russians, including scientists, faced food shortages.[58] A professor in Prague, Bohuslav Brauner, agreed with Mellor's anonymous colleague. Citing the experiences of an unnamed Russian friend who had just "escaped" from the USSR, Brauner said that scientists there "have practically no rights, and there is no possibility of free scientific work."[59] A Russian émigré named Vladimir Korenchevsky wrote to *Nature* to solicit donations for the American Relief Administration, which promised to use the funds to send food to struggling Russian scientists.[60] A few months later, Richard Gregory and C. Hagberg Wright, colleagues at the British Science Guild, appealed to British scientists to donate scientific literature for their Russian colleagues.[61] Notably, many of the contributions to *Nature* from Russian émigrés came from scientists who had moved to the United States, not to Britain—further evidence of *Nature*'s growing international reach.

Commentary on scientific affairs in the USSR abated somewhat in the mid-1920s but resumed in the 1930s with two controversial episodes: the detainment of the Russian-born physicist Peter Kapitza in 1934 and the cancellation of the International Genetics Congress in 1937. Kapitza was a former student of Rutherford's who had spent many years in the United Kingdom. On a trip to the USSR in September 1934, Soviet officials informed him that he would not be permitted to return to the United Kingdom because the Soviet government needed his scientific talents.[62] This development alarmed Kapitza's friends, and *Nature* reported the episode with concern in its News and Views column, writing "This commandeering of Kapitza's services on behalf of the USSR ignores the personal and psychological factors involved.... It comes as no surprise to his friends to learn from reliable sources that his health has already been seriously impaired by anxiety and strain."[63] The complaints had little effect; Kapitza would spend the rest of his life in the USSR.

The suppression of genetics in the USSR provided another reason for *Nature* contributors to criticize the Soviet government. In the mid-1930s, Trofim Lysenko—now one of the most infamous figures in the history of science—gained the ear of Soviet leader Joseph Stalin with his theory that Soviet agriculture could be improved by using environmental factors to alter crop heredity (an idea he drew from the work of the Russian horticulturalist Ivan Michurin). Lysenko was an opponent of "theoretical" genetics, and Stalin was attracted to the idea of science that valued "practice" and results over abstract theories. By the late 1930s, opponents of Lysenko's ideas were subject to political scrutiny and even censorship. One of Lysenko and Stalin's main targets was the respected geneticist Nikolai Vavilov, whom they accused of promoting Western, bourgeois theories that offered no useful applications for the USSR.[64]

Lysenko's ideas had previously been discussed in *Nature*; a 1936 article by the botanist V. H. Blackman discussed and dismissed Lysenko's theories but did not treat them as obviously unscientific.⁶⁵ The attitude in *Nature* toward Lysenko became much more negative after the Soviet government canceled the 1937 International Congress of Genetics, which was supposed to be held in Moscow. The Soviet government justified the decision by declaring that geneticists were anti-Marxist and promoted ideas hostile to the Soviet government.⁶⁶ *Nature* also reported—incorrectly, as it turned out—that Vavilov had been arrested after the cancellation of the Congress.⁶⁷

The News and Views column of 30 January 1937 began with an extremely critical item on "Genetic Theory and Practice in the U.S.S.R." The piece scornfully described Michurin and Lysenko's work as "Lamarckian" and expressed astonishment that the Soviet government had elevated Lysenko's "vernalization" above genetics.⁶⁸ Notably, this was followed by an item titled "Scientific Freedom," which praised a resolution by the American Society of Naturalists condemning political interference in scientific progress.⁶⁹ In August 1937, after receiving translations of several speeches at the Lenin Academy of Agricultural Sciences in Moscow, *Nature*'s staff devoted a news article to refuting Lysenko's theories as described in the speeches. The article concluded that "on the whole, the discussions [of Lysenko's theories] appear to be of very limited theoretical interest. They are, however, of outstanding significance in revealing the atmosphere in which scientific investigators in totalitarian countries have to live and work."⁷⁰

Nature's contributors also had much to say about controversies in the United States in the mid-1920s over the teaching of evolution in public schools. Twelve days after the governor of Tennessee signed a bill forbidding public schools from teaching any theory that conflicted with the Biblical creation story in Genesis, *Nature*'s lead editorial took the bill and its principal supporter, William Jennings Bryan, to task for their "futile" attempt to refute Darwinism.⁷¹ The editorial's author argued that Southern racial beliefs were fueling the fight against Darwinism because the theory of evolution suggested that Africans and Caucasians might have a common ancestor. The editorial criticized the "educational backwardness" of these racial beliefs but said that given the low standard of living in the American South, "it is not surprising that they retain beliefs which, according to British notions, are decades out of date." The editorial's author predicted that Bryan's fight to refute Darwinism "will probably fail in the end as completely as his famous appeal to the United States to adopt bimetallism."⁷²

While the initial editorial was dismissive of Bryan and his allies, J. T. Scopes's trial for teaching evolutionary theory at his Tennessee high school occasioned greater alarm.[73] On July 11, the day after Scopes's trial began, *Nature* printed a fifteen-page special supplement on "Evolution and Intellectual Freedom." The supplement led with an editorial asserting that politics and religion were never acceptable justifications for the restriction of science. "We cannot help being astonished," the editorial said, "that there should be States in the United States of America which deliberately adopt a policy of scientific stagnation."[74]

The journal's staff asked several prominent British professors and religious thinkers to write short letters explaining their views on the Tennessee law. While there was some disagreement about the seriousness of the Tennessee bill and the Scopes trial—E. Ray Lankester said that so long as the universities remained unaffected, scientific inquiry would not be seriously hindered, and D'Arcy Wentworth Thompson of the University of St. Andrews considered evolution too advanced to be taught in secondary schools at all—the contributors to the supplement all said that there was no doubt about whether evolution was a real phenomenon. The religious figures who wrote letters for *Nature*'s supplement declared they did not consider the theory of evolution to be in conflict with their religious beliefs. The Reverend J. Scott Lidgett of London went so far as to say that the Fundamentalist point of view was "injurious" to Christianity, as it assumed that God only acted on the world from outside it instead of acknowledging "the deeper teaching of Scripture as to the organic relation of God to His World."[75] Several follow-up letters from other contributors continued on this theme: the Tennessee ban was wrongheaded, and the belief that Genesis superseded evolutionary theory was backwards.[76]

Perhaps the most notable feature of the material on the Scopes trial is the way *Nature*'s contributors characterized Britain as a model of a society that had moved beyond religious control over science—a rhetorical strategy in strong contrast to the journal's older pattern of comparing Britain unfavorably to other nations. At least one prominent figure in American science took offense at *Nature*'s superior tone. J. McKeen Cattell, the editor of *Science*, wrote a letter to the editor in late July stating,

> Not intending to bite the hand that feeds us, I still venture to express a doubt as to whether the strength and courage of American men of science in their efforts to attain the intellectual freedom established in Great Britain will be greatly forwarded by the series of little articles published in NATURE and by the editorial comments.[77]

The most visible commentary on foreign science in *Nature* during the interwar period, however, was about the academic policies of the National

Socialist government. Shortly after Adolf Hitler assumed emergency powers in March 1933, Albert Einstein was declared a traitor. On April 7 the Nazi government passed the Law for the Restoration of the Career Civil Service, which declared that government employees who lacked the "proper" qualifications, who had unreliable political affiliations, or who were of non-Aryan descent would be fired. This included university professors and scientists working for government laboratories. Over one-fifth of German scientists lost their positions between 1933 and 1935 because of the Nazi civil service rules.[78]

Nature's editor and contributors were highly critical of the firings and of Nazi views in general. A report on a psychology conference in Germany noted the absence of Jewish leaders in the field and questioned the scientific validity of the conference's goal, which was to promote "a psychology which expresses the genuine German spirit."[79] In August 1935, the Nazi government removed Dr. Arnold Berliner from his position as the editor of the journal *Die Naturwissenchaften* because of his Jewish birth. *Nature*'s News and Views column expressed "much regret" over Berliner's firing and printed an excerpt from a letter by an anonymous friend of Dr. Berliner's praising the work he had done with *Die Naturwissenschaften*.[80] Contributors to *Nature* stridently questioned the supposed scientific evidence for the superiority of the Aryans and the degeneracy of the Jews. A 5 August 1933 item in News and Views described Hitler as "ill-balanced, fanatical and otherwise abnormal" and said the German leader's racial views "belong to a 'science' which would be out of date even if it had not failed to justify itself when submitted to the test of scientific analysis."[81]

The insufficient scientific basis for Nazi racial theories was a theme *Nature*'s editors and contributors would return to again and again in the years following Hitler's coup. The lead editorial for 17 February 1934 claimed that the German leadership "has glorified and idealised war" in the name of "Aryan virtue." But the entire concept of an Aryan race was scientifically ridiculous, said the editorial's author: "neither physical nor cultural anthropology endorses the exclusive ideal of 'Aryanism' as having a basis in historic fact; and a patriotism which pursues its end without regard to considerations of logic or common sense may in the long run be as destructive of the Fatherland as treason."[82] Six months later, another leader on "The Aryan Doctrine" stridently questioned the claim that Germans were descended from a highly cultured Nordic racial line and described Aryan racial theories as either "untenable or discredited."[83] Julian Huxley revisited the scientific bankruptcy of Aryan racial doctrines in March 1936 in a discourse at the Royal Institution titled "The Race Problem." *Nature* reported on Huxley's talk in their News and Views column, writing that in scientific terms, there was no such thing as an "Aryan race."[84]

Nature contributors also criticized the German government's attacks on intellectual freedom, especially scientific freedom. An editorial about the physiologist A. V. Hill's 1934 Thomas Huxley Memorial Lecture praised Hill's defense of scientific internationalism but expressed concern about science in certain national contexts. The editorial's author was especially worried about science in Germany, which had "evicted some of her greatest scientific investigators ... content, as it had been put, that her science should lag behind that of the rest of the world, provided that it were German."[85] *Nature*'s staff also expressed great alarm over the teaching of history in Germany after the release of a circular instructing German teachers to emphasize the heroic history of the Aryan race. They accused the Nazi Party of conspiring to "abandon all standards of intellectual honesty in pursuit of a political ideal."[86] The Nazi Party, according to *Nature*, cared nothing for the truth unless it served their political ends.[87] An editorial from February 1936 declared that although many countries had compromised scientific progress for nationalistic goals, "the major threat to academic freedom has come from Germany."[88] Similarly, a June 1936 editorial on "The Protection of Scientific Freedom" declared that "the devastation of the German universities continues, and in Russia and Italy freedom of study and teaching in large portions of the field of learning are still proscribed,"[89] and it urged British scientific workers to take a more active role in politics to avoid befalling a similar fate. Following a report on the opening of the Philipp Lenard Institute at Heidelberg, at which several leading figures in the German academic world had given speeches decrying "Jewish physics" or "Jewish influence" on learning, a British correspondent writing under the initials "P. F. F." suggested that German scientists were facing the same situation as Galileo during the Inquisition.[90] A lead editorial for 5 June 1937 put these sentiments the most bluntly: *Nature*'s front page declared that Nazism "is the form of totalitarianism that has been most destructive of science" and said that German Minister of Science and Education Bernhard Rust had "degraded" science in Germany.[91]

Notably, the reaction to National Socialism in *Nature* was more negative than the reactions in Britain as a whole. There were a wide range of responses to Nazism in Great Britain, ranging from moral outrage to indifference to outright support.[92] The statements about Nazism in *Nature*, on the other hand, were unreservedly negative, possibly because British scientists knew many colleagues in Germany who had been affected by the civil service firings. Some of *Nature*'s antifascism may also have stemmed from the presence of several leading Marxists in the British scientific community, including J. B. S. Haldane, J. D. Bernal, and Lancelot Hogben. The Nazi Party was extremely hostile to Communists, and the sentiment was returned.[93]

These criticisms did not go unnoticed in Germany. The Nobel Prize-winning German physicist Johannes Stark led the charge in *Nature* in defense of German academic policies.[94] Stark, a member of the Nazi Party since 1930, was one of the new regime's most visible scientific advocates. Despite his Nobel Prize, Stark had long felt that he had been pushed to the side of the German physics community. He had resigned a professorship at Würtzberg in 1922 in order to pursue commercial interests; when this proved unsuccessful, he tried to reenter academic physics but could not find a position. Stark came to believe that a Jewish conspiracy was responsible for his inability to secure an academic post. His infamous 1922 book *The Present Crisis in German Physics* offered a shrill indictment of relativity, quantum theory, and the perceived preference for theory over experiment in German physics. By 1933 he had secured a position at the Imperial Institute of Physics and Technology through the support of fellow Nazi sympathizer Philipp Lenard.

Although Stark had not previously contributed to *Nature*, he took it on himself to defend Nazi policies in the journal. Stark first wrote to *Nature* in response to extracts from A. V. Hill's Huxley Memorial Lecture. Stark called Hill's allegations against the Nazi government "inaccurate" and gave the following account of the civil service laws in Germany:

> The National-Socialist Government has introduced no measure which is directed against the freedom of scientific teaching and research.... Measures brought in by the National-Socialist Government, which have affected Jewish scientists and scholars, are due only to the attempt to curtail the unjustifiable [sic] great influence exercised by the Jews. In Germany there were hospitals and scientific institutes in which the Jews had created a monopoly for themselves and in which they had taken possession of almost all academic posts.... Only a very small part of the 600,000 Jews who earn their living in Germany have been affected by the National-Socialist measures. No Jewish civil servant was affected who had been in office before August 1, 1914, or had served at the front for Germany or her allies or whose father or son had fallen in the War.[95]

Two months later, Stark wrote in again to correct English "misunderstandings" of what was happening in Germany. He insisted that the effect of the new laws was actually quite small and that Jews were not the only employees who had been removed from their posts as the result of the new laws. Stark finished his letter by suggesting that *Nature* refrain from future commentary on the German government: "The withholding of criticism of the new regime in Germany, or at least a conscientious regard for the truth in scientific circles, will be to the advantage not only of international cooperation but also of the Jewish scientists themselves."[96]

A somewhat more measured and less anti-Semitic defense of the German regime came from Professor R. Woltereck, a German professor of agriculture. Woltereck said he understood the reasons for English alarm but that the measures were necessary to restore Germany's independence and sovereignty. If Britain had suffered catastrophe on the level of Germany after World War I, Woltereck insisted, the British would understand the need to "postpone everything, including scientific research (objective science), in order to strengthen the mental forces of the people, especially of its future leaders." He expressed confidence that Germany would "reinstate the full academic freedom of its universities and science, as soon as political sovereignty in our own country is assured."[97]

Nature's readers and editor were unmoved by such defenses. A. V. Hill, who had been sent an advance copy of Stark's remarks about his Huxley lecture, responded in the same issue as Stark's letter. He called Stark's anti-Semitism "absurd" and declared that it was a fact that a significant number of German Jews had lost scientific posts. "No doubt in Germany, after this reply, my works in the *Journal of Physiology* and elsewhere will be burned,"[98] he concluded wryly. Hill's fellow biologist J. B. S. Haldane was equally dismissive of Stark's justifications, writing, "The fact that non-Aryans have been expelled from other posts does not necessarily justify their expulsion from scientific positions unless the premise that 'two blacks make a white' has first been conceded."[99] Gregory himself wrote a short response to Woltereck, praising his "restrained and courteous letter" but insisting that men of science could not but deplore "the acceptance of a policy which teaches that to attempt to find and hold truth is but a secondary and subordinate activity of the human mind."[100]

Criticism of Nazi academic policies grew more strident as Germany prepared to celebrate major milestones for its two leading universities, Heidelberg and Göttingen, in 1936 and 1937 respectively. On 22 February 1936, an anonymous correspondent writing from Germany noted that the celebration of Heidelberg's 550th anniversary had been scheduled for June 30, the same date Hitler and the Nazi leadership had executed a violent purge of a paramilitary organization called the Sturmabteilung (SA) in 1934.[101] The correspondent lamented that the university where the great Jewish-born philosopher Spinoza had once held a professorship had now expelled forty-five professors "on the ground either of their opinions or of their Jewish descent."[102] The next week, the News and Views column wondered how many British scientific societies or universities would send delegates to the Heidelberg celebration; the University of Birmingham had already voted not to accept the invitation, and *Nature* predicted that Oxford and Cambridge would follow suit.[103] A week later, the Germans revoked all of the invitations to British universities with the sole exception of Cambridge.[104]

Nature's anonymous German correspondent continued to send the journal updates on the status of science in Heidelberg. In January 1937, he wrote that only 99 of the 215 faculty members who had held their jobs before the Nazi rise to power remained at Heidelberg and that there was rapid turnover among the new teaching staff. Furthermore, all students matriculating at Heidelberg were now required to provide proof of the religion of all four grandparents. "Science has been abolished in the German universities and its spirit has abdicated from the Reich,"[105] the correspondent declared. The same correspondent wrote again as the bicentennial of the University of Göttingen approached, naming a list of twelve eminent scientists who had either left or been fired from Göttingen. He concluded the article in pessimistic terms: "Göttingen ceased in 1933 to be a scientific centre. On June 30, visitors to Göttingen will celebrate a unique series of losses of learning, liberty and life."[106]

NATURE'S PLACE IN THE WORLD

The blunt criticisms of National Socialist academic policies and the state of German science proved too much for the Nazi government to ignore. With his 12 November 1937 order, Bernhard Rust swept *Nature* from "general usage" at all German universities and research libraries. Rust's concise order did not specify which articles had prompted the ban, although the timing suggests that the anonymous correspondent's articles about Göttingen and Heidelberg were the proverbial straw that broke the camel's back.

A Munich professor, H. Rügemer, elaborated on National Socialist feelings about *Nature* in a 1938 article for the *Zeitschrift für die gesamte Naturwissenschaft*, a scientific journal that had been founded in 1935 and taken over by Nazi Party officials in 1936 with the aim of promoting the "German spirit" of science.[107] Rügemer's article, "*Nature*, an abominable magazine," was part essay on the "Aryan foundations" of science and the importance of excluding Jews from scientific institutions, part refutation of *Nature*'s charges against National Socialism. Rügemer accused the British publication of being influenced by Germany's Jewish enemies. "Since 1933, *Nature*'s foreign science correspondence service has been built on an antifascist spy organization in Germany and Italy," Rügemer wrote. "We hope, that in the fondness of the English for fairness and understanding the rabble-rousing writings of a Jewish scribbler will fail to have their desired effect."[108] The comment about the "Jewish scribbler" appears to have been directed at *Nature*'s anonymous correspondent.

Despite the prominent placement Gregory and the *Nature* staff had given to criticism of the Nazi government, the journal's reaction to Rust's decree was fairly subdued. Rather than devote a lead editorial to the defense of the journal or print a lengthy article castigating the censorship of *Nature* in Germany, the staff relegated the restrictions on *Nature* to their News and Views column of 22 January 1938. The German government's new policy on *Nature* did not even lead the column. Instead, News and Views opened with an item on a recent American Association for the Advancement of Science resolution in favor of creating a World Association for the Advancement of Science that would protect scientific freedom. An item on "Prohibition of NATURE in Germany" followed this report.[109] The News and Views notice quoted Rust's order removing *Nature* from general use in libraries and then responded to Rust's claims:

> We welcome the opportunity of recording worthy additions to the literature of science or to natural knowledge from any country or any race; but we should be false to the traditions of science if we failed to condemn any influence which would make scientific research subservient to political or theological domination. The misrepresentation of our attitude contained in the announcement of the German Minister of Education is hard to bear, but we regret that the penalty involved in the withdrawal of *Nature* from libraries and other institutions will be felt more by some of our readers in Germany than by ourselves.[110]

Nature's condemnation in Germany illustrates both that country's increasing isolation and tensions within the German scientific community during the Third Reich. But it also shows the growth in *Nature*'s international influence—*Nature*'s importance as a publication site for recent experimental results made it a necessary subscription for German research libraries. Rust would not have bothered restricting the journal, and Rügemer would not have railed against it so vehemently, if they thought no one was reading it. And indeed, Otto Hahn's close reading of *Nature*'s coverage of the fission discovery shows that at least some German scientists continued reading it despite the restrictions. A *New York Times* article on the Nazi ban of *Nature*, which described the journal as the "manual of research workers," further illustrates *Nature*'s reach.

> London, Jan. 29—British scientists were shocked at the German Government's ban on Nature, internationally famous scientific weekly edited by Sir Richard Gregory.
>
> Sir Richard said today that he expected the ban would continue indefinitely.
>
> "Copies of Nature have not been displayed openly in German bookshops for some time," he said. "Now I suppose the universities will keep it in a secret drawer as if it were an obscene publication. Of course we shall continue to publish news of German scientific work and reviews of German publications. But we shall go on fighting for academic freedom."
>
> Research workers here recognize that no publication in the world tries to be more objective in its news columns. Week after week it publishes news of the latest laboratory discoveries in letters from scientists everywhere, including Germany.... Yet German scientists no longer will be able to keep abreast of discoveries announced in Nature except by special permission of the authorities.[111]

By 1938, *Nature* was internationally newsworthy.

CHAPTER SIX

"It Almost Came Out on Its Own": *Nature* under L. J. F. Brimble and A. J. V. Gale

At the same time that Richard Gregory was dealing with the German ban on his journal, he was also facing a more pressing *Nature* question. Like Lockyer before him, Gregory had decided it was time to find a successor. After "frank and friendly" talks with Daniel Macmillan over the editor's salary in May of 1938, Gregory chose to step down as of the end of the year. "'Nature' is now recognised as the leading international organ of science, and my association with this development makes me both proud & glad," he told Macmillan in a letter announcing his plans.[1] Sir Richard's claim about *Nature*'s international preeminence was perhaps a bit exaggerated (the Germans and the Soviets would certainly have disagreed with his assessment), but its centrality to the British scientific community could not be questioned, and the journal's international influence and readership had indeed grown significantly during Gregory's editorial tenure.

Unlike his predecessor, Gregory had no qualms about recommending his editorial assistants Arthur J. V. Gale (1895-1978) and Lionel J. F. "Jack" Brimble (1904-1965) as his replacements. Although neither man had Gregory's public profile (and, indeed, neither man seems to have been interested in acquiring such a profile), Gregory assured Macmillan that "Mr Gale & Mr Brimble are quite capable of carrying on and of maintaining the high reputation of the journal."[2] But Brimble and Gale's capabilities would soon face a major challenge. On 1 September 1939, Germany invaded Poland. Two days later France and Britain declared war on Germany. For the second time in twenty-five years, Europe was at war.

Brimble and Gale's editorship was shaped in the crucible of these wartime years. They successfully guided *Nature* through wartime paper shortages, postage restrictions, and last-minute editorials about the Manhattan Project, but their habit of running a stripped-down, low-maintenance version of *Nature* would carry on well past the Second World War. As a result, Brimble and Gale's postwar *Nature* lost much of the lively, controversial character that had distinguished *Nature* from its competitors in earlier years. But paradoxically, the relatively staid Brimble and Gale period was responsible for some of the most famous articles ever printed in *Nature*, including the 1953 DNA papers and some major contributions to the development of plate tectonics. And even though Brimble and Gale's *Nature* showed minimal interest in science outside of Britain's borders, the percentage of international contributions to *Nature* steadily increased after the war.

While Brimble and Gale's *Nature* made for less exciting reading than Lockyer's or Gregory's, their tenure gives us an opportunity to assess *Nature*'s place in the postwar scientific publishing landscape—and an opportunity to think about how editorial styles might affect the reception and reputation of a journal. While Brimble and Gale's laissez-faire approach to editorship made *Nature* comparatively dull, their editorship did not remove *Nature*'s desirability as a site for publishing new research findings or theories. Throughout their tenure *Nature* remained a widely read journal that attracted a solid number of submissions (in fact, rather more submissions than either of the editors might have liked). Perhaps most interestingly, *Nature* was still seen as scientifically respectable even though its editorial staff often eschewed outside refereeing—which in turn suggests that the history of peer review at scientific journals is more complicated than many observers have assumed.

NATURE AND THE SECOND WORLD WAR

Gale and Brimble had both been at *Nature* for a number of years when they assumed their joint editorship in January 1939. Gale had graduated from Selwyn College, Cambridge, with a degree in agriculture, and spent the First World War in military service. In 1920 he accepted a position as Gregory's assistant, with a salary of £200 per year for two days' work per week.[3] Brimble, nine years Gale's junior, came to the journal in 1931. He had earned his BS degree from the University College of Reading and spent several years as a lecturer on botany, first at the University of Glasgow, then at the University of Manchester. Brimble had never been a passionate researcher, however, and after Brimble wrote a book review for *Nature* that Gregory admired, the Macmillans offered Brimble the opportunity to work for the journal as Gregory's second assistant.[4]

In November 1938 Gale wrote to Gregory to say that he had discussed the joint editorial appointment with Daniel Macmillan, who had officially confirmed the position and offered him and Brimble raises.

> I expect the directors discussed the financial aspects of the "Nature" changes with you, but in case they did not, you will be glad to know that both Brimble and I are to have substantial rises. Mr. Dan sent for me this afternoon, formally confirmed the new appointments and offered me an increase of £200. Directly or indirectly I know you are responsible and I am grateful—but you understand.
>
> Mr. Dan also promised not to interfere with Nature editorially—that was a useful point I think.[5]

Daniel Macmillan likely felt that there was little risk in promising Brimble and Gale editorial autonomy. Not only was the pair well versed in *Nature*'s workings, but unlike their predecessors, neither Gale nor Brimble brought much of a personal agenda to *Nature*, and neither man was much interested in leveraging his position as editor of the journal into wider public influence. In the absence of an editorial archive, it is difficult to determine exactly how Brimble and Gale divided their responsibilities, but it is clear that the new editorial team was invested in maintaining the status quo and not in remaking Macmillan's successful weekly.

Nothing Brimble and Gale had done under Gregory, however, had prepared them for managing *Nature* during a period of crisis such as the Second World War. At first the war had minimal impact on *Nature*'s management and was only reflected in the journal's editorials. Interestingly, given the contributors' strong anti-Nazi sentiments, *Nature*'s first editorial on the war focused not on any German wrongdoing but on an optimistic vision of what the world could be like after the war:

> If the fruits of victory are ultimately to be reaped, we must bring to the struggle not merely the full force of our moral and material resources, but also constructive and imaginative statesmanship.... Science, at least, has given men a vision of the world that might be when man's moral and spiritual development is in keeping with his material advance. If that world is ever to be realized, scientific workers, amid the stress of the present emergency, must guard zealously their loyalty to truth, not less than their belief that science transcends national frontiers.[6]

The possibilities of the postwar world would be a recurring theme of *Nature*'s wartime editorials, almost all of which were written by Rainald Brightman, the chief librarian for the Dyestuffs Group at Imperial Chemical Industries.[7] The wartime leaders were particularly concerned with how the postwar rebuilding process could benefit the development of Britain's empire; Brightman offered many suggestions on how science and technology could improve quality of life in Britain's colonies once the war was over.[8] (These suggestions did not include decolonization.)

Despite the optimistic tone of its editorials, the war left *Nature* far from unscathed. The *Nature* offices themselves escaped damage during the London Blitz, but Brimble was injured in a midwar bombing raid and would never fully recover. Wartime paper restrictions cut *Nature*'s average number of pages and, although Brimble and Gale maintained consistent dating of *Nature*'s issues, printing and mailing of many wartime issues was delayed.[9] In June 1940, Brimble and Gale announced that they would eliminate summaries of the Letters to the Editor, would stop issuing supplemental sections of short book reviews, and would limit all Letters to five hundred words.[10] A year later, subscribers were notified that *Nature*'s size would be further decreased and that the price of an issue would go up 2 pence, to 1s 6d.[11] The following volume eliminated the masthead to save space. In 1939 an issue of *Nature* averaged about forty pages; by 1943 this had fallen to thirty pages.

The war also affected *Nature*'s correspondence with its contributors. In May 1940 the News and Views column informed readers that the receipt of manuscripts could no longer be acknowledged and that authors would only receive a single proof of their articles.[12] In March 1941, Brimble and Gale ceased issuing proofs to authors outside Great Britain.[13] They also discontinued the practice of sending multiple copies of *Nature* to authors of Letters to the Editor, instead restricting authors to a single copy.[14]

Recovery came slowly after the war with Germany ended in May 1945. Attempting to clear a significant backlog of unpublished letters, Brimble and Gale shrunk the type of the Letters to the Editor column in February 1946; just nine months later they restored the original type, not because the backlog had been cleared but because subscribers were complaining that the smaller type was too difficult to read.[15] For years after the war's end, government restrictions on printing and electric power occasionally delayed issues of *Nature*.[16] It was not until mid-1948 that *Nature*'s issues returned to their average prewar length of forty-plus pages.

The wartime experience permanently affected the way Brimble and Gale edited *Nature*. Wartime paper restrictions and mail problems led them to publish a stripped-down version of *Nature* that contained little high-maintenance debate or discussion. Instead, wartime *Nature* ran many reprints of lectures, book reviews, and letters to the editor detailing new theories and recent experimental findings. These characteristics carried forward into the rest of Brimble and Gale's editorship. The controversies and debates that had made prewar *Nature* distinctive among specialist scientific periodicals in Britain all but vanished during the Brimble-Gale era. Avoiding heated discussions seems to have been an explicit policy rather than an inadvertent omission. In 1950, for example, the News and Views column informed readers that a recent article by Julian Huxley on Soviet genetics "could clearly not be allowed to become the subject of debate in the correspondence columns of this journal"—a sentiment that would surely have seemed odd to anyone who remembered *Nature* under Lockyer or Gregory.[17]

Brimble and Gale's low-maintenance editorial style affected other areas of the journal as well. Unlike their predecessors, Brimble and Gale wrote very few of *Nature*'s famous lead editorials. Instead, almost every editorial from 1939 to 1966 came from the pen of Rainald Brightman. Brightman had begun writing editorials for *Nature* in 1931, when Richard Gregory's attention was increasingly split between *Nature*, his work for Macmillan's scientific books division, and other organizations such as the British Association for the Advancement of Science and the British Science Guild.[18] When Brimble and Gale took over in 1939, Brightman was already *Nature*'s most prolific editorial writer; he would soon become almost its only one. Brightman's articles in the 1950s and early 1960s generally summarized the contents of a recent government report or a lecture (often one that would be reprinted later in that week's issue) and offered a few points of commentary at the end. The tone tended toward straightforward reporting rather than provocative editorializing, and Brightman's preferred subjects were uncontroversial ones, such as the state of technical education in Great Britain.[19]

A typical example of a Brightman leader is "University Expansion in Britain," the editorial for 22 June 1957.[20] The editorial was based on a three-part *Manchester Guardian* article by Sir James Mountford, the vice chancellor of the University of Liverpool. Mountford, as Brightman explained, had expressed concern over whether British universities would be able to train enough scientists and engineers in the coming decades. Brightman summarized some of the debate about the article that had taken place in the House of Lords, explained various recommendations that the University Grants Committee was currently considering, and closed the article by adding that expansion in postgraduate scientific education would have to be accompanied by an expansion in the number of technicians and craftsmen qualified to build laboratory equipment and new buildings. The piece demonstrated impressive familiarity with the university expansion debate both in the public sphere and in Parliament, but—like most Brightman editorials—it did not exactly make for gripping reading. The opening paragraph of "University Expansion in Britain" gives a good overall impression of Brightman's style:

> Sir James Mountford, vice-chancellor of the University of Liverpool, has examined the problems with which universities in the United Kingdom will be confronted in the next ten years in a recent series of three articles in the *Manchester Guardian*. Foremost is the difficulty of providing enough places to meet the demands for scientists and technologists. Numerical expansion is, of course, no new problem to the universities, and Sir James rightly pays tribute to the magnificent performance of the universities since the end of the Second World War. In 1938-39 there were 50,250 students; by 1949-1950 this figure had risen to 85,400, mainly because of the influx of ex-Service men. It was assumed that this remarkable achievement was but a temporary operation and that numbers would settle at about 70,000. In 1946, however, the Barlow Committee had wisely assessed the country's needs and had called for a considerable increase in science and technology graduates, together with a substantial expansion in students of humanities. The universities' response to the challenge is well known; their numbers have now risen to 88,700.

Mary Sheehan, who joined the journal in 1966 as the assistant to the new editor John Maddox, recalled that each week a box of material on the latest scientific news would be sent out to the then elderly Brightman so that he could write the leader and send it back to the London office. Gale himself would tell David Davies, a later editor of *Nature*, that he had written only one editorial during his entire tenure at the journal: the one that ran after the bombing of Hiroshima, when there simply was not time to send the usual materials out to Brightman.[21]

Brimble and Gale's reliance on Brightman for their leaders suggests that the two saw editing *Nature* as a content-management job, not a content-generation job. Indeed, one of the only active changes Brimble and Gale made to *Nature* was the introduction of a new masthead in 1958 (see fig. 7). When Sheehan arrived at *Nature* in 1966, her impression was that "it almost came out on its own in a funny sort of way."[22] Brimble and Gale (and, after 1961, just Brimble) had remade the journal so that it required little active editorial management. Arguments were avoided, Brightman wrote the leaders, News and Views was put together from institutional press releases, and any research article or letter that looked reasonable was likely to be accepted, especially if it came from a well-known laboratory. As a result, *Nature* lost much of the liveliness and sense of immediacy that it had pos-

FIGURE 7 *Nature*'s new masthead, introduced in 1958. Reprinted by permission of the Nature Publishing Group.

sessed under Lockyer and Gregory. Davies would aptly call *Nature* of the 1950s "worthy but dull."[23]

A CHAT DOWN AT THE CLUB: EDITORIAL POLICY UNDER BRIMBLE AND GALE

On 1 April 1950, *Nature* opened with an editorial on Letters to the Editors, likely written by Brimble instead of Brightman. The item was reminiscent of Richard Gregory's notes from the 1930s on the popularity of the Letters to the Editor column, but the 1950 editorial seemed more exasperated than pleased. "For the past three or four years," the editorial began, "there has been great congestion of the portion of *Nature* appearing under the title of 'Letters to the Editors.'" The author said this congestion was "perhaps inevitable" given that "these 'letters' have come to be regarded as the usual mode of announcement of the results of new work," but the sheer number of submissions was clearly causing problems for *Nature*'s editorial offices. *Nature*'s correspondence column, which had once been able to print most submissions within a week of receipt, now held submissions for six months or more.[24] The editorial offered suggestions for contributors eager to see their letters in print as quickly as possible. The author asked contributors to submit clean copies of their letters rather than assuming they could edit infelicitous phrases or mistakes when the letter's proofs were printed and to put their references in the journal's style instead of expecting *Nature*'s editorial staff to rework their references for them.

More interestingly, the editorial criticized "men of science" for being so eager to claim priority over their colleagues. Brimble pointed to a letter in that same issue of *Nature* that had been coauthored by two research groups investigating the riboflavin concentration of sow's milk, one at the National Institute for Research in Dairying in Reading, England, the other at the Oklahoma Agricultural Experiment Station.[25] The Oklahoma team had realized that their work overlapped with the Reading team's work and had contacted their potential rival to discuss a collaboration. "This seems to us essentially the right way for men of science to act; *o si sic omnes!* [if only they were all like that!]" the editorial lamented.

Brimble and Gale's exhortation for contributors to *Nature* to collaborate rather than compete for the right to be first forms an intriguing contrast with the journal's prior history. Ernest Rutherford, for example, would certainly have balked at the suggestion that he should collaborate with the Curies in order to reduce the strain on his preferred venue for establishing his priority claims. It also stands in contrast to the story behind the journal's most famous paper. In 1953, James Watson (b. 1928) and Francis Crick (1916-2004), of Cambridge's famous Cavendish Laboratory, submitted "A Structure for Deoxyribonucleic Acid" to *Nature*.[26] It was an achievement with a backstory very much at odds with Brimble and Gale's idea that men of science should curb their competitiveness. As Watson would tell the story in his famous autobiography *The Double Helix*, nothing mattered more than being the first to solve the puzzle of DNA's structure—especially if it meant beating Cal Tech's Linus Pauling (1901-1994), who was also working on the same question. An account by Watson's collaborator Crick supports this recollection; Crick recalled that Cavendish supervisor Lawrence Bragg had been "quite cast down" when Pauling beat him to the structure of the alpha-helix (an important structural feature of proteins) and that "this failure on the part of my colleagues to discover the α helix made a deep impression on Jim Watson and me."[27]

Watson and Crick were also "racing" against another group of researchers in England: a team at King's College London that included the biophysicist Maurice Wilkins (1916-2004) and the crystallographer Rosalind Franklin (1920-1958). In their quest to be first, Watson and Crick famously went so far as to make use of Franklin's results without her knowledge or permission. Franklin and Wilkins had a tense working relationship and had not been sharing results for some months. In January 1953, however, Franklin's graduate student Raymond Gosling gave Wilkins a copy of an x-ray photograph he and Franklin had taken of the DNA "B" form, apparently without Franklin's knowledge.[28] Wilkins showed the photograph to Watson a few days later. As Watson would describe it, "The instant I saw the picture my mouth fell open and my pulse began to race.... The black cross of reflections which dominated the picture could arise only from a helical structure."[29] Watson and Crick would subsequently request a copy of a nonconfidential report from King's College to Britain's Medical Research Council (MRC), a government funding body for biological research. The report contained much of Franklin's numerical data and proved helpful as Watson and Crick refined their model.[30] Franklin died just five years later and likely never knew how important her photograph of the B form of DNA, or the data from the MRC report, had been to Watson and Crick's model.[31]

Watson, Crick, and Wilkins have each written personal recollections of the events leading up to the publication of the DNA papers. Interestingly, none of them mention why they chose to send their findings to *Nature* instead of another journal. In *The Double Helix*, the choice of journal appears obvious to the authors; Watson simply says that "*Nature* was a place for rapid publication," and a week after the crucial insight, "the first drafts of our *Nature* paper got handed out." Wilkins recalls that Watson and Crick had already decided to submit "a short paper to be published quickly in *Nature*" but that "consultation and negotiation with the editor gave King's a week or two" to write accompanying papers detailing their own results.[32] On 18 March 1953, Wilkins wrote to Crick to say that he wanted to publish a note alongside theirs to *Nature* since King's had done the experimental work, adding that Franklin (still unaware that Wilkins had shown Watson the photo of the DNA B form) and Gosling were insisting on adding a piece of their own about their crystallographic work on the DNA molecule.[33] In their retrospectives, the scientists portray submission to *Nature* not as a decision, but as a foregone conclusion.

Given Watson's interest in priority for the findings, *Nature*'s weekly publication schedule was probably the most important reason behind the DNA team's choice of journal. Wilkins's comment about Watson and Crick's desire for a note that could be "published quickly" supports this interpretation, as does Watson's observation about *Nature* being a place for fast publication. Watson's conviction that DNA was "dynamite,"[34] and its solution a surefire path to a Nobel Prize, might tempt us to assume that Watson and Crick also chose *Nature* because they considered it internationally prestigious. Certainly by the early 1950s *Nature* had become one of the major publications for scientists working in the field of nucleic acid research. Scientists from not just Britain but the United States, Belgium, Sweden, and India contributed dozens of articles on nucleic acid research to *Nature* between

1945 and 1953.³⁵ Both Crick and Wilkins had already published in *Nature* when they submitted the DNA papers to that journal.³⁶ *Nature*'s centrality to publication strategies in molecular biology was probably due to the fact that the emerging discipline did not yet have its own publication apparatus; the *Journal of Molecular Biology* was not founded until 1957. Furthermore, Britain, not the United States, was regarded as the world leader in protein and nucleic acid research, which probably also fueled the choice of *Nature*.³⁷

The assumption that results would be sent to *Nature*, however, was not just about *Nature*'s international reach in the field of molecular biology. In fact, local scientific networks seem to have been equally (if not more) important in determining that the DNA papers would be printed in *Nature*. Sir Lawrence Bragg (1890–1971), the head of the Cavendish Laboratory, had longstanding connections with Brimble and Gale. Although Bragg himself did not send his own research articles to *Nature*, preferring more specialized physics journals, he participated in *Nature* in other ways. Beginning in the early 1940s, Bragg wrote a number of articles on the history and accomplishments of the Cavendish Laboratory for *Nature*, as well as essays on science in Cambridge or Britain more generally.³⁸ Watson's recollection of Bragg's first look at their famous paper suggests that Bragg both approved of the choice of *Nature* and felt that his own endorsement was likely to improve its chances of acceptance: "After suggesting a minor stylistic alteration, [Sir Lawrence] enthusiastically expressed his willingness to post it to *Nature* with a strong covering letter."³⁹ Personal connections with *Nature* also helped the King's College group. John Randall, head of the King's College London laboratory, was a member of the Athenaeum along with Brimble. This social connection prompted Brimble to alert Randall to the forthcoming Cambridge publication; Brimble wanted to make certain that King's was aware of the Cambridge paper and had the opportunity to publish their work as well.⁴⁰

Whether *Nature* would accept their articles apparently did not worry Watson, Crick, or Wilkins much. In their retrospective accounts, none of them recalls any anxiety over the manuscript's fate or excitement when news of forthcoming publication came from the *Nature* editors. The articles were all submitted in early April and were not sent out for further review. On April 25, all three were in print.⁴¹ As Wilkins had expected, the Watson-Crick model quickly overshadowed the two papers from King's.⁴²

The famed Watson-Crick paper gives us an opportunity to assess *Nature*'s place in scientific publishing in the 1950s. Scholars have disagreed on whether the Watson-Crick DNA paper had a large immediate effect on the field of molecular biology, although it is worth noting that the Watson-Crick paper was *Nature*'s most cited article throughout the 1950s and 1960s.[43] A more interesting question for this chapter than when Watson and Crick achieved their widest fame, however, is how citations of the Watson and Crick paper compared to citations of major biology papers published in other journals in the early 1950s.[44] We can learn much about *Nature*'s reach in the early 1950s by comparing the Watson-Crick paper with influential papers from field journals.[45] Alfred Hershey and Martha Chase's 1952 article "Independent Functions of Viral Protein and Nucleic Acid in Growth of Bacteriophage," published in the *Journal of General Physiology*, provides one useful point of comparison.[46] The Hershey-Chase paper described the famous "Waring blender experiment," which demonstrated that the reproductive abilities of bacteriophage were due to their nucleic acids, not to their proteins. Most molecular biologists—including James Watson—believed that the Hershey-Chase paper confirmed that nucleic acids, not proteins, were the genetic material.[47] Alfred Hershey would share the Nobel Prize for Physiology or Medicine with Max Delbrück and Salvador Luria in 1969 largely on the basis of this work.[48]

Notably, short-term citations for the Hershey-Chase paper were comparable with those for the Watson-Crick paper. Between 1952 and the end of 1955, the Hershey-Chase article was cited 117 times. Between 1953 and 1956, Watson and Crick's double helix paper was cited 132 times.[49] The closeness of the two numbers suggests that in the 1950s, *Nature* was fairly widely read among practitioners in the biological sciences (and was likely to be particularly significant to molecular biologists for reasons discussed above). However, the numbers also suggest that being published in *Nature* did not garner a paper significantly more notice than immediately sending results to a respected field journal. Having one's work in *Nature*, in other words, appears to have been neither a particular advantage nor a disadvantage in terms of obtaining scientific colleagues' recognition.

"I FELT FRUSTRATED WITH THE SYSTEM": *NATURE* AND SEAFLOOR SPREADING

As we have seen, personal relationships between Brimble and Gale and laboratory heads played a major role in the story of the DNA papers. Recommendations from Lawrence Bragg and John Randall were enough, in Brimble and Gale's view, to justify printing the papers without further review. Personal relationships may also account for another striking editorial decision made under Brimble in 1963 (two years after Gale retired): the acceptance of Frederick Vine and Drummond Matthews's paper on magnetic "stripes" on the seafloor and the rejection of a very similar paper by the Canadian geophysicist Lawrence Morley. In the 1960s, *Nature* was one of the major publication sites for the revolution in the earth sciences that produced modern plate tectonic theory. The Vine-Matthews paper and Morley's letter both proposed that alternating "stripes" of normal and reversed magnetic polarity over oceanic ridges might be evidence in favor of seafloor movement—and in favor of continental drift.

Continental drift was far from a new idea. It had first been proposed at the 1912 meeting of the Geological Association of Frankfurt, where the German geophysicist Alfred Wegener gave a paper suggesting that continents could move and that some continents now separated by oceans might once have been connected.[50] But Wegener's ideas met with significant skepticism. His evidence was considered circumstantial, and his mechanism for explaining how continents could drift was unsatisfactory to his contemporaries. Although many earth scientists rejected Wegener's theory, continental drift never vanished entirely. Two-thirds of the geology textbooks printed between 1930 and 1960 contained at least a brief mention of the theory; most of the textbooks portrayed drift as an interesting but controversial and unproven explanation for the current shape of the earth's land masses.[51] It was a textbook, in fact, that first introduced the teenaged Fred Vine (b. 1939) to the idea that the continents had drifted apart over time.[52] Vine later said that his interest in drift theory bemused his undergraduate supervisors at the University of Cambridge; he recalled submitting an essay to the respected palynologist Norman Hughes, who "crawl[ed] up the wall because he didn't believe in continental drift."[53] Marie Tharp (1920–2006), one of the oceanographic cartographers who created the first map of the entire ocean floor, encountered this skepticism in the early 1950s as well. She would later recall a colleague at Columbia's Lamont Geological Observatory telling her that her profiles of mid-oceanic ridges looked too much like continental drift to be accurate. The colleague, said Tharp, "dismissed my interpretation of the profiles as 'girl talk.'"[54]

While Vine and Tharp were horrifying their teachers and colleagues with their talk of continental drift, across the Atlantic the Princeton geologist Harry Hess (1906–1969) was working on a new theory about the ocean floor. Unlike Wegener, who had believed that the oceanic substrate acted like a viscous fluid, Hess believed that the seafloor was a constantly spreading solid. Hess's theory outlined a model of seafloor spreading in which molten basalt rose up from the earth's upper mantle through oceanic ridges and solidified to form new ocean floor. As the molten basalt flowed up through the oceanic ridge, older ocean floor was pushed further and further away from the ridge, eventually being pushed into a "subduction zone" and rejoining the upper mantle. Hess circulated a preprint of his ideas, but the theory was first formally published by Robert Dietz (1914-1995), a geologist at the US Navy Electronics Laboratory. Dietz proposed the theory in a 1961 letter to the editor in *Nature*.[55] Hess and Dietz's idea won few immediate converts. Hess himself described the theory as more "geopoetry" than scientific theory—an elegant explanation that had, as yet, little empirical evidence to validate it.

On the other side of the continent—at the US Geological Survey in Menlo Park, California—geologists Richard Doell (1923-2008) and Allan Cox (1926-1987) were conducting research on the magnetization of the earth's crust. Cox and Doell were among a growing number of paleomagnetists—scientists studying the earth's magnetic structure—who believed that the earth's magnetic field had reversed polarity several times over its history. In a 1960 review article for the *Bulletin of the Geological Society of America*, Cox and Doell strongly suggested that their developing timeline could be used to investigate drift theory.[56] The two stressed, however, that more extensive and reliable data were needed. Ultimately they would collect much of that data themselves. Between 1959 and 1963, Cox and Doell, working with a University of California, Berkeley, graduate student named Brent Dalrymple, collected data on the magnetization of rocks of different ages and used their data to construct a timeline of the reversals of the earth's magnetic field.[57]

In October 1962, Vine joined the Department of Geodesy and Geophysics at Cambridge as a PhD candidate. He began developing computer-based methods for reconstructing the possible effects of reversing magnetization on the ocean floor. When another member of the department, Drummond Matthews (1931-1997), returned from an expedition to the Carlsberg ridge in the Indian Ocean, Vine used his computer model to interpret the magnetic data and found magnetic "stripes" of normally and reversibly magnetized oceanic floor running parallel to the ridge. From there, Vine and Matthews began to develop the idea that would be referred to as the Vine-Matthews hypothesis: that Hess-Dietz seafloor spreading combined with reversible magnetization of the oceanic crust would produce "stripes" of normally and reversibly magnetized ocean floor at oceanic ridges. In the spring of 1963, Vine began to write up the Carlsberg ridge data and his interpretation for publication. (Matthews was then on his honeymoon.)[58] They chose to submit their paper to *Nature*. Maurice Hill, one of Cambridge's senior geologists, read the paper and felt that *Nature*'s editor would want at least some physical evidence in favor of their theory. He gave Vine and Matthews permission to publish magnetic data from ridges in the North Atlantic and northwest Indian oceans to provide an empirical underpinning to their arguments.[59]

But unbeknown to Vine and Matthews, another geophysicist had already submitted a paper very similar to theirs. In February 1963, Lawrence Morley (b. 1920) at the Geological Survey of Canada sent a letter to *Nature* suggesting that magnetic patterns around oceanic ridges could support a model of seafloor spreading. Going a step further than Vine and Matthews, Morley also suggested that better knowledge of the magnetic reversals chronology could help geologists calculate the rate of seafloor spreading.

Two months later Morley received a rejection letter from the editor of *Nature*. According to Morley, the letter simply said that the editor "did not have room to print" his communication.[60] Morley would go on to submit his piece to the *Journal of Geophysical Research*, where it was again rejected. When Morley saw the Vine and Matthews piece in September, he knew his paper would no longer be considered novel within the geological community; he even worried that if he persisted with trying to publish his own version, he might be accused of plagiarizing the Cambridge geologists. In 1970 Morley moved out of geophysics and accepted a position managing the Canadian Centre for Remote Sensing.

In a 1979 interview with the historian Henry Frankel, Morley seemed skeptical of the editor's explanation for his rejection. He noted that his article would have taken up three-quarters of a page of *Nature*.[61] However, Morley submitted the piece as a letter to the editor. For that section of the journal, three-quarters of a page was not a trivial amount of space—most letters were less than a page, and many took up half a page or less. As we have seen, the backlog of submissions to the popular Letters to the Editor column was quite substantial by 1963. Given the volume of correspondence Brimble was facing, it is possible that he was intrigued by Morley's ideas but not enthusiastic enough about the piece to create room for it. But if that were the case, why would the *Nature* staff have accepted Vine and Matthews's very similar article?

Because *Nature*'s archives have not survived and we do not have access to referee reports or in-house communications about various papers, it is difficult to determine the exact rationale behind the Morley and Vine-Matthews editorial decisions. However, there are some possible explanations. The first, offered by Vine himself, is that there were significant differences between Morley's rejected letter and Vine and Matthews's accepted piece.[62] Morley's article was almost entirely speculative and theoretical, offering no new data. Vine and Matthews's paper, on the other hand, did contain new data—and in fact, their colleague Maurice Hill at Cambridge had urged them to make this change because he thought it would strengthen the article's chances of acceptance. In later interviews, Vine indicated that he thought he and Matthews had benefited from Hill's good instincts about what would impress the editorial staff at *Nature*.[63]

Furthermore, although Morley's rejection letter from *Nature* was not particularly illuminating, a second rejection letter from another journal suggests that the theoretical aspects of Morley's letter did strike some readers as problematic. When the *Journal of Geophysical Research* rejected Morley's paper, the editor enclosed a note from a referee that made it clear the anonymous reviewer considered the letter too speculative, more appropriate "over martinis" than in a communication to the *Journal of Geophysical Research*.[64] And yet the speculation-versus-data explanation seems unsatisfactory, or at least incomplete. Robert Dietz's article on seafloor spreading was also theoretical and speculative, and yet *Nature* printed his piece.

Given Brimble and Gale's reliance on local networks of scientific authority, and in particular their strong ties to Cambridge, we must ask whether there was institutional or national bias at work in the rejection of Morley's piece. The idea that the Cambridge-affiliated Vine and Matthews received preferential treatment from *Nature* would certainly be in line with North American geologists' impressions of *Nature* under Brimble and Gale. Bruce Heezen, a geologist at the Lamont Geological Observatory at Columbia University, once suggested that Brimble and Gale were favorably disposed to speculative papers from Cambridge or Oxford but regarded "speculation from a redbrick university in the United States [as] bullshit"—an especially striking statement given that Lamont's charismatic director Maurice Ewing had a collegial relationship with Brimble and Gale.[65] (In fact, the relationship between Ewing and Brimble and Gale was so close that if Ewing did not personally approve of a paper written by a Lamont scientist, he would call *Nature* and ask that it be rejected—a request Brimble and Gale apparently honored.)[66] Morley's case has been cited in Canada as an example of an apparent lack of international respect for Canadian science, an accusation applied equally to the British *Nature* and the Americans who ran the *Journal of Geophysical Research*.[67]

Morley, certainly, seems to believe that his British competitors had an advantage. In his interview with Frankel, Morley claimed that the Cambridge connection had worked against him and that the person who read magnetic geology articles for *Nature* had worked to ensure priority for the Cambridge geologists:

> I found out that the reason it was rejected was that the reader for *Nature* on magnetic methods knew at that time through verbal communications that Vine and Matthews were hoping to publish their paper and for that reason he did not want my letter to scoop their paper.[68]

Morley softened his take on his rejection for a 2001 collection of retrospectives on the development of plate tectonics, saying only,

> I felt frustrated with the system. I knew that when a scientific paper was submitted to a journal, the editors choose reviewers who are experts on the topic being discussed. But the very expertise that makes them appropriate reviewers also generates a conflict of interest: they have a vested interest in the outcome of the debate. We could call this the "not invented here syndrome": scientists may be biased against good ideas emerging from someone else's lab. In retrospect, that is exactly what happened.[69]

Morley ultimately seemed to conclude that his paper had been the victim of a silent bias against researchers from nonelite institutions. Again, Dietz's 1961 paper provides a useful counterexample. Dietz's article was a highly speculative piece from an American geologist working at the US Navy Electronics Laboratory in San Diego, not an Oxbridge don or a professor at one of the elite American universities. Morley's more serious charge, that an unnamed person hindered his letter because he knew about Vine and Matthews's work, is difficult to substantiate, especially in the absence of that person's name or institutional affiliation. However, given the *Nature* staff's strong reliance on institutional and personal connections, Morley's complaint seems uncomfortably plausible.

The stories behind the DNA papers and the Vine-Matthews and Morley papers show that Brimble and Gale placed great power in the hands of influential laboratory heads when deciding what to print and what to reject. Perhaps more significantly, it is clear from the two stories that under Brimble and Gale, pieces with the right institutional affiliations or recommendations could reach *Nature*'s pages without going through external peer review. This was not inconsistent with how Lockyer and Gregory had run the journal; neither of the previous editors had solicited outside opinions systematically. However, it seems unlikely that the combative Lockyer or the debate-loving journalist Gregory would have rejected an interesting piece because a laboratory head said so, as Brimble and Gale did for Maurice Ewing. Furthermore, as we saw in chapter 3, Gregory was extremely cautious of showing favoritism even to his closest friend, H. G. Wells. But Brimble and Gale were more retiring than their predecessors and more willing to be influenced. As a result, Brimble and Gale are usually seen as affable but low energy, the heads of a regime in which editorial decisions were as likely to be made over dinner at the Athenaeum as they were in the *Nature* offices.

LOSING AND GAINING GROUND: *NATURE'S* COMPETITORS AND CONTRIBUTORS DURING THE BRIMBLE-GALE EDITORSHIP

Accounts from contemporaries suggest that *Nature* was not considered a particularly important or prestigious journal under Brimble and Gale. Walter Gratzer, who became *Nature*'s molecular biology correspondent in 1966, recalled that when he was a PhD student in the early 1960s, *Nature* was "widely seen" but "it wasn't regarded as a high-grade journal."

> I published a few things in *Nature* when I was a PhD student and almost anything could get into it at the time, if it wasn't actually wrong. Refereeing was pretty erratic and I think they took more notice of where it came from than the content. And it wasn't that important a journal. There were of course enormously important things in it from time to time, but it was commonly regarded as the thing where you put your first preliminary report and then a complete description would be written up in a more serious journal.[70]

Similarly, in 1979 Fred Vine recalled a lighthearted conversation with his colleagues about his now famous 1963 *Nature* paper.

> It must have been in June or July '63.... We were sitting around at coffee, and not an awful lot was being said. Somebody said, "Do you know if *Nature* gets their articles reviewed, or do they publish almost anything?" I said, "Well, we're just about to find out because, you know, I just put my paper in, and if they publish that they'll publish anything."[71]

Like Gratzer, Vine and his colleagues saw *Nature* as a journal that might publish "almost anything." Furthermore, both Matthews and Vine also recall being disappointed at the lack of response to their ideas after *Nature* printed their paper on 7 September 1963.[72] Matthews would later say that "the paper dropped into a sort of vacuum, as we expected it to.... Teddy Bullard [a prominent Cambridge geologist] used to proselytize for it a bit, but American labs wouldn't hear anything of it—thought it was all nonsense."[73] Vine recalled that by 1964, "I was getting pretty discouraged and beginning to lose faith myself. It went over like a lead balloon; in some ways there was no response. People just sort of turned away."[74] Between 1963 and 1966, the Vine-Matthews paper was cited just 28 times. Publication in *Nature* was no guarantee of immediate respect, or even attention, for a scientific paper.

Just as the DNA papers gave us an opportunity to assess *Nature*'s place in the biological literature, the development of the seafloor spreading model gives us a useful opportunity to examine *Nature*'s place in the earth sciences literature. It was not until Vine began collaborating with the University of Toronto geologist J. Tuzo Wilson (1908–1993) in 1964 that the Vine-Matthews hypothesis began winning more converts. Vine and Wilson soon found a strong test case: the Juan de Fuca Ridge, an oceanic ridge off the coast of northern Washington State and southern British Columbia. After collaborating on an analysis of the Juan de Fuca data, Vine and Wilson published back-to-back articles in the 22 October 1965 issue of *Science*. The first article, authored by Wilson, outlined how the Juan de Fuca data supported his theory of transform faults, a type of fault running perpendicular to an oceanic ridge that connected oceanic ridges or tectonic plates at its ends.[75] The second, co-authored paper argued that the pattern of magnetic anomalies over the Juan de Fuca ridge closely matched the symmetrical model predicted by the Vine and Matthews theory.[76]

Two more articles in *Science* converted most geophysicists to the seafloor spreading theory. On 2 December 1966, *Science* ran an article coauthored by two Lamont scientists, James Heirtzler and Walter Pitman, that described results from the Reykjanes and Pacific-Antarctic ridges. Pitman and Heirtzler concluded that their results "strongly support the essential features of the Vine-Matthews hypothesis and of ocean-floor spreading as postulated by Dietz and Hess."[77] The Heirtzler-Pitman paper was quickly overshadowed, however, by a longer paper that appeared in *Science* just two weeks later: Vine's "Spreading of the Ocean Floor: New Evidence," which used Lamont's data from the Pacific-Antarctic ridge to not only support his previous work with Matthews and Wilson but also to posit a general model of seafloor spreading.[78] In later interviews, many prominent geophysicists cited Vine's 1966 paper as the work that won them over to the theory of seafloor spreading.[79] Over the next four years, many scientists worked to explain the implications of seafloor spreading for continental geology, eventually combining continental and seafloor geology into an overarching theory of plate tectonics.

The development and acceptance of seafloor spreading gives us a window onto *Nature*'s relationship with geophysics journals. Unlike the structure of DNA or the discovery of radioactivity, there was no single paper that revealed the "discovery" of seafloor spreading; it was a theory developed through the work of many researchers, and many papers were an essential part of the theory's construction. In 1972 Allan Cox, whose work on magnetic chronologies had been an important part of the development of the seafloor spreading model, put together a volume of major papers that had contributed to the development of seafloor spreading and plate tectonics.[80] Cox's choices are not uncontroversial—Robert Dietz's 1961 *Nature* article is excluded in favor of a later essay by Harry Hess, and Cox's own field of geomagnetic reversals is heavily represented—but if we combine Cox's table of contents with the papers historian John A. Stewart identifies as the most influential ones in the development of plate tectonics,[81] we gain a list of twelve major articles published between 1954 and 1966 that made significant contributions to the development of the seafloor spreading model.[82] Out of these twelve articles, five were published in *Nature*, five in *Science*, and two in the *Bulletin of the Geological Society of America*.

The reasons authors would choose the *Bulletin of the Geological Society of America* are fairly clear—it was a major field journal in geophysics and published lengthy articles averaging roughly twenty pages. A more direct and more interesting comparison can be drawn between *Nature* and *Science*, the two weekly journals of general science. *Science*, as we saw in chapter 2, had been founded in 1880 as an explicit attempt to give American scientists a *Nature*-style weekly publication where they could discuss the latest scientific developments. However, for much of its history *Science* had been in *Nature*'s shadow, both in terms of international readership and, arguably, quality of papers. Despite his concern for priority, for example, in the early 1900s Ernest Rutherford did not submit articles to *Science* even though *Science*'s New York editorial offices were closer to Montreal geographically than *Nature*'s London ones. In 1936, J. McKeen Cattell, who had then been editor of *Science* for forty-one years, wrote to Gregory to invite him to come to America for a lecture tour. He frankly admitted to admiring and envying *Nature*: "It has been one of the trials of my life that Nature is better than SCIENCE."[83] Cattell's words may have been simple flattery designed to entice Sir Richard into accepting the proposed lecture tour, but it seems likely that he also genuinely admired *Nature* (although he could not resist pointing out that his own journal was offered at a much lower subscription price).[84]

In the development of seafloor spreading, however, *Nature* and *Science* appear to be on equal footing. Each weekly journal ran short articles containing a mix of new geophysical data and provocative suggestions for the implications of those findings. There are a few subtle differences between the articles published in *Nature* and the ones published in *Science*. The geophysics articles in *Science* were slightly longer, averaging 7,000 words each versus an average of roughly 6,000 words each for *Nature*. The slightly more speculative articles, such as the Vine and Matthews article and Robert Dietz's 1961 article, were more likely to be published in *Nature*, while data-heavy articles such as the ones Vine and Wilson wrote were more likely to be published in *Science*. Generally speaking, however, there were far more similarities than differences when it came to *Nature*, *Science*, and the articles each ran on plate tectonics. The material on geophysics from the 1950s and 1960s strongly suggests that in the postwar world, *Science* and *Nature* were on increasingly equal footing in terms of prestige and ability to attract influential papers.

Another American competitor was also gaining prominence in the 1960s and would have a significant influence on *Nature*'s content: *Physical Review Letters*, founded in 1958.[85] In 1951, Samuel Goudsmit became the managing editor for the American Physical Society and took on responsibility for the Society's flagship journal, *Physical Review*. Goudsmit became frustrated by the ever-increasing size of *Physical Review*, fearing that its length dissuaded physicists from actually reading the journal. *Physical Review* had published *Nature*-like letters to the editor since 1929. Goudsmit hoped that a journal devoted entirely to short letters would enable readers to keep abreast of the most recent findings in physics even if they could not read entire issues of the *Physical Review*.

Physical Review Letters debuted in 1958. It did not do much to relieve the pressure on its parent publication; the size of *Physical Review* continued to grow. But *Physical Review Letters* did become an important competitor to *Nature*. Ultimately, it attracted many of the short pieces in the physical sciences that might once have been directed to the British journal. Notably, Brimble and Gale both had backgrounds in the life sciences—Gale in agriculture, Brimble in botany. Many members of the *Nature* staff under Brimble and Gale's successors felt that they had inherited a journal that was strong in biology but weak in chemistry and physics, and several saw *Physical Review Letters* as a reason for the shift. As Mary Sheehan put it,

> Brimble was a biologist and knew nothing about physics, from what I've gathered, so people stopped submitting, well didn't completely stop, but you know, there wasn't a great flow of physics papers, which John [Maddox] tried to remedy because he was a physicist at heart. But things like *Phys Rev Let* really did most of the business.[86]

David Davies concurred, saying that while *Nature* remained extremely strong in his own field of the earth sciences, competition from *Physical Review Let-*

ters had noticeably weakened *Nature*'s content in physics by the time he stepped into the editor's chair in 1973.[87]

CONTRIBUTORS' INTERESTS AND PEER REVIEW UNDER BRIMBLE AND GALE

With Brimble and Gale so embedded in British academic culture, Brightman's leaders focused on domestic issues, and with important foreign competitors gaining prominence, we might expect *Nature* to have lost some of its appeal to non-British contributors during the 1950s. Postage restrictions and wartime mail difficulties led Brimble and Gale to draw on a heavily British contributor base during the 1940s. In 1937, Gregory had boasted that half of the letters to the editor came from outside Britain; by 1950, this number was down to 40 percent. Furthermore, *Nature*'s longer articles about new theories or experimental findings (pieces that ran from three to five pages in contrast to just half a page for a letter to the editor) were overwhelmingly British—more than 70 percent of *Nature*'s research articles came from British laboratories in 1950. Interestingly, however, despite Brimble and Gale's reliance on personal connections within the British scientific community and despite increased competition for the types of short experimental articles that *Nature* was known for printing, between 1950 and 1965 the percentage of non-British experimental content in *Nature* steadily *increased*. By the time of Brimble's death in 1965, roughly 60 percent of articles and letters to the editor in *Nature* came from outside Great Britain—an even higher percentage than under Gregory.

These figures raise two obvious questions. First, why did *Nature*'s supply of international contributions continue to rise even when the journal's editors were focused on science in their home country? As with the journal's nineteenth-century transition to a specialist publication and with the initial stirrings of internationalism during the radioactivity boom, it was *Nature*'s contributors who drove the trend. Brimble, Gale, and the rest of the *Nature* staff did not seek to expand their reach outside of Britain's borders, but contributors outside Britain continued to find *Nature* useful for announcing a forthcoming paper or publishing a short piece to claim priority for their theories and findings. Brimble and Gale's low-maintenance approach to editing meant that while they certainly did not work to recruit international contributions, they also did not work to discourage them, and they would usually print articles that seemed legitimate if they could find the room.

Second, why did contributors outside Britain find *Nature* a desirable place to publish their results—especially when, as we have seen, fast publication was no longer a guarantee? Some of *Nature*'s international growth during this period was almost certainly due to continuing British strengths in fields such as molecular biology and geophysics, where making one's results known in Great Britain would have been an essential part of ensuring they reached the "right" groups of scientists. Another factor was growing international mobility among scientists, who were increasingly likely to accept jobs and fellowships outside their native countries. Fred Vine, for example, moved from Cambridge to Princeton for a postdoctoral fellowship, and James Watson was an American who came to Cambridge after finishing his PhD in Illinois. As more and more scientists came to Britain for a period and more and more British scientists spent part of their careers abroad, more international research groups were likely to contain a member familiar with *Nature*'s submission process—a trend similar to the one we saw in radioactivity in chapter 4. Finally, it is important to note that much of *Nature*'s increasing international content came from the United States, another English-speaking nation that was experiencing a large boom in its number of scientific researchers and, therefore, an expansion in the number of its scientific papers. As an English-language journal, *Nature* was able to attract papers from American laboratories with far more ease than other international competitors.

Nature's continuing appeal among researchers reveals something extremely striking: in the 1950s and 1960s, *Nature* did not have to employ systematic external peer review in order to remain desirable as a venue for announcing new findings. Many observers have considered peer review to be one of the most important features of the scientific journal. In 1969, the physicist John Ziman described peer review, along with an editor's approval, as the crucial reason why scientific journal articles were considered trustworthy:

> An article in a reputable journal does not merely represent the opinions of its author; it bears the *imprimatur* of scientific authenticity, as given to it by the editor and the referees whom he may have consulted. The referee is the lynchpin about which the whole business of Science is pivoted.[88]

In most accounts of the history of the scientific journal, it was Henry Oldenburg, the legendary secretary of the Royal Society of London, who introduced this essential feature of scientific publication to the newly created *Philosophical Transactions* in the late seventeenth century. Oldenburg, the story goes, wisely saw that he needed to consult experts in order to judge the quality of manuscripts, and thus peer review was born and was ever after a crucial feature of any reputable scientific publication.[89]

However, the history of peer review—like the history of the scientific journal—is not nearly this simple.[90] Far from springing full grown from the head of Henry Oldenburg, peer review did not become a consistent feature of scientific journals until well after journals became the scientific community's site for establishing knowledge claims during the nineteenth century. *Nature* itself stands as a clear illustration of this fact: as we have seen, Lockyer, Gregory, Brimble, and Gale all felt perfectly comfortable making in-house editorial decisions. Some manuscripts were sent out for external opinions, but this was by no means a necessary condition for publication in *Nature*.

The retrospectives from Gratzer and Vine make it clear that British contributors knew (or at least suspected) that the refereeing process at *Nature* was somewhat lax, and prominent laboratory heads such as Lawrence Bragg and Maurice Ewing would certainly have realized how much influence they wielded over which papers made it into print. Morley's 1979 comments about "the reader for *Nature*" who rejected his article suggest that North Americans, too, knew that *Nature* relied on a small number of local opinions. And yet, *Nature*'s contributors and readers do not appear to have considered *Nature*'s unsystematic peer-review process a reason to distrust the scientific claims made in the journal. While journals affiliated with scientific societies often employed refereeing procedures in the nineteenth century, this was not seen as a special guarantee of scientific accuracy.[91] As we saw in chapter 2, publishing in *Nature* became a sign of scientific legitimacy in Britain not because Lockyer's staff employed peer review—they did not—but because its *readers* were considered "the right people" (to borrow Crookes's 1895 phrase). It was assumed that the editorial staff would filter out anything obviously inferior, but ultimately *Nature*'s readership would assess an article's credibility. Well into the second half of the twentieth century, *Nature* could still be considered a legitimate place to publish scientific findings even in the absence of systematic external peer review—and even when headed by an editor with no claim to scientific expertise, like Gregory, or editors who were scientifically low profile, like Brimble and Gale.

Notably, in the mid-twentieth century *Nature* was not considered a "high-grade" journal, but this seems to have had more to do with the journal's perceived selectivity than with its peer-review system. As Gratzer put it (and as Vine's retrospective confirms), contemporaries saw *Nature* in the early 1960s as a journal that might publish "almost anything" that "wasn't actually wrong." Indeed, Brimble and Gale seem to have been keen to accept as many pieces as they could, especially ones that came recommended by trusted sources. Brimble's exasperation with the demands on the Letters to the Editor column strongly suggests that he and Gale were unwilling to solve their problem by rejecting more papers. The desire to accept as many manuscripts as possible, in turn, created longer and longer delays between submission, acceptance or rejection, and publication. When Brimble died in 1965, he left behind a backlog of manuscripts that stretched back fourteen months.

THE COSTS AND BENEFITS OF INTERNAL REVIEW

Under Brimble and Gale, *Nature*'s editorial processes relied heavily on connections within the British scientific community. A personal recommendation from a prominent British scientist meant an article was far more likely to be printed; a letter from an unknown scientist was more likely to languish in a growing pile of papers in the *Nature* editorial office.

And yet, *Nature*'s contributor base in the 1950s and early 1960s was far more international than *Nature*'s editorials and news content might have suggested. Despite the limited ambitions of its editors during this period, publishing in *Nature* still had some advantages—the journal accepted short articles and reached a fairly wide readership even if the journal itself was not considered a place for particularly noteworthy results. During this period serious competition arose from journals such as *Physical Review Letters* and *Science* that also published short pieces, but *Physical Review Letters* was only a competitor in physics, and for some sorts of papers *Nature* had a significant advantage over *Science*. The comments from Vine about how *Nature* might "publish anything" suggest that, *Nature* had something of a reputation for publishing articles and letters that might not make it through the review process in other journals.

Peer review is now an expected part of scientific publishing, and when we look back, it is tempting to see only the pitfalls of Brimble and Gale's system. For instance, Brimble and Gale's trust in top British laboratory officials may have helped obscure Rosalind Franklin's contributions to the DNA model; their successor John Maddox would later claim that he "would have smelled a rat" when he read Watson and Crick's sentence about being "stimulated by a general knowledge" of the crystallographer's unpublished findings.[92] Lawrence Morley's case suggests that a journal whose editor relied on personal connections to choose articles might give submissions from researchers outside the editor's network less thorough consideration than pieces from those inside it.

But the Brimble and Gale style gave *Nature* scope to print papers that might have faced skepticism from external reviewers—and some of those papers turned out to constitute major advances in their fields.[93] Maddox was also fond of saying that the Watson-Crick DNA paper would never have made it into print if *Nature* had employed peer review in 1953: "It is only necessary to imagine what people would say if it reached them in the mail: 'It's all model-building, just speculation, and such data as they have are not theirs but Rosalind Franklin's!'"[94] Notably, after he came to the journal in 1966, Maddox changed much about *Nature* but retained the editor's absolute right to decide which articles should be printed whether or not referees had been consulted. As we shall see, Maddox found a great deal of value in having such wide scope to print unusual, controversial, or speculative articles based solely on his own authority.

CHAPTER SEVEN

Nature, the Cold War, and the Rise of the United States

In February of 1970, Macmillan and Company and the *Nature* staff took a step that would have been almost unthinkable five years earlier: they opened a *Nature* office in Washington, DC. John Maddox, *Nature*'s editor since 1966, announced the new office's opening in a leader on 13 December 1969. Although the editorial staff would temporarily relocate to Washington to oversee the new office, Maddox firmly stated that *Nature*'s core staff would return to Britain after no more than six months. "*Nature* will cherish its British accent," Maddox assured readers. He added that *Nature* would also seek to "strengthen links with other centres in Europe" in the months and years ahead.[1]

Nature's Britishness has been a recurring theme in this book. Even as their contributor base became increasingly international, Brimble and Gale kept their sights on scientific opinions and scientific issues within their own nation. But by 1980, *Nature*'s news and opinion had undergone a significant shift under the leadership of two editors: Maddox and David Davies, both of whom actively sought to improve *Nature*'s news gathering outside Britain and to expand commentary on scientific issues around the globe. Furthermore, the number of international contributors—in particular, American contributors—continued to increase under Maddox and Davies. Just four years after Brimble's death, Macmillan and Maddox felt it not only desirable but essential for *Nature* to open a Washington office.

This chapter will examine the tremendous expansion in international news coverage and international contributors that took place under Davies and Maddox. The rise of the United States as the world's scientific powerhouse fundamentally changed *Nature*'s place in the scientific community. This British scientific journal gradually transformed into an international journal with, as Maddox put it, a "British accent." However, *Nature*'s internationalism had an important limitation. Between 1966 and 1980, scientists in the USSR represented a gap in *Nature*'s contributor base. This gap reflected larger trends in Western and Soviet scientific publishing and gives us the opportunity to examine how the Cold War affected international scientific communication. *Nature*'s example illustrates that far from being a trivial consequence of Cold War hostilities, the Cold War publishing divide had significant consequences for the development and spread of scientific theories in the late twentieth century.

MADDOX AS EDITOR, 1966–1973

Much like Richard Gregory before him, John Maddox (1925–2009) came from a working-class background; his father, Arthur Maddox, worked in the furnaces of an aluminum plant in the Welsh town of Penllergaer. At fifteen, Maddox won a scholarship from the British Crown that enabled him to attend Oxford's Christ Church College. After finishing his Oxford undergraduate program, Maddox went on to obtain a doctorate in physics at King's College London. In 1949 he accepted a position as a lecturer in theoretical physics at the University of Manchester.

Six years later, Maddox decided that the academic life was not for him and changed careers. He accepted a position with the *Manchester Guardian*, becoming the newspaper's first-ever science correspondent—a position that paid him a much larger salary than his Manchester lectureship.[2] Maddox also contributed to the *New Scientist*, a weekly popular science magazine founded in 1956 and published by Reed Elsevier, and served as a science correspondent for the *Washington Post* for a period in 1960. The former academic distinguished himself quickly in his new profession, twice serving as the chairman of the Association of British Science Writers. Maddox wrote between two and five articles per week during his time as the *Guardian*'s science correspondent. He drew on *Nature*, *Science*, and other scientific journals to find interesting results to share with the *Guardian*'s readers; he also drew on corporate press releases to report on industrial advances. He reviewed books about scientific subjects and contributed many in-depth articles about ongoing controversies between scientists, the state of science education, and potential health crises in Britain. As Cold War tensions developed, Maddox frequently reported on the intersection of science and politics with articles about nuclear power, nuclear weapons, and the space race between the United States and the USSR.³

In 1964 Maddox accepted a job as an administrator with the Nuffield Science Teaching Project, an organization that sought to improve British science curricula in primary and secondary schools. Two years later, after Brimble's death, Maddox received an unexpected visitor: Maurice Macmillan MP.⁴ Maddox described the meeting in a 1995 article:

> Macmillan had said that he had wanted to talk about *Nature*, which we did.... But then the conversation became more serious, and Macmillan seemed to be asking whether I would be interested in Brimble's job. I remember a flush of ambivalence.... It had been more than a year since *Nature* had abandoned the practice of appending to research articles and notes the dates on which it had received them. That seemed worse than merely bad, which I explained to Macmillan. "How big is the backlog?", I asked.... When we next met, he produced the answer: 2,000-odd.⁵

Despite this unpleasant information about the size of the manuscript backlog, Maddox agreed to take the job. He also persuaded his highly efficient assistant, Mary Sheehan, to leave the Nuffield Science Teaching Project and move with him to *Nature*. A genteel note in the News and Views section of *Nature* on 22 January 1966 announced that Macmillan had found a replacement for L. J. F. Brimble. The notice described Maddox's qualifications and assured readers that the forty-year-old Maddox would not assume full duties at *Nature* until his obligations with the Nuffield Science Teaching Project were concluded.⁶

It was arguably the last time that the word *genteel* could be used in connection with Maddox's work at *Nature*. He quickly stripped the editorial "we" from the journal, and letters from the editorial office no longer "begged to inform" contributors of the status of their manuscripts. He also instructed contributors that they were no longer to refer to themselves in the third person when writing articles for *Nature*.⁷ Most importantly, Maddox overhauled *Nature*'s methods of choosing which manuscripts would make it into print. Under Gale and Brimble, as we saw in chapter 6, manuscript refereeing for *Nature* relied heavily on personal connections within the British scientific community. In the absence of those connections, a manuscript was more likely to be rejected—and, perhaps even more frustratingly for the authors, more likely to languish forgotten in a file cabinet. When Maddox arrived, Brimble's office was piled high with old manuscripts still awaiting final judgment on acceptance or rejection. As Maddox described it,

FIGURE 8 John Maddox in 1996. Reprinted by permission of the Nature Publishing Group.

[The *Nature* office] was an open-plan space without much of a plan. A window facing West ran 10 metres along the room and the broad window-ledge supported the famous backlog. That was arranged in piles, one for each month, providing a histogram of Brimble's problem, soon to be mine. There were fourteen monthly piles when I first saw them.[8]

Maddox began restoring *Nature*'s reputation for speedy publication by working to clear the backlog.[9] He had to give three months' notice at the Nuffield Science Teaching Project before starting at *Nature*, but even before leaving Nuffield, Sheehan recalled, Maddox "used to go into the *Nature* office every day and pick up a suitcase full of manuscripts and take them home and take them back the next day."[10] Once he officially began the job at *Nature*, Maddox began holding daily editorial meetings about manuscripts sent for consideration.[11] Sheehan replaced the old system of piling papers by month of submission with a more streamlined index card system that made it easier to look up the status of any submitted manuscript.[12]

Under Maddox, some unsuitable papers were rejected outright, others were sent out for referee opinions, and some Maddox simply accepted on his own authority without sending them out for referees' comments. Walter Gratzer, who became the journal's molecular biology correspondent in 1966, thought that Maddox "didn't worry too much about refereeing" early in his tenure; his priority was quick publication and a compelling journal. He even tried an innovative experiment to speed up the refereeing process: he collected a group of referees around a table piled high with manuscripts, hoping they would come to "instant decisions" about the submissions. The process worked less well than he had hoped. It was difficult to get all of the referees together at once, and many were unhappy making decisions with the speed Maddox expected. Gratzer recalled that one colleague "would immerse himself in the first paper and couldn't be shifted until the whole thing was over."[13] The experiment was ceased after only half a dozen referee meetings.

Maddox would also solicit contributions from laboratories and scientists he thought were doing interesting work, sometimes going so far as to encourage an author to withdraw an exciting paper from another journal and submit it to *Nature* instead.[14] Gone were the days when an influential laboratory director might stop *Nature* from printing a compelling piece. Lamont's Maurice Ewing, who had been able to stop publication of papers from his laboratory with a phone call to Gale or Brimble, found that the new regime was not nearly so accommodating. Maddox, in the words of one anonymous geophysicist, "just wouldn't put up with pressure from the establishment to stop something."[15] By December 1966 Maddox was pleased enough with *Nature*'s improved turnaround time that he began printing the date of submission at the end of each scientific paper, pointedly reminding the scientific community that *Nature* could get their work into print much more quickly than other publications, often within a month of the initial submission.[16]

Unsurprisingly, given his background in journalism, Maddox was especially eager to overhaul *Nature*'s newsier sections. Later commentators frequently remarked that Maddox brought a "newshound sensibility" to *Nature*. Maddox sought to make *Nature* the journal of choice for scientists who wanted to learn about and discuss the world's most important—and most current—scientific news. His wife, Brenda Maddox, a noted historian and journalist, believed that her husband's approach to the leaders and News and Views was shaped by his experience on the *Guardian*.

> I think he wanted to beat the *Guardian* at its own game. He wanted the news pages at *Nature* to be ... as newsy as the science pages of the *Guardian* and probably even leading the way. He did always see [*Nature*] ... [as] a newspaper ... as up-to-the-minute as it possibly could be.[17]

Within a few months Maddox had pushed Brightman out so he could write the editorials himself.[18] He also changed the journal's organization. Brimble and Gale had mixed experimental articles in among news articles; Maddox put all of the news writing at the front of the journal, created a single section for longer research articles, and placed the Letters to the Editor immediately after the research articles. In effect he divided the magazine into two sections: the front section was news and opinion and the back section contained new experimental findings and book reviews (a structure that brings to mind *Nature* in its early years under Lockyer). Determined to make the front end of the journal as current as possible, Maddox often wrote the leaders at the last minute—sometimes at the printer's office while the typesetters waited for his text. In many ways, Maddox's focus on keeping *Nature* current hearkened back to the immediacy and controversy that had made *Nature* distinctive in the competitive nineteenth-century science publishing market.

Under Maddox, *Nature* continued its tradition of commenting on science and scientific issues within Great Britain; in particular, educational reform was a frequent topic between 1966 and 1973.[19] Maddox, however, expanded the range of editorial commentary well beyond Britain's borders. He wrote editorials on Governor Ronald Reagan's battles with the University of California system, funding for graduate students in the United States, and the Cultural Revolution in China.[20] The Maddox regime was unafraid to court international controversy. For example, under Maddox, *Nature* took a provocative stance on American and British environmentalism, cautioning against environmentalist "hysteria" and urging "moderation" in the regulation of DDT.[21]

Maddox also took steps to improve the scientific reporting at *Nature*. Under Brimble and Gale, the News and Views column had been a list of recent promotions, scientific society appointments, and policy changes at various universities and laboratories. After Gale stepped down in 1961, Brimble attempted slightly more ambitious news coverage, but the articles were largely focused on matters within Britain and were heavily weighted toward institutional news, such as anniversaries of other publications, summaries of recent government reports, or the opening of new laboratories. The News section under Maddox still contained some institutional intelligence, such as a budget crisis at Euratom (the European Atomic Energy Commission) or what was happening within a task force on science policy in the United States.[22]

But Maddox's front section included far more information about recent research, including papers published in other journals, than Brimble's had. He recruited a network of correspondents who reported anonymously on what was happening in their discipline. *Nature* thus became an important source of scientific news not just because of the scientific papers it printed but also because Maddox's correspondents could be relied on to share news of important findings in other publications—and, in part because of the anonymity, they felt little need to pull their punches. According to Walter Gratzer, the anonymous molecular biology correspondent, he had "free scope" over the topic of his column.

> [Maddox] wanted me to write about anything.... He was also quite pleased whenever there was a bit of controversy. And I'd try and make it a principle to get at people—talk about papers that were bad as well. And then occasionally there was a controversy and some bad-tempered exchanges in the Correspondence columns, and he always backed me up on those sorts of things.[23]

The changes made an impression on *Nature*'s readers. David Davies, a geosciences correspondent who would later succeed Maddox as editor, recalled that Maddox "transformed [*Nature*] almost overnight.... Scientists of all sorts who were reading *Nature* in the mid-1960s noticed a sudden change." It went from a "very serious and rather dull journal" to one that "started to be interesting."[24]

"THE INEVITABLE ROW WITH MANAGEMENT"

Maddox and *Nature* seemed to be an ideal pairing—Brenda Maddox said that her husband "loved it right away." The job fulfilled both his passion for science and his love of news and gave him tremendous opportunities to write opinion pieces and news articles himself. Being the editor of *Nature* also provided many chances for travel and networking, which the gregarious Maddox enjoyed. Macmillan and Company, impressed with the work he had done, gradually expanded his position to include responsibilities with other Macmillan journals such as *Education and Training* and Macmillan's science books division—a position very similar to the one both Lockyer and Gregory had held at Macmillan.[25]

When combined with his determination to have the final say on what would appear in *Nature*, however, Maddox's other activities began to have consequences for the journal. In the early 1970s some contributors noticed a change in *Nature*'s correspondence with its authors. Davies, by then the director of the Seismic Discrimination Group at the Massachusetts Institute of Technology's Lincoln Laboratory, recalled,

> I published a few papers in *Nature*, and I noticed, and I think other scientists noticed, that the standard of dealing with authors was in decline. I think that the attention to the detail of scientific publication—getting everything spot-on correct... proofs corrected properly and that sort of thing—it seemed not to be as good as it used to be. I would get proofs days after the journal had actually published. And that didn't go down well with American scientists.... For me it was a minor irritant, but it was not a minor irritant for others.[26]

When Philip Campbell, the current editor of *Nature*, began at *Nature* in 1979, he was told that Maddox had been "overinvolved" and that the result had been "sometimes whimsical decision-making and delays in the handling of scientific papers."[27]

In 1971, Maddox tried an ambitious experiment. He divided *Nature* into three publications: *Nature*, *Nature New Biology*, and *Nature Physical Science*. On 1 January 1971, Maddox used the first page of News and Views to announce the change. "From today," he explained, "*Nature* will be published in three editions each week, on Monday, Wednesday and Friday." On Monday, subscribers would receive *Nature Physical Science*, on Wednesday *Nature New Biology*, and on Friday the original *Nature*. Maddox admitted that splitting *Nature* by discipline might run the risk of "rob[bing] *Nature* itself of its interdisciplinary character," but he suggested that *Physical Science* and *New*

Biology would enable *Nature* itself to be more accessible to a wider range of readers:

> Care will be taken to ensure that the Friday edition ... intended to be read as widely as possible by specialists as well as those with a general interest in science, will contain a wide selection from the original records of research submitted for publication. More rigorously than in the past, however, care will be taken to ensure that the material published on Friday is not merely of general interest but that it is accessible to a general audience as well.[28]

It is tempting to view Maddox's tripartite experiment as an idea ahead of its time, an early version of *Nature*'s current publishing empire that includes prestigious sister publications such as *Nature Genetics* and *Nature Medicine*. However, Maddox's *New Biology* and *Physical Science* were different from the current sister publications in an important way: they did not have their own editorial regimes. "They weren't separate entities really, they were just sort of offcuts," Davies would later recall. "And in fact they didn't have separate editors, they were just done by members of staff [at the original *Nature*]."[29] Furthermore, there was no separate submission process for the three journals—authors would send their manuscripts to *Nature* and manuscript editors would choose pieces to print in the satellites.

Some of those who knew Maddox suggested that the three-journals plan was an attempt to move toward a daily *Nature*, a scientific newspaper that would not only print the most exciting scientific research findings but would also beat the *Guardian* and other scientific news organizations to the punch.[30] Maddox wrote that he planned for *New Biology* and *Physical Science* each to be "a pleasure in its own right to read."[31] However, the satellite publications had very little in the way of the opinion pieces and strong editorial voice that made their parent publication distinctive. The satellite journals contained some news reporting—largely focused on matters of interest within a discipline—and a few book reviews, but the content of both was heavily weighted toward research articles. In fact, many issues contained only articles and then a handful of letters, with no news or book reviews at all. Both *Physical Sciences* and *New Biology* ultimately bore a stronger resemblance to quarterly scientific society journals than to Maddox's lively and newsy *Nature*.

According to Sheehan, Maddox's decision also stemmed from a desire to print more papers than could be bound in a single *Nature*, and initially, *Nature* did print more papers under the three-journals plan. The total number of letters per week increased from an average of about 35 to an average of about 42, although the average number of articles per week stayed relatively static at about 7. Despite the increased space for letters, *New Biology* and *Physical Science* proved unpopular with authors. Would-be contributors to *Nature* who found their papers printed in *New Biology* or *Physical Science* felt jilted by the original *Nature* rather than grateful for the extra space. As Davies recalled, "people who sent their stuff to *Nature* and found it was being published in one of the satellites felt they'd been demoted to the second division and this wasn't really a *Nature* paper."[32] Maddox's three-journals plan was also expensive—he had wished to keep the cost of subscribing to all three *Nature*s the same as the pre-1971 subscription to *Nature* alone, but printing and mailing three journals was naturally more costly than printing just one.

The perceived failure of the three-journals plan came at an unfortunate time for Maddox. Maddox's superiors at Macmillan, particularly managing director Nicholas Byam Shaw, had generally been supportive of Maddox's busy schedule; they admired his energy and recognized how much he had done to resurrect *Nature* from its stagnant pre-1966 state.[33] Initially, *Nature* under Maddox had almost complete editorial autonomy with very little interference from the rest of the publishing company.[34] In 1970, however, Macmillan hired Jenny Hughes, a former member of the Foreign Office, as a new director in the Macmillan journals division. Hughes and Maddox were both driven individuals with strong personalities, and the two did not see eye to eye about Maddox's management of *Nature*.[35] The result was what Brenda Maddox described as "the inevitable row with management."[36] Hughes became convinced that Maddox's domineering editorial style and his busy schedule were simply not acceptable for Macmillan's most prestigious journal. Armed with complaints from scientists who said that *Nature*'s standards of contributor correspondence had declined, Hughes began arguing that Macmillan and Maddox should part ways. When *Nature* showed a financial loss at the end of 1972, Hughes received approval to find a new editor.

DAVID DAVIES AND THE 1970S AT *NATURE*

David Davies (b. 1939) first learned of the search when he noticed a small advertisement in *Nature* inviting applications for the position of editor. Davies had been reading and contributing to *Nature* throughout almost his entire scientific career. After finishing his doctorate in Cambridge's famed geophysics department, Davies had accepted a fellowship at Cambridge's Peterhouse College. In 1970 he left Cambridge to lead the Seismic Discrimination Group at the Massachusetts Institute of Technology (MIT). Davies's research at MIT focused on detecting nuclear explosions, with the specific goal of monitoring compliance with the 1963 Partial Nuclear Test Ban treaty signed by the United States, the USSR, and Great Britain. His résumé suggested that he might be a good match for the job: he was an editor of the *Geophysical Journal of the Royal Astronomical Society* and had been *Nature*'s geophysics correspondent since 1968. Maddox had recruited Davies after reading a report Davies compiled about a conference held by the Stockholm International Peace Research Institute.[37]

Despite his familiarity with *Nature* and his editorial experience, Davies initially decided not to apply for the position. He was happy at MIT and was also concerned his research background might not be wide ranging enough. However, Hughes personally solicited his application. Maddox—despite his unhappiness with Hughes's decision to replace him—also called Davies to encourage him to apply. Davies submitted his application and flew to London for an interview. After a second interview, which included a meeting with former prime minister Harold Macmillan (a member of Macmillan's owning family), Davies was offered the position.

Davies came in to *Nature* with a number of goals. He sought to overhaul *Nature*'s appearance by eliminating the orange, advertisement-dominated covers and introducing interesting cover images, new fonts, and cartoons. "The hardest thing, I think, was getting the technical side of it right," he later recalled.

> I spent a lot of time at the typesetter's. . . . It was all done with hot metal in those days, and I was really keen that they should get out proofs that were in really good condition. . . . So I spent I think a lot of the first year or two worried about getting the technical side there so we didn't have misprints.

Unsurprisingly, Davies eliminated the unpopular *New Biology* and *Physical Science* and restored *Nature* to a unitary publication. Davies's other major goal was, as he put it, "getting the refereeing system beyond reproach." Unlike Maddox, who had felt perfectly comfortable accepting a paper because he found it interesting, Davies admitted nothing without reports from at least two referees, even in his own field of geophysics.

In addition to eliminating the satellites and overhauling the referee system for experimental papers, Davies made a distinct mark on the front section of the journal. Mindful of complaints about anonymous correspondents, particularly from American readers, Davies removed the anonymity of News and Views—after his own experience as one of those anonymous

FIGURE 9 David Davies in the late 1960s. Personal collection of David Davies.

correspondents, he said, he felt that the anonymity was not particularly useful to the writers or readers. Articles were now run under the author's byline instead of being labeled as "By our molecular biology correspondent" or "From our physics correspondent."

The tone of the editorials changed as well. Although Davies was less

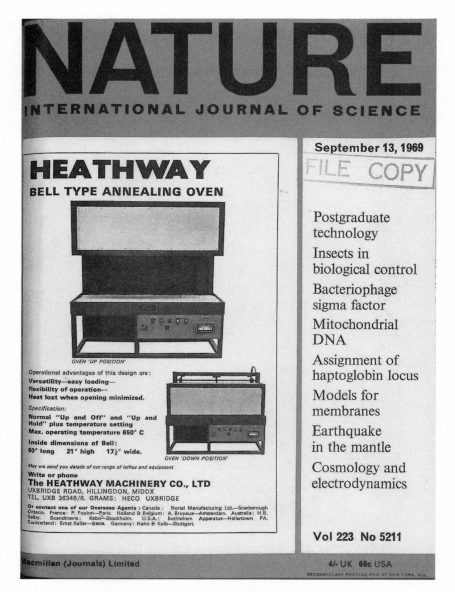

FIGURE 10 1969 cover of *Nature* from the Maddox era. Reprinted by permission of the Nature Publishing Group.

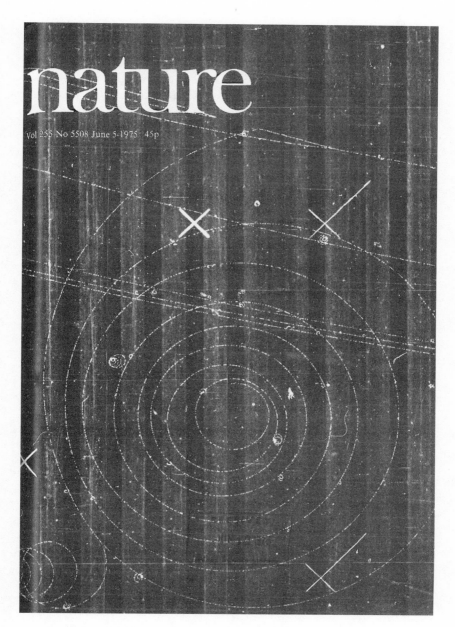

FIGURE 11 One of Davies's overhauled covers for *Nature* from June 1975. The cover image is a cloud chamber pattern. Reprinted by permission of the Nature Publishing Group.

enthusiastic about writing the leaders than Maddox, he estimated that he wrote approximately 85 percent of them during his editorship. He delegated the occasional piece about the life sciences to other staff members but wrote the bulk of the rest himself. "I would have liked to have other leader writers, but the staff were really stretched in all sorts of other ways," he explained. "I used to have to churn them out every Sunday; it was a real burden."[38]

From the beginning of his editorship, Davis sought to engage readers and contributors as active participants in shaping *Nature*'s path. In his first editorial, appropriately titled "*Nature* in the Future," Davies wrote that

> *Nature* must be an open journal, reflecting the sense of community which is still very strong amongst scientists. In the long run much of *Nature* is simply scientists talking to scientists about things which have a broad interest. Opinions differ about what constitutes "broad interest" and we can make no claim that the exposition and opinion sections of the journal give a uniform and totally balanced coverage. But the remedy to that is in the hands of readers. If you think there is more to tell or another side to a story, let us hear it.[39]

Three months later, with the elimination of the satellite journals nearing, Davies printed *Nature*'s revised style manual as part of the leader and clarified the criteria that would be used to select *Nature* papers, including length, whether the piece would be of wide interest, the plausibility of the conclusions, and the clarity of the writing.[40]

In another leader Davies told contributors "It's your journal" and invited readers to share their opinions on *Nature*'s form and content.[41] Were there ways that *Nature*'s acceptance rate (currently at approximately 35 percent) might be made more generous? Should the distinction between articles and letters be eliminated? Should length limits be raised or lowered? And should *Nature* continue to strive to be accessible to a general scientific readership or admit more specialized pieces that might only be understood by other scientists in their discipline? While the piece seemed to have little immediate effect on *Nature*'s format, Davies's choice of topic reflected his belief in editorial transparency, his conviction that a journal should serve the interests of its contributors, and his determination to admit as many high-quality papers as possible. "[I had] very much an academic scientist's view that the key thing was getting as many papers into print [as possible]," he explained.[42]

Davies also brought a deft sense of humor to *Nature*'s leaders. One pre-Christmas piece poked fun at convoluted academic language, concluding with the declaration that "At this time the personnel associated with this information-dissemination mode convey to its recipients appropriate greetings for the festive period upcoming momentarily."[43] Another leader consisted of a series of made-up rejection letters from increasingly obscure grant sources.[44] An April 1975 editorial satirized grant accounting reports; the imaginary professor began by declaring, "The work reported in this paper was supported by the Science Research Council who paid £15,000 for a project, the title of which can only be said to bear a remote resemblance to the title of this paper." The piece went on to describe expenditures such as

> a highly sophisticated microscope which turned out to be unnecessary ... a trip to America to give a ten-minute lecture at 8.20 on the first day of a conference, numerous train journeys to London to attend committee meetings the value of which the entire committee agrees privately is nil and a research assistant whose use has been minimal but whom the professor could find nobody else to pay for.[45]

Davies lent a playful sensibility to the journal, but he was also concerned that *Nature* should fulfill its readers' and contributors' expectations for a major scientific publication. In his view, *Nature* should be a voice for scientists across disciplines—and those scientists themselves had a responsibility for holding *Nature* to account.

INCREASING INTERNATIONALISM UNDER MADDOX AND DAVIES

As we saw in chapter 6, even under the British-focused Brimble and Gale, the percentage of experimental articles and letters to the editor submitted from international laboratories steadily increased. Under Maddox the trend toward an increasing number of international contributors continued. In 1966, the experimental articles and letters in *Nature* came from approximately 40 percent British sources and 60 percent international; by 1973, that figure stood at approximately 33 percent British sources and 67 percent international. As mentioned previously, Maddox also greatly expanded the international scope of the front half of the journal; *Nature*'s news reporting and opinion contained much more about science outside Britain's borders than it had under Brimble.

The increasing internationalism of *Nature*'s editorial purview was accompanied by a managerial shift: the opening of *Nature*'s first North American office in Washington, DC, in February 1970. *Nature*'s new Washington office was, in some ways, confirmation of a new scientific world order. The Second World War had brought massive destruction to scientific centers in Europe, the USSR, and Asia. After hostilities ceased, the United States—which had emerged from the war with its infrastructure intact and was riding high on wartime scientific successes such as the Manhattan Project—emerged as the unquestioned world leader in scientific research.[46] In the postwar period, America experienced tremendous growth in the number of practicing scientists and produced an increasing number of papers as a result.[47] As a well-known English-language journal, *Nature* was an appealing periodical for American scientists, especially after Maddox improved the journal's turnaround time. During Maddox's first editorship, American laboratories were behind roughly 35 percent of the articles and letters that appeared in *Nature*, and the Americans were inching ever closer to outnumbering British contributors. Americans had also outstripped the British in the number of subscriptions to the journal. Mary Sheehan explained that the new office was designed to facilitate contact with the increasingly large American readership base:

> We had an enormous number of American subscribers . . . more American subscribers than English ones. And because Americans tend not to like ringing out of country, or, you know, they think very internally, John thought it was a good idea to have an office in Washington.[48]

Maddox and Sheehan travelled to Washington in January 1970 to set up the new office; they stayed in Washington for six months. Maddox continued writing editorials, putting them on a freight plane every week so they could be in London for the printing of the journal.[49] When Maddox and Sheehan departed, Nicholas Wade, later the science editor of the *New York Times*, became *Nature*'s first Washington editor. Mary Scallan, a longtime *Nature* employee, moved to Washington to take on the secretarial duties, and in 1972 *Nature*'s assistant editor Mary Lindley moved from London to Washington to become the Washington manuscripts editor. The Washington office had the power to arrange refereeing for manuscripts and communicate with American authors about the status of their submissions; editorial operations and printing, however, remained centralized in London.

However, the opening of the Washington office seems to have had a limited effect on American scientists' view of *Nature* as an essentially British publication. Before he left MIT, Davies, accompanied by Mary Lindley, embarked on a series of visits to scientists in the Boston area to discuss their impressions of *Nature*. "I thought of it as sort of a lap of honor, but it turned out to be exactly the opposite," he later recalled. "They all were complaining." The complaints, he said, were partly about the sometimes barbed anonymous commentary in News and Views, partly about flaws in their proofs, but largely about *Nature*'s perceived British bias: "They all said ... it's a very British establishment journal. I bet all your referees are London and Cambridge."[50]

Maddox's three-part *Nature* may have been responsible for some of the perception that the British scientific establishment ran the journal. While the ratios of British and American contributors remained steady after the journal was split into three, articles by British contributors were more likely to appear in the "regular" Friday *Nature* than articles by contributors outside Britain. In 1971 and 1972, British laboratories were responsible for, on average, over half of the articles in *Nature* but only about a third of the articles in the satellites. By contrast, in 1971 and 1972, American laboratories contributed approximately 30 percent of the articles in classic *Nature* but 45 percent of the articles in the satellites. The international breakdown of the Letters section was fairly consistent across the three journals, but given the longer length and prominent placement of the articles, it is not difficult to imagine that American readers—especially contributors who had been unwillingly relegated to the satellites—took note of the discrepancy.

Mindful of the complaint that *Nature* was a "British establishment journal," Davies and his subeditors built an increasingly international network of referees who reviewed papers for *Nature*.[51] By 1980, *Nature*'s scientific content reflected Davies's goal of drawing in more contributors from outside Britain. Roughly one in five experimental articles came from Britain, a third came from the United States, and the rest came from laboratories in other parts of the world. He also pushed to make *Nature*'s news even more international. In his first leader, Davies wrote that *"Nature's news gathering facilities around the world must grow in the next year or two, as it is increasingly necessary to understand the scientific scene away from the trans-Atlantic axis."*[52] Davies felt that *Nature* had strong coverage of Britain and the United States but lagged in coverage elsewhere, and he added a number of new correspondents who worked in countries other than the United States and the United Kingdom.

We can put *Nature*'s contributor base in perspective by comparing it to another major scientific weekly: *Science*. In 1962, *Science* had acquired a transformative editor of its own, former *Journal of Geophysical Research* editor Philip Abelson.[53] Abelson recruited new staff writers to enhance *Science*'s News and Comment section and added a new section, Research News. He also reformed *Science*'s peer-review process, allowing scientists to suggest referees who would be qualified to review their papers (although as Abelson would tell his editorial board in 1973, "Authors complain equally about prejudice or incompetence of referees, whether the referees are their choice or ours or a mix"[54]). By the late 1970s the number of subscribers to *Science* dwarfed the number of subscribers to *Nature*: in 1979, *Science* had 152,000 subscribers worldwide, while *Nature* had only about 25,000.[55]

Interestingly, the data show that Abelson's changes to *Science* did not extend to recruiting more contributions from outside the United States. Like *Nature*, *Science* published two types of experimental articles. At the front of the journal *Science* printed its Articles section, which contained two to five pieces that were a mix of reviews of recent work, lecture reprints (often from Nobel laureate speeches), and new experimental findings in the physical, biological, or social sciences. At the back of the journal came the Reports column, which contained short experimental articles that bore a strong resemblance to *Nature*'s Letters to the Editor. Between 1960 and 1980 the percentage of experimental articles and reports in *Science* that came from non-American laboratories held steady at about 15 percent. In contrast, 80 percent of *Nature*'s experimental content was coming from outside the United Kingdom by 1980. Furthermore, only 9 percent of *Science*'s subscribers lived outside the United States, whereas more than 50 percent of *Nature*'s subscribers lived outside the United Kingdom. Between *Nature* and *Science*, there was no question as to which was the more international publication—a fact that illustrates both *Nature*'s international reach and the extent to which Americans dominated the production of English-language scientific papers.

Nature's impact factor gives us another means of examining the journal's growing reach during the 1970s. Many observers have pointed out that impact factors are a problematic way to assess a journal's readership or its importance; however, the numbers do show that *Nature* became more frequently cited during the late 1970s.[56] Eugene Garfield, the president of the Institute for Scientific Information (ISI), first compiled impact factors (a number measuring the average number of times recent articles in an academic journal have been cited in the rest of the literature) for journals in 1974. Garfield hoped that the numbers would help libraries assess which publications would be most useful to their researchers. In 1975, the first year Garfield and the ISI published the Science Citation Index Journal Citation Report, *Nature* had the 109th highest impact factor among the journals ranked; by 1980, *Nature* ranked 49th. *Science*'s rank, in contrast, remained relatively stable during the same period: in 1975 *Science* was 48th; in 1980, it was 56th.

LIMITED INTERNATIONALISM? *NATURE* AND SOVIET SCIENCE AFTER WORLD WAR II

As we have seen, both Maddox and Davies worked to expand *Nature*'s international scope during the 1960s and 1970s. If we mapped the origins of the articles and the letters to the editor during the Maddox and Davies editorships, however, we would see a significant gap in *Nature*'s contributor base: very few submissions came from the USSR. Between 1960 and 1980, only approximately 1 percent of letters and articles in *Nature* came from the USSR—a strikingly small number when we consider the USSR's vastness and scientific productivity. To put that number in perspective, Japan—a country far smaller than the USSR—contributed approximately 2.5 percent of *Nature*'s articles and letters. *Nature*'s content in the 1960s and 1970s provides a window onto the consequences Cold War scientific divides had for scientific communication and collaboration and once again shows that *Nature*'s editors and contributors were not only aware of intersections between science and politics, they sought to comment on and engage with them.

The low proportion of Soviet contributors is unsurprising when we consider some of the USSR's policies on collaboration between international scientists. In the years after World War II, the Soviet government under Stalin became increasingly concerned that Soviet science had fallen behind Western science and began pushing the Soviet scientific community to catch up and overtake Western science in productivity. Stalin felt that cultivating international ties could damage this program.[57] Soviet scientists who collaborated too closely with Western colleagues or were believed to have sent Soviet research abroad could face prosecution. In 1947, for example, two Soviet scientists, the cytologist Grigorii Roskin and the microbiologist Nina Klyueva, were convicted of sharing Soviet scientific secrets after sending some of their work to the United States for publication.[58]

Roskin and Klyueva were reprimanded and allowed to continue working, but in the aftermath of their trial, Stalin declared that all experimental findings and technical improvements would be regarded as state secrets and that divulging them to anyone outside the USSR without the government's permission would be punishable by eight to twenty years in a "reformatory labor camp."[59] Furthermore, Stalin halted publication of Soviet academic journals published in foreign languages and instructed all Soviet scientific, medical, and technical journals to cease the practice of translating their abstracts and tables of contents into foreign languages.[60]

Interestingly, although Stalin had forbidden the printing of foreign-language journals in the USSR, scientists still occasionally published abroad (with the government's permission) and expected that their discoveries would be recognized in the West. In May 1949, for example, the Soviet astrophysicists A. I. Alichanian and A. I. Alichanow wrote a letter to the editor of *Nature* titled "Concerning New Elementary Particles in Cosmic Rays."[61] Alichanian and Alichanow were writing in response to an October 1947 *Nature* article by the Bristol astrophysicists C. H. G. Lattes, G. P. S. Occhialini, and C. F. Powell.[62] Alichanian and Alichanow agreed with the authors' conclusions but objected to the authors' failure to cite their work, which had been published in the *Proceedings of the Academy of Sciences in the Armenian Soviet Socialist Republic* (also called the *Doklady Armenian S.S.R.*) in December 1946 and in English in the *Journal of Physics* in January 1947.[63] Alichanian and Alichanow argued that their results were not only relevant to the British group's work but were more complete than their British colleagues' and supported their conclusions about the relative masses of elementary particles more effectively than the British experimental results.[64]

The Bristol physicist C. F. Powell, one of the criticized paper's coauthors, was given the opportunity to respond in a letter to the editor directly following the Soviet letter. Powell wrote that he had not learned of Alichanow's work until May 1948, when a colleague in Paris had mentioned the Soviet group's research. In subsequent publications, Powell said, he and his colleagues had indeed referred to Alichanow's work. Furthermore, he had invited Alichanow to a symposium in Bristol on cosmic ray research in September 1948 but Alichanow had not attended. Powell wrote that the absence of citations to Alichanow's work was not the result of malice or intent to deny his Soviet colleagues credit but was "a consequence of the difficulties of communication and intercourse between us which exist at the present time." Powell went on to call for more international conferences and more readily available translations of Russian papers.

> I believe that similar difficulties are likely to occur in the future unless steps are taken to remedy this situation. . . . It would be of considerable assistance to physicists in Britain if reliable translations of important papers in Russian could be made available, with a minimum of delay, to interested workers in different fields of study. The responsibility for such translations could, perhaps, be undertaken by one of the learned societies.[65]

It is unclear whether Powell's comments were made in ignorance of the strict Soviet prohibition against foreign translations or whether he was aware of the policy and sought to call attention to its drawbacks while avoiding potentially inflammatory criticism of the Soviet government. But the Alichanow-Powell episode illustrates some of the consequences of the Soviet scientific rules for scientists on both sides of the Iron Curtain. British researchers like Powell might miss an opportunity to read relevant research in their field; Soviet researchers like Alichanow might find themselves denied international credit for their work. Even though Alichanow had published his work in the English-language *Journal of Physics*, a single English-language publication was not enough to ensure that foreign colleagues would be aware of his latest research, especially given the high volume of scientific papers most researchers had to contend with. Furthermore, Alichanow's letter was dated February 1949—more than a year after the October 1947 *Nature* papers. It seems likely that the delay was due to, at least in part, Alichanow's need to obtain permission before sending a letter containing some of his results.

Stalin's death in 1953 removed one of the major obstacles for Soviet scientists who wished to increase their connections with researchers in other nations.[66] Reforms enacted during the 1950s and 1960s loosened restrictions on foreign translations and foreign publishing by Soviet scientists, and papers by Soviet scientists were printed in Western journals, including *Nature*. However, there continued to be publishing restrictions in Communist nations, as Czechoslovak scientists Jiri Zemlicka and Stanislav Chladek discovered when they attempted to contribute an article to the *Collection of Czechoslovak Chemical Communications*. In a letter to *Nature*'s Correspondence column, the two wrote that their paper had been rejected on the grounds that they were "currently living abroad [in the United States] without the approval of the Czechoslovak government" and were therefore prohibited from publishing in any Czechoslovak journal.[67]

Furthermore, restrictions on academic travel from the USSR made it more difficult for Soviet scientists to participate in international conferences or accept places as visiting researchers than it was for their Western colleagues.[68] Academics who were considered loyal to the USSR would be allowed to visit the United States or the United Kingdom, but less politically reliable researchers would often be denied permission to attend professional conferences or become visiting professors outside the USSR. The ability to travel between departments and countries was not a trivial academic perk. The Alichanow-Powell episode illustrates how difficult it was for Alichanow to ensure awareness of his work or forge intellectual connections with his Western counterparts when he was not given permission to attend events such as the conference on cosmic rays. Similarly, in the case of plate tectonics, conferences, seminars, and visiting faculty appointments were instrumental in the acceptance of the new theory. When the crucial profiles of the Reykjanes and Pacific ridges were published in December 1966, most of the major figures in American and British geophysics already had some familiarity with the data. From the perspective of the British-American network, the published articles and their striking conclusions were the culmination of years of discussion at professional events and over coffee in department break rooms. From the perspective of those outside the network, however—including Soviet geophysicists—the crucial articles seemed a more sudden and isolated development. As a result, Soviet earth scientists, who tended to focus on continental geology rather than oceanic geology, were much less enthusiastic about plate tectonics than their Western colleagues. Soviet geophysicists were not represented in the *Nature* papers about seafloor spreading.

Indeed, Soviet scientists faced barriers to obtaining copies of *Nature* at all. In the USSR, *Nature* largely circulated as a collection of photocopies distributed from Moscow and unauthorized by Macmillan. This photocopied version was occasionally edited to remove potentially objectionable material.[69] Soviet academic libraries subscribed to the official version of *Nature*, but on occasion, pages containing anti-Soviet material would be removed before the issue was made available. Some issues—such as the one for 19 September 1970, which contained excerpts from Zhores Medvedev's book *The Medvedev Papers* about science in the USSR—never arrived there at all.[70]

NATURE AND THE USSR DURING THE COLD WAR

As we saw in chapter 5, *Nature* was highly critical of the USSR during the 1930s, giving voice to Soviet émigrés' charges that Soviet scientists lacked intellectual freedom and expressing outrage over Joseph Stalin's promotion of Lysenkoism over genetics. During the Second World War, the alliance between Great Britain and the USSR led to several *Nature* articles praising Soviet achievements in science. However, as plans for postwar Europe began to take shape, *Nature*'s contributors became more skeptical about the USSR. As early as March 1945, a lead editorial for *Nature* expressed concern over whether the members of the alliance could find common ground in the postwar world.[71]

Rainald Brightman, who wrote the majority of the leaders under Brimble and Gale, was not an admirer of the USSR and occasionally used *Nature*'s leaders to express distrust of the Soviet system. For example, in the fall of 1946, Brightman wrote a two-part lead editorial for the journal titled "Conditions of Survival" about the burgeoning international efforts to regulate atomic energy and weapons. Brightman claimed that honesty, integrity, and internationalism in science "cannot be realized under communism as we see it in Soviet Russia" and argued that "respect for human personality, freedom of thought and utterance, freedom of worship, freedom of investigation" were "denied by communism and cherished by the Christian ethic."[72]

The crystallographer J. D. Bernal, one of Britain's most vocal Communist Party members and an unabashed supporter of the USSR, wrote a letter to the editor strongly objecting to the editorial.[73] Bernal argued that the USSR had proven itself an able defender of intellectual freedom and that *Nature*'s divisive language might harm the attempt to find common ground over atomic energy. However, another article in the same issue of *Nature*, coauthored by John R. Baker and A. G. Tansley of the Society for Freedom in Science, arguably undercut Bernal's argument. In "The Course of the Controversy on Freedom in Science," Baker and Tansley wrote that the USSR had initiated a "movement against pure science and against freedom in science."[74]

In an unsigned leader, the editorial staff tried to chart a middle course. The leader suggested that perhaps Baker and Tansley were "jumping to conclusions" about the state of science in the USSR and noted that *Nature* itself had frequently advocated for more government planning of science in Britain because of science's importance for society. However, the piece was far more pointed in its criticism of Bernal's letter: "We think it is Bernal who is allowing politics to intrude upon his scientific views, and this is the type of attitude which we feel must be checked."[75]

This brief exchange serves as a useful illustration of *Nature*'s general editorial attitude toward the Soviet Bloc following the Second World War. The prolific Brightman favored leaders that focused on British domestic issues, not on international relations, but when the subject of the Cold War came up, the general tenor of editorials and news coverage under Brimble and Gale was pro-Western and skeptical of the USSR.[76] The most pointed commentary came when *Nature* discussed Lysenkoism and genetics in the USSR. Many prominent British critics of Lysenkoism published anti-Lysenko articles in *Nature*, and the journal's editorial staff decried the Soviet government's support of Lysenko's career.[77] But compared to *Nature*'s content about the USSR under Richard Gregory, the material about the USSR during the Brimble-Gale regime seems fairly mild, verging on indifferent.

Maddox's editorials were far more outspoken. Most visibly, under Maddox *Nature* publicized the plight of Zhores Medvedev, a Soviet biologist best known for his work on aging. Medvedev was a vocal critic of the Soviet regime. In 1969, he published *The Rise and Fall of T.D. Lysenko*, which criticized Stalin for supporting Lysenkoism. Although Lysenko had fallen from power in 1965, Medvedev was subsequently dismissed from his position at Timiryazev Academy. Undeterred, Medvedev wrote another book criticizing the Soviet government's treatment of scientists; Macmillan published the English translation, titled *The Medvedev Papers*. The publication of this second book in 1971 led to Medvedev's detainment at a Soviet psychiatric hospital. Following his release, Medvedev came to London to take a one-year visiting fellowship at the National Institute for Medical Research in 1973. The one-year fellowship turned into a permanent post when the USSR revoked Medvedev's citizenship, confiscated his passport, and told him he could not return home.

Nature reported all of this in careful detail, and Maddox frequently used the editorial page to call attention to Medvedev's situation. In September 1970 Maddox printed an excerpt from the forthcoming *Medvedev Papers* and devoted the leader to praising Medvedev's "fearless" advocacy for increased freedom of travel from the USSR.[78] In July 1972, Medvedev was arrested in Kiev and sent back to Moscow shortly before he was due to give a paper at the International Congress on Gerontology. Maddox's leader the following week recounted the incident and called on Western scientific societies to "seriously consider" canceling any future plans to participate in conferences held in the USSR, at least until they had obtained "formal assurances that the Russian police will not, in their own special way, help in choosing the participants."[79] When the USSR revoked Medvedev's passport, Maddox decried the USSR's treatment of Medvedev and suggested that the Western scientific community should make its displeasure known.

> Is this the time perhaps when invitations to Soviet scientists to attend conferences in the west should be withdrawn and participation in exchange schemes temporarily halted? Drastic measures indeed, which might damage several years of careful nurturing of Western-Soviet relations, but this is a drastic situation.[80]

It was under Davies, however, that *Nature* would take its most overtly political Cold War editorial stance. Davies would become known for his impassioned *Nature* editorials about nuclear test bans. (The Maddox regime also commented on Cold War issues such as nonproliferation treaties, although this was not a major theme of *Nature*'s editorials in the 1960s and early 1970s.) Davies's work at MIT had focused on using seismic data to detect subterranean nuclear tests to enforce compliance with a partial test ban. However, he came to believe that a comprehensive test ban—not merely the 1963 partial ban—was the only way to stem nuclear proliferation and prevent a nuclear war. His editorials were sharply critical of partial test ban treaties and especially of British nuclear policy, which he argued was too closely tied to American interests.[81] In his second week as editor, Davies criticized the British government for relying too heavily on American opinions for their nuclear defense policy.[82] Ten months later he accused the British government of outright "nuclear hypocrisy" for testing an atomic bomb in Nevada while pressuring other countries to restrict their own tests.[83] Interestingly, under Davies, *Nature*'s leaders were far less anti-Soviet than they had been under Maddox; the editorial page under Davies was more interested in commenting on policies west of the Berlin Wall.

Nature's news section in the 1970s, however, reinforced the impression of a divide between Soviet and Western scientists. The coverage of geology in the USSR frequently argued that Soviet refusal to accept continental drift isolated Soviet earth scientists and prevented them from collaborating with their Western colleagues. A 1976 article on recent scientific advances in the USSR by associate editor Vera Rich said that "One of the obstacles to communication between Soviet geologists and their foreign colleagues in recent years has been the unwillingness of Soviet geologists officially to admit the possibility of continental drift." Rich attributed this not to any "political considerations" but rather to "personal opposition from certain leading geologists, notably Professor V. V. Belousov of the Institute of Earth Physics in Moscow."[84] Four years later, Rich wrote a piece titled "Russian Plate Tectonics: Drift of Change" and led the article with the same sentiments:

> "Plate tectonics" and "continental drift" have been dirty words to orthodox Soviet geophysics for many years—orthodoxy being, in this case, laid down not by Party doctrine but by the personal views of Academician Belousov, the acknowledged head of the Soviet geophysical establishment.[85]

Many of Rich's articles on other areas of Soviet science played on a similar theme of division between the USSR and the West—and on the idea that Soviet science was deviant from acceptable scientific practice. In a 1974 article about a delayed international celebration for the 200th anniversary of the USSR Academy of Sciences, Rich wrote that "delay" was a "euphemism for 'scrapped'" and that the Soviet government had never intended to allow international scientists to be invited to celebrate the academy's milestone.[86] In an article on Soviet psychiatry, Rich wrote that psychiatrists who used their science to attempt to "cure" political dissidents were engaging in "a deliberate and cynical abuse of professional skill."[87] Rich also frequently reported on the treatment of refusenik scientists—Soviet researchers who wanted to emigrate from the USSR (mostly, but not exclusively, Jews who sought permission to emigrate to Israel). Rich's articles chronicled both the initial denials of the visas and the subsequent harassment and professional dismissals of scientists who had sought to leave the USSR.[88] Under Davies, criticism of the USSR was relocated from the editorials to the News section—shifting the USSR's political and scientific deviance from a matter of opinion to be argued to an objective fact to be reported.

DAVIES MOVES ON, MADDOX MOVES BACK

Unlike predecessors such as Lockyer, Gregory, or Brimble (who has the rather grim distinction of being the only editor of *Nature* to die while holding that position), Davies did not see editing *Nature* as a job for life. He was a firm believer in changing jobs—in fact, one of his *Nature* editorials criticized the British tendency to stay in the same position for decades, which Davies argued led to stagnation.[89] By 1979 Davies felt it was time to move on. He accepted a position with the Dartington Hall Trust, an organization in North Devon that was working to provide opportunities for employment, business growth, and rural development in a relatively remote part of Britain.

Davies and the others in the *Nature* office were surprised to learn the identity of his successor: none other than John Maddox, who had left *Nature* seven years earlier under somewhat contentious circumstances.[90] By then Jenny Hughes's star at Macmillan had dimmed and, despite her objections, Maddox was brought back.[91] In the interim, Maddox and Mary Sheehan had run a small publishing business called Maddox Editorial; when that folded, he and Sheehan returned to the Nuffield Science Teaching Project. Maddox seems to have had few qualms about returning to *Nature* despite what had happened in 1973. Brenda Maddox thought the Nuffield Project was "peaceful" but "slow-paced"; her husband missed journalism and had always regarded *Nature* as his ideal match.[92] Sheehan, who found the Nuffield Foundation dull, also welcomed the opportunity to return to *Nature*. Davies recalled that Macmillan's directors told him that this time, Maddox had agreed not to divide his attentions—he would work "twenty-four seven for *Nature*."[93]

It appears that Macmillan, at least at first, also attempted to exercise more control over Maddox's work. In a letter dated 3 October 1980, Maddox told the cell biologist Sir Michael Swann that Macmillan was "looking for [more] detailed control of the way in which *Nature* is written and compiled than I think good for the journal."[94] Indeed, Maddox was frustrated enough to apply for a position at Oriel College, Oxford, where Swann was provost; however, he and Macmillan evidently managed to work out their relationship to their mutual satisfaction. Maddox would edit *Nature* for the next fifteen years.

By the time Maddox returned to *Nature*, the Washington office was well established, and four-fifths of the articles and letters to the editor came from outside Britain. But there is a great deal of evidence to suggest that *Nature*'s readers still viewed *Nature* as essentially a British journal—Davies's colleagues' complaint that *Nature* was a "British establishment" publication indicates that *Nature* maintained its "British accent" (as Maddox had put it). At first glance, it is somewhat surprising that so many scientists from other countries sought to contribute to what they saw as a British journal. The 1960s and 1970s were precisely the period when the United States overtook Britain—and every other country—to become the West's unquestioned scientific powerhouse. Furthermore, Britain's colonial empire had dissolved. Great Britain was no longer the scientific and political power it had been in the early twentieth century. What drew scientists from other parts of the globe to contribute to a publication in a country whose scientific influence had waned?

Some of *Nature*'s continued influence can be attributed to the large number of foreign subscribers, particularly Americans. Even if Britain was no longer the scientific or political power it had been, *Nature* reached a wide audience in the new scientific powerhouse. Indeed, the comparison with *Science* suggests that *Nature*'s move toward internationalism was also helped by the fact that it was *not* an American journal. In fact, Britain's decline as a world power was exactly what paved the way for a British journal to position itself as a spokesman for a worldwide scientific community in the 1980s. *Nature* was no longer the leading journal from the seat of the world's most extensive colonial empire; it was a journal published in yet another country whose scientific influence had been eclipsed by that of the United States. Contributors from France, Germany, or Japan could see themselves as part of *Nature*'s community in a way that might not have been possible in the nineteenth century. Under Maddox's leadership in the 1980s, *Nature* opened new offices in Tokyo, Paris, Munich, and Hong Kong, reflecting the growing internationalism of both the front and back sections of the journal.

Nature's transition from a British scientific journal to an international one is somewhat a story of its editors—Brimble and Gale were focused on Britain while Maddox and Davies sought to expand the scope of *Nature*'s coverage. But as with *Nature*'s transition from a lay periodical to a specialist journal in the 1870s, the transition from a national to an international publication was ultimately driven by *Nature*'s contributors. Even under Brimble and Gale, the percentage of international contributors to *Nature* steadily increased, and after Maddox restored *Nature*'s reputation for speedy publication, the journal became an appealing publication site for an even larger number of scientists. The growth in submissions came with a corresponding decline in the acceptance rate. In 1974, *Nature* printed approximately 35 percent of submitted papers; by the end of the 1980s, the acceptance rate would be down to 1 in 8 papers, or 12.5 percent. *Nature* could no longer be considered a journal that "might print anything," as it had been in the early 1960s. As *Nature* became more selective, publication in *Nature* became increasingly prized as a sign of scientific success—and the more that researchers sought to publish in *Nature*, the more difficult that distinction became to win.

Furthermore, as scientific careers increasingly crossed national borders, the kinds of national publishing strategies that drove journal selection in the nineteenth and early twentieth centuries became less and less important—with the significant exception of the Cold War divide. As we have seen, the internationalism of *Nature*'s community in the 1960s and 1970s had a major limitation: it was largely restricted to non-Communist countries. This gap in *Nature*'s contributor base was not a conscious creation of *Nature*'s editorial staff but rather a reflection of divisions between scientific practitioners. While there were occasions such as the International Geophysical Year when Soviet and Western scientists collaborated and some exchange via programs that brought Soviet scientists abroad and sent Western scientists to the USSR, Cold War researchers faced obstacles to collaborating or communicating with colleagues on the opposite side of the Berlin Wall. Even if a Western researcher had wanted to publish a piece in one of the USSR's most prominent scientific journals, few Western scientists would have been able to speak or write enough Russian to do so, and the language barriers would have meant that an article in a Russian journal would be unlikely to yield many career rewards for an American or British researcher. Similarly, as had been the case for French or German scientists in the early twentieth century, Soviet scientists had little career incentive to publish in *Nature* in the 1960s or 1970s. Their advancement depended more on recognition from fellow Soviets than on recognition from Western scientists. Contributors' interests, not editorial policies, were the cause of the gap in the contributor base. As the examples of plate tectonics and astrophysics show, however, this publishing and communications gap had significant consequences for scientists who found themselves excluded from discussions that led to major theoretical changes or who would have liked their work to receive more international recognition. Scientific communication shaped—and separated—Cold War scientific communities.

CHAPTER EIGHT

"Disorderly Publication": *Nature* and Scientific Self-Policing in the 1980s

Nature's subscribers opened their copies of the journal on 30 March 1989 to find a leader on difficulties in the recent elections for the Soviet Academy of Sciences.[1] Not a few readers, however, probably found their eyes drawn to a second, smaller headline tucked into the right-hand corner of the first page: "Cold (con)fusion."

Just days before, Martin Fleischmann and Stanley Pons, two chemists at the University of Utah, had made a startling claim at a press conference: by submerging a palladium electrode in a solution of heavy water and running a current through it, they had produced nuclear fusion at room temperature. The accompanying heat and energy had, Pons and Fleischmann said, caused one of their cells to explode, leaving a small crater in the cement floor of their laboratory. The economic and environmental implications were remarkable—had these two chemists discovered a process that could provide cheap, clean energy to the world?

Pons and Fleischmann told their audience that they had submitted a full report to *Nature* and that their paper would soon be published alongside an article by their colleague Steven Jones at Brigham Young University. But according to the "Cold (con)fusion" article, the *Nature* staff was not yet ready to endorse the Pons-Fleischmann findings. "Reports that an account of cold nuclear fusion is soon to appear in this journal are premature," the article crisply declared. "No one was more surprised than the editors of *Nature* to learn, on reading last Friday's *Wall Street Journal* that two papers on room-temperature nuclear fusion would probably appear simultaneously in this journal, perhaps in May."[2]

The editorial raised no objection to the science of cold fusion—indeed, aside from what had been said at the press conference, little was known about the methods that had enabled Pons and Fleischmann to obtain such extraordinary results. The editorial did take issue, however, with the way in which the discovery had been announced. "When scientists find themselves reading about their colleagues' discoveries in newspaper columns before anything has been submitted, let alone accepted or published, in a research journal, there is cause to be worried," the article stated.

> The procedure of peer-review, slow and irksome as it sometimes can be, has evolved to protect not only journals but scientists and science itself.... Patents that turn out to be worthless and investments that disappear may demonstrate the value of cautious peer-review to those who now think of it as a fusty institution much loved by pedants.[3]

The article closed with a caution to science reporters, reminding them that just two years ago, "reports of superconductivity at increasingly incredible temperatures were appearing daily."[4] The subtext of this reminder was obvious: until an article appeared in a specialist journal, cold fusion was an inherently unreliable claim that should be treated with extreme caution.

Although Pons and Fleischmann had indeed submitted an article to *Nature*, that journal never printed it: only Steven Jones's article, with its far more modest claims about neutron production and excess heat from the reaction, would be published in *Nature*. Instead of being the forum where a new era of energy was declared, *Nature* quickly became a major center of cold fusion skepticism. By 29 March 1990, a year to the week after the first mention of cold fusion in *Nature*, Maddox felt secure enough to declare "Farewell (Not Fond) to Cold Fusion" in the magazine's leader.[5]

Maddox's strong public stance on the cold fusion findings was part of a larger discussion about *Nature*'s role in the scientific community in the late twentieth century. Another significant episode from the 1980s, Maddox's investigation and denunciation of the French immunologist Jacques Benveniste's *Nature* paper on highly dilute antisera, reveals that Maddox saw *Nature* not merely as a passive forum for the printing of scientific papers but as an institution that could, if necessary, take dramatic steps to defend the scientific community from "careless" research. However, *Nature*'s readers and contributors rejected Maddox's efforts to claim an investigator's role for the journal; they decried the Benveniste episode as a "circus" and criticized *Nature* for printing a paper that was apparently so questionable. During the cold fusion controversy, Maddox shifted his tactics. He and the rest of the editorial staff cast the cold fusion episode as a battle between careful, peer-reviewed, properly conducted science and sloppy science revealed through press conferences in hopes of wealth through patents. Maddox wrote editorials criticizing Pons and Fleischmann's methods, associate editor David Lindley wrote news articles forecasting the death of cold fusion, and the journal's editorial staff gave significant space to cold fusion's most prominent scientific critics. Where *Nature* led, science reporters followed. News outlets such as *Time*, the *Economist*, and the *Wall Street Journal* all covered *Nature*'s role in the cold fusion controversy and portrayed the journal's skepticism as proof that the scientific community was rejecting the Pons-Fleischmann claims. Ultimately, the cold fusion episode convinced many observers of the scientific journal's continued importance to the scientific community and illustrated *Nature*'s influence among both scientists and laymen at the end of the twentieth century.

WHAT SHOULD BE IN A JOURNAL?

Controversies over questionable science were not new to *Nature*'s pages in the late twentieth century. In 1904, for example, *Nature* printed a letter to the editor from the American physicist Robert W. Wood in which Wood described a visit to the laboratory of the French physicist René Blondlot.[6] Wood had visited Blondlot at the request of the German physicist Heinrich Rubens, one of many researchers who had been unable to duplicate Blondlot's reported results.[7] After his visit, Wood publically declared Blondlot's signature discovery, N rays, a delusion.

According to Wood's account in *Nature*, he had asked Blondlot to show him an experiment meant to isolate and refract the N rays through an aluminum prism. During the experiment (which was conducted in a dark room), Wood secretly removed the aluminum prism from the apparatus. This experimental sabotage had no effect on Blondlot's results. Wood proposed several experiments that he said would "settle the matter beyond all doubt," but wrote that he "left with a very firm conviction that the few experimenters who have obtained positive results have been in some way deluded." The "deluded" scientists did not instantly abandon N rays following Wood's critique (Blondlot, who remained in his position at Nancy until his retirement in 1910, believed in their reality until his death in 1930), but by 1906 even Blondlot had ceased to study or publish about them.[8]

Brimble and Gale were not particularly interested in these kinds of debates; as we have seen, under their guidance *Nature* was generally staid and unobjectionable. John Maddox, in contrast, was unafraid to court controversy during his first editorship—his provocative editorials on DDT and science in the USSR stand as obvious examples. But it was David Davies who began testing the boundaries of what might be acceptable as a scientific paper. In October 1974 *Nature* printed a paper about extrasensory perception by two researchers in the Electronics and Bioengineering Laboratory at Stanford University, Russell Targ and Harold Puthoff.[9] Targ and Puthoff concluded that some tested individuals (including the claimed telekinetic Uri Geller) did indeed have the ability to "see" information about a remote, unknown location. In a leader about the paper, Davies published a summary of the referees' reports and explained that despite some reservations among the referees, he had printed Targ and Puthoff's paper because

> *Nature*, although seen by some as one of the world's most respected journals cannot afford to live on respectability. We believe that our readers expect us to be a home for the occasional "high-risk" type of paper. This is hardly to assert that we regularly fly in the face of referees' recommendations. . . . It is to say that the unusual must now and then be allowed a toe-hold in the literature, sometimes to flourish, more often to be forgotten within a year or two.

Davies then challenged readers to look into Targ and Puthoff's claims:

> Publishing in a scientific journal is not a process of receiving a seal of approval from the establishment; rather it is the serving of notice on the community that there is something worthy of their attention and scrutiny. And this scrutiny is bound to take the form of a desire amongst some to repeat the experiments with even more caution.[10]

Nature's readers, for the most part, declined the challenge. Only one research group responded to the Targ-Puthoff article in *Nature*.[11] Davies later said that given the chance he might not print the paper again because it had not incited the discussion he had expected: "I regretted it afterwards, but we took a lot of advice on it and thought let's get it in there and see whether anybody follows up on it. But people just said no, I'm sure there's cheating going on there."[12]

The Davies regime was also well known for printing a piece by Sir Peter Scott and Robert Rines about the Loch Ness Monster, including photographs from Loch Ness and a discussion of how the creature might acquire an appropriate taxonomical name (Scott and Rines suggested *Nessiteras rhombopteryx*).[13] According to Davies, *Nature* staffers accepted the piece while he was at *Nature*'s Washington office.

> I said I think it should go in the front end of the journal, not as a scientific paper but as an opinion, Comment and Opinion. So we put it in the front end of the journal. . . . And it was an amusing piece really, I don't think there was anything in it, and it was just before Christmas.

The article, however, generated a great deal of attention. Davies was even summoned before Harold Macmillan (the former Prime Minister and member of Macmillan's owning family) to explain the piece. "He just wanted to know that we weren't going soft-headed," Davies explained. "And I told him no, it's absolutely a one-off."[14]

Davies, however, was not speaking for John Maddox, who returned to *Nature* in 1980 with his "newshound" sensibility and appetite for controversy fully intact. During Maddox's second editorship, he wanted *Nature* to pursue not only scientific preeminence but journalistic triumphs. Maddox's choices of journalistic coups reveal his very real interest in *Nature*'s role as a *scientific* journal, an institution that wielded great influence over which new scientific knowledge claims would be accepted or rejected. Maddox took strong editorial stances and relished direct engagement in scientific controversies, believing that as the editor of *Nature*, he had a responsibility to weigh in on the most important scientific issues of the day.

A series of scientific scandals in the late 1980s throws Maddox's views about *Nature*'s role into particularly strong relief. In 1988, the Nobel Prize–winning immunologist David Baltimore became embroiled in a government investigation into a 1986 paper that he coauthored for the journal *Cell*.[15] Massachusetts Institute of Technology (MIT) biologist Thereza Imanishi-Kari was the lead author of the paper, but Baltimore's status as a Nobel Prize winner quickly ensured that the investigation became known as the "Baltimore affair."[16] The paper had found that transgenes (segments of DNA transferred from one organism to another) were capable of affecting antibody production in mice—suggesting that transgenes might one day be a method of strengthening the body's defenses against disease. However, Margot O'Toole, a postdoctoral fellow working in Imanishi-Kari's laboratory, was unable to reproduce the results Imanishi-Kari's team had reported in *Cell*. The project had received federal funding from the National Institutes of Health (NIH). After O'Toole contacted the NIH's Office of Research Integrity (ORI) to express concerns about Imanishi-Kari's work, Michigan congressman John Dingell created a committee to investigate whether Imanishi-Kari had forged her data—and, by extension, if Baltimore had signed off on fraudulent results. The initial ORI report found Imanishi-Kari guilty of nineteen counts of research misconduct based largely on a Secret Service investigation of the biologist's laboratory notebooks. The Baltimore affair would stretch out over nearly a decade until a 1996 report by the Department of Health and Human Services exonerated Imanishi-Kari of any wrongdoing in her experiment.

The Dingell investigation unsettled many within the scientific community, including John Maddox. In a lengthy article for *Nature*'s News and Views section on 30 June 1988, Maddox described the growing controversy as a "Greek tragedy" and expressed deep concern over the US government's involvement in the case.

> The remedy in this sad and sometimes shabby case does not reside with [NIH investigators] Ned Feder and Walter Stewart, with Margot O'Toole, David Baltimore or even, with great respect, John Dingell. It is a communal problem, that of how the scientific community should learn to live with the evident imperfections of the scientific literature.... The remedy rests with the scientific community.... The snag is that, even in the United States, the community does not appreciate what it has to lose. And that the Dingell committee does not understand what damage it may do.[17]

It was almost certainly not a coincidence that Maddox's article on the Baltimore affair appeared in the same issue as a paper that seemed to provide a test case for the scientific community's ability to evaluate problematic results. The paper was "Human Basophil Degranulation Triggered by Very Dilute Antiserum against IgE" from a team of researchers led by the immunologist Jacques Benveniste (1935–2004) at Paris's Institut national de la santé et de la recherché médicale (INSERM) laboratory. *Nature* warned its readers that this was no ordinary article. The leader for the 30 June 1988 issue was titled "When to Believe the Unbelievable," and the subheadline alerted readers that "An article in this week's issue describes observations for which there is no present physical basis." Despite the exhortations for readers to approach the paper with caution, the editorial portrayed Benveniste and his work in a generally positive light. *Nature*'s staff described the French immunologist as a humble seeker of truth, as perplexed by his results as anyone else and eager to receive the scientific community's feedback and advice on his work.[18]

The claims were indeed astonishing. Benveniste's paper described a series of experiments showing that highly dilute antiserum solutions—solutions so dilute, in fact, that statistically they no longer contained antiserum molecules—still produced the same chemical results as antiserum solutions at traditional dilutions. The data also displayed an unusual "rhythmic" pattern to the activities of the different dilutions—antiserum activity would drop off at one dilution, rebound dramatically at another even more dilute concentration, then drop off and rebound again at increasing dilutions. Furthermore, all dilutions had to be accompanied by vigorous shaking (or "vortexing" in the laboratory's terminology) in order for the effects to be observed.

The paper had implications that went far beyond the field of immunology. Although the paper did not explicitly mention possible medical applications, the Benveniste team's findings seemed to be laboratory evidence for the effectiveness of the alternative medical practice of homeopathy. Homeopaths treated (and still treat) illnesses by having the patient ingest highly dilute solutions of various substances, often ones toxic in higher doses such as mercury or arsenic.[19] Traditional medical practitioners had long scoffed that there was no theoretical or empirical basis for believing that a solution of what was, statistically speaking, plain water could have any therapeutic effects. But if the Benveniste team's findings were correct and a solution of antiserum that had been diluted by a factor of 10^{120} could still produce the same chemical actions as a solution of antiserum diluted by a factor of 10^2, suddenly homeopathy had, if not a satisfying theoretical explanation, at least laboratory-based evidence in favor of its efficacy.

Maddox and the *Nature* staff warned against such an interpretation in their editorial, admitting that among homeopaths "there will be a natural inclination to welcome Benveniste's article as aid and comfort, but that would be premature, probably mistaken." Maddox further cautioned against embracing the Benveniste results in an "Editorial reservation" printed after the article, where he wrote that the paper was likely to trigger "incredulity" among readers and declared there was "no physical basis" for the team's results. The final sentence of the "Editorial reservation" hinted at what would come next in *Nature*: "With the kind collaboration of Professor Benveniste, *Nature* has therefore arranged for independent investigators to observe repetitions of the experiments. A report of this investigation will appear shortly."[20]

A scientific journal sending a team of investigators to a contributor's laboratory to observe their work was, to say the least, an unusual step. Maddox appears to have had the idea in mind as early as the spring of 1987, when Benveniste first submitted the paper. Maddox and the staff at *Nature* sent Benveniste's paper out for peer review, as the journal had done for all research articles since 1973. Walter W. Stewart, an NIH employee famous for his audits of scientific fraud—in particular, his investigation of the Baltimore affair—acted as one of the referees.[21] In the cover letter to his (largely negative) referee report, Stewart wrote, "If you do send a team to Paris to check on the laboratory, please keep me in mind," indicating that Maddox was considering a visit to the Paris laboratory long before he agreed to publish the heavily revised Benveniste paper.[22] Walter Gratzer, the former molecular biology correspondent who had maintained connections with Maddox and *Nature*, believes that Maddox had the 1904 N rays episode in mind when he conceived the idea and that he expected a similarly quick and clean explanation for the Benveniste team's results.[23] Indeed, it seems likely that the chance to make such a visit was a key reason that Maddox agreed to print the paper. Significantly, the Benveniste paper ran under the special heading of "Scientific Paper." According to Charles Wenz, then the coordinating editor of *Nature*, none of Maddox's subeditors would accept responsibility for printing the paper in their own sections.[24] This was clearly the dynamic editor's personal project.

The result was one of the most astonishing—and bizarre—episodes in the history of the modern scientific journal. The three independent investigators *Nature* chose to observe the experiments were Stewart; James "The Amazing" Randi, a former laboratory technician who had made his name as a stage magician and debunker of alleged psychic phenomena; and Maddox himself.[25] The three men arrived at Benveniste's Paris laboratory on 4 July 1988. Meanwhile, controversy was already bubbling as correspondents wrote to weigh in on the Benveniste article and *Nature*'s unusual approach to its publication.[26]

Nature withheld publication of letters critiquing or explaining Benveniste's article until the three-man team of Randi, Stewart, and Maddox was ready to release their report. On 28 July 1988, the Benveniste findings once again took over *Nature*, but the French immunologists had little reason to celebrate their return to *Nature*'s pages. Randi, Stewart, and Maddox titled their report "'High-Dilution' Experiments a Delusion."[27] The article that followed was a damning account of what the three had found at the Benveniste laboratory. The team declared that the high-dilution experiments were "statistically ill controlled," claimed that "no substantial effort has been made to exclude systematic error, including observer bias," and further claimed that all data in conflict with the team's hypothesis had been excluded from Benveniste's analysis. "The phenomenon described," said Maddox, Randi, and Stewart, "is not reproducible in the ordinary meaning of the word."[28]

Maddox, Randi, and Stewart based their condemnation of the French team's experimental methods on three "double-blind" experimental runs in which coauthors Dr. Elisabeth Davenas and Dr. Francis Beauvais were asked to measure antiserum activity without knowing which dilution they were examining. At one point Randi, concerned about the possibility of fraud, had taped the decoding key to the laboratory's ceiling. The results of the double-blind experiments, said the team, showed no antiserum activity at high dilutions and no evidence of the "rhythmic" phenomenon described in Benveniste's initial paper.

The investigative team did not accuse the French laboratory of fraud; instead, they said, "We have every reason to believe that Dr Benveniste was (and, perhaps, still is) convinced of the reality of the phenomenon reported in his article."[29] Rather, the report portrayed the team as enthusiastic but incompetent researchers who had carried out sloppy experiments and ignored negative results in their zeal to find antiserum activity at high dilutions. "The climate of the laboratory is inimical to an objective evaluation of the exceptional data," the report said. "The folklore of high-dilution work pervades the laboratory, as epitomized by the suggestion that decanting diluted solution from one tube to another might spoil the effect and the report that the repeated serial dilution by factors of three and seven (rather than ten) always yields negative results."[30] The report further noted that Benveniste's research had been heavily funded by the French company Boiron et Compagnie, a supplier of homeopathic pharmaceuticals.

Benveniste was given the chance to write his own article in response. He portrayed Walter Stewart as a one-man scientific Inquisition with no significant research to his name, accused Maddox of being nothing more than Stewart's dupe, and called the investigation "a mockery of scientific inquiry" and a "Salem witchhunt."[31] Despite the "lip service" paid to his research team's honesty, he said, the presence of stage magician Randi, famous as a debunker of alleged psychics such as Uri Geller, indicated that the investigators assumed fraud was involved from the very beginning.[32] Furthermore, Randi had been the investigator who had caused his research team the most agony: during a crucial point in one of the experiments, "Randi [was] playing tricks, distracting the technician in charge of its supervision!" Benveniste hinted strongly that the entire episode, from the acceptance of his original paper to the July 28 article dismissing his results, had been a setup designed to discredit his work: "Why then accept a paper on 13 June to publish June 30th to destroy on 8 July data so easily spotted as wrong or made up? Is it a display to the world of the almighty anti-fraud and heterodoxy squad?"[33]

The controversy continued in the next issue of *Nature*. The Correspondence column for 4 August 1988 was filled with letters about the Benveniste episode; many suggested possible explanations for the phenomenon described in the original article.[34] But the longest and most prominent letter came from two scientists from the NIH, Henry Metzger and Stephen C. Dreskin, who wrote to say that they had repeated the Benveniste experiments and "observed no results such as theirs. We therefore see no basis as yet for concluding that the chemical data accumulated over two centuries are in error." Despite their conclusion that the high-dilution article had been flawed, Metzger and Dreskin expressed disappointment in the way *Nature* had handled the episode:

> It is reasonable to ask whether the observations of Davenas *et al.* should have been published by *Nature*. We think not. One of us (H.M.) reviewed their paper at the request of *Nature* in April 1987, and urged that the findings be checked by one or more laboratories chosen by the editor. Instead Dr Benveniste made his own choice, and *Nature* decided to publish the report and then to despatch [*sic*] an international investigative team consisting of the editor, a magician and a scientist, none of whom has experience in the relevant field. Their report provides no support for the published claims and will dismay serious scientists: it adds to the circus atmosphere engendered by the publication of the original paper.... We believe that the approach chosen by *Nature* is regrettable.[35]

Discussion of the Benveniste data continued in *Nature* for eight more weeks. Some letters came from homeopaths protesting the treatment of the Benveniste paper; others came from scientists trying to explain what known chemical or biological processes might have caused the phenomenon observed in the data.[36] A few letters came from immunology laboratories where the scientists had not been able to reproduce the INSERM team's results.[37] Other letters continued to question the accuracy of the data presented in the Benveniste team's original paper; a particularly scathing letter from Dr. P. M. Gaylarde at the Royal Free Hospital in London explicitly accused the Benveniste team of inventing the data, claiming that the Benveniste team's small standard errors could not have been obtained in a real experiment.[38]

But a substantial number of the printed letters, like the Metzger and Dreskin letter, directed their criticisms not at Benveniste but at *Nature* and its editor. Maddox had tried to push the boundaries of his role as *Nature*'s editor; the journal's readers and contributors pushed back. Gregory Petsko, from the Massachusetts Institute of Technology Department of Chemistry, wrote that "simple human error" was the most likely cause of irreproducible results. He argued that *Nature*'s approach to the Benveniste paper had been counterproductive, even harmful to science: "Fraud is a very serious matter, but I think it is more apt to occur in a climate where mistakes are treated too harshly."[39] Other correspondents echoed these sentiments, lamenting the "elements of farce, witch hunt and arrogance"[40] in the Maddox team's report and expressing concern that *Nature*'s "rushed and evidently prejudiced attempt to discredit" the Benveniste paper had only served to make the results impossible to disprove convincingly.[41] G. J. Neville from the Institute of Laryngology and Otology in London suggested (tongue in cheek) that perhaps the scientific community ought to create a "hit squad" that would "descend unannounced upon unsuspecting laboratories, ruthlessly checking routines for the inclusion of relevant sampling and statistical errors." This approach, he said, "would provide the suitable climate of moral fear and financial accountability under which basic science is expected to operate."[42]

Maddox shut down the discussion on 27 October 1988 with two final articles about the controversy: one by Benveniste and one by himself. Benveniste took the occasion to reiterate his charges of sloppiness and witch-hunting on the part of the investigative team, implying that Nature had only agreed to print his article as part of a "plot" to discredit his results. "Unfortunately, facts are stubborn and so are we," he declared. "The numerous truth-seeking scientists all over the world, some of them prompted by our paper and the obviously biased inquiry, have intellectual and technical means either to understand the error or to establish this new field. There is more to come."[43]

Maddox's four-page article on the controversy followed Benveniste's final response. The embattled editor adopted a generally self-assured yet bemused and slightly regretful tone to explain his decision making during the Benveniste affair, suggesting that the readership's strong reaction had indeed made an impression. "I have learned a great deal from the controversy, as have many of my colleagues, although I do not pretend fully to understand why such great passions have been aroused," he began.[44]

Maddox argued that his journal's conduct had been basically sound. He defended the participation of both Walter Stewart and James Randi in the investigation, describing Stewart as a deft laboratory scientist who also happened to possess "a flair for spotting inconsistencies in intricate arguments" and noting that Randi was not only a Macarthur Fellow but a former laboratory technician. Although his previous writings had depicted Benveniste as humble and genuinely convinced of his results, Maddox now portrayed the French scientist as a bully who demanded that Nature print his work even after negative referee reports:

> I should also have been more cautious when, having rejected the paper for what my colleagues hoped would be the last time, Dr Benveniste telephoned indignantly to protest that Nature was proposing to suppress news of one of the greatest discoveries of the twentieth century. I forget whether he compared his dilemma to that of Galileo on that occasion or in a conversation during the later visit to Paris. . . . But the accusation of suppression by the head of a government-supported laboratory cannot be dismissed lightly.[45]

Asking the Benveniste team to agree to a visit before the paper was published, said Maddox, would have bestowed all of the power on Benveniste: "given a favourable or even noncommittal report, he would have claimed publication, but otherwise could have withdrawn."[46] Maddox also defended his investigation from the charge that it had been overzealous and unfair. "*Nature* has no ambition to lead a pack of vigilantes seeking to rid the scientific literature of error," he insisted.[47] While Maddox maintained that *Nature* would continue to be on the lookout for "erroneous" or "careless" science, he was clear that the independent inquiry into Benveniste's laboratory was an experiment that would not be repeated.

Maddox's unprecedented actions during the Benveniste controversy reveal much about his view of *Nature*'s role within the scientific community. First, they illustrate Maddox's belief that *Nature* should be an active voice for his vision of good science. The decision to personally evaluate the INSERM laboratory's methods rather than rejecting the paper provided an opportunity for Maddox to not merely print the most exciting scientific papers but to discredit striking results that he believed had not been obtained through careful experimentation.[48] Second, the episode reveals Maddox's conviction that science should police itself and that *Nature* was one of the scientific institutions that ought to be responsible for a critical evaluation of results that seemed too good to be true. Overall, Maddox sought to use the Benveniste affair to illustrate that existing scientific institutions could detect and discredit findings from laboratories that failed to meet the standards for quality research—and to expand the role that journals such as *Nature* might play in such detection. However, Maddox's vision for *Nature* did not go unchallenged. The journal's readers and contributors took the editor to task for his unusual actions; they made it clear that they did not think a laboratory "hit squad" fell within the mission of a scientific journal. The next time Maddox encountered a chance to discredit problematic results, he changed his tactics dramatically.

COLD FUSION AND *NATURE*

About the same time that *Nature*'s readers and contributors were debating the journal's conduct in the Benveniste affair, Steven E. Jones (b. 1949) of Brigham Young University's physics department was reviewing a Department of Energy funding proposal from two chemists at the University of Utah. To Jones's surprise the chemists, Martin Fleischmann (1927–2012) and Stanley Pons (b. 1943), were working on a project extremely similar to his own: an attempt to produce nuclear fusion using a palladium cathode submerged in heavy water. Palladium has the unusual property of being able to absorb large amounts of hydrogen and deuterium ions into its metal lattice.[49] Jones, Pons, and Fleischmann were all attempting to use electrolysis of heavy water to force so much deuterium into a palladium cathode that the resulting pressure would force the deuterium to undergo nuclear fusion. The expected fusion would release neutrons, gamma rays, and much larger amounts of energy than would be observed from a standard electrolytic cell.

Jones, who was preparing to announce his findings at a scientific conference in March 1989, contacted the Utah group to discuss the overlap in their research projects. Jones's revelation appears to have unsettled Pons, Fleischmann, and their employer, the University of Utah. If cold fusion proved to be a viable source of energy, billions of dollars in patent rights would be at stake; if Jones published first, he would establish an indisputable priority claim no matter how long Fleischmann and Pons had been working on the same problem. The two research groups, with some assistance from their universities, began negotiating an arrangement that would allow all three men to share equally in the credit for palladium-cathode fusion. The Brigham Young and Utah teams eventually agreed that on 24 March 1989, representatives of both research teams would meet at the Federal Express office at the Salt Lake City airport to simultaneously mail their papers to *Nature*.

In accordance with their agreement for simultaneous publication, Jones canceled his planned talk at a March conference and instead submitted an abstract to the American Physical Society (APS) for their May 1 meeting. However, the APS published abstracts in advance. When Pons and Fleischmann saw Jones's abstract in the APS program in mid-March, they felt the Brigham Young team had broken their agreement. Pons and Fleischmann submitted their paper to the *Journal of Electroanalytical Chemistry* and secured a promise that the piece would be published on 10 April 1989. The University of Utah then scheduled a press conference for 23 March 1989. Well-placed advance leaks to reporters at the *Financial Times* and the *Wall Street Journal* ensured that the Utah team had the attention of science reporters around the globe when Pons and Fleischmann took the stage. A furious Jones dispatched his paper to *Nature* on March 23. The following day—the agreed-on deadline for sending the two papers to *Nature*—Marvin Hawkins, a graduate student from the University of Utah, waited at the Salt Lake City airport in vain for his Brigham Young counterpart to arrive. He eventually mailed his team's paper by itself.[50]

Most of the reporters who attended Pons and Fleischmann's press conference knew nothing of the shattered Utah–Brigham Young publication deal. They saw two respected electrochemists take the stage and announce, in unemotional language filled with technical terms, that they had produced nuclear fusion at room temperature using an apparatus so simple that it seemed almost any undergraduate chemistry major could replicate it.[51] Unsurprisingly, the announcement created a media sensation. The oil crisis of the 1970s was a not too distant memory; it was three years after the Chernobyl disaster had further tarnished the view that nuclear fission might solve the energy crisis; and the very next day the *Exxon Valdez* oil tanker struck a reef off the coast of Alaska, causing one of the worst environmental catastrophes in history. Could the two Utah electrochemists have found a cheap, clean way out of any future energy crises? In the absence of a published scientific paper, laboratories from around the world began trying to glean exactly what Pons and Fleischmann had done from newspaper reports and television broadcasts. A preprint of the April 10 *Journal of Electroanalytical Chemistry* paper soon became one of the most prized objects in the physical sciences.[52]

As newspaper articles around the globe reported the Utah announcement and as scientists began their first attempts to replicate the experiment in their own laboratories, *Nature* demurred on substantial discussion of the Pons-Fleischmann claims. A brief news article on April 6 listed laboratories trying to confirm the Utah team's findings but explicitly said the phenomenon "lack[ed] confirmation." On April 13, Maddox used the leader to explain why *Nature* had been relatively quiet on the subject of cold fusion. In a piece titled "Disorderly Publication," he confirmed that both the Jones and Pons-Fleischmann teams had submitted articles to *Nature*. "Both articles have been sent to referees," said Maddox; "each is now being revised in the light of the many comments that have been made." While *Nature* was abiding by "standard procedures" with respect to the cold fusion articles, the media furor surrounding the claims seemed to call for a scientific publication to make the relevant data as widely available as possible. Therefore, said Maddox, "revised versions of one or other or of both articles will be published later in the month. In these exceptional circumstances, they will be accompanied by such comments of the referees as remain pertinent."[53]

Twice, Maddox's editorial drew an explicit parallel between the Pons-Fleischmann-Jones claims and the Benveniste affair. Both featured intense media interest in the claims, both conclusions "[flew] in the face of orthodox belief," and both teams drew criticism that "the data available are insufficient for a careful judgment of its validity."[54] The article closed by chiding Pons and Fleischmann for revealing their findings at a press conference rather than waiting for the peer-review process to admit their research into an academic journal. "It is naturally difficult to bottle up exciting news, but impatience is a poor guide to action. The greater the importance of a discovery seems, the longer it should be worthwhile waiting to see it properly established."[55] Maddox thus positioned himself and *Nature* as the defenders of the scientific journal's relevance in an age of commercial concerns.

The following week, *Nature* opened with yet more cold fusion news: Maddox announced that the April 27 issue of *Nature* would include Steven Jones's piece but not the article Pons and Fleischmann had submitted. The publication of the Brigham Young team's article, said Maddox, "should not be taken to imply that all those who have seen it are persuaded to its chief conclusion."[56] The reason Jones's paper had been accepted while the Pons-Fleischmann piece had not, said Maddox, was that Jones had responded to his reviewers' comments while Pons and Fleischmann had declined to change their article.[57] In that same issue of *Nature*, IBM physicist Richard Garwin contributed an article to the News and Views section analyzing Fleishmann and Jones's presentations at an April 12 forum on cold fusion research. Garwin argued that both teams had "insufficient evidence" for their claims of nuclear fusion and bluntly concluded, "I bet against its [cold fusion's] confirmation."[58]

By the time "Observation of Cold Nuclear Fusion in Condensed Matter" reached *Nature*'s pages on 27 April 1989, no article in the magazine's history, not even the Benveniste piece, had received a comparable amount of buildup or experienced such careful distancing from its claims before a single word from its authors had been printed. The Jones et al. piece contained little that would have surprised those who had been following the cold fusion saga. The Brigham Young team reported "the observation of deuteron-deuteron fusion at room temperature during low-voltage electrolytic infusion of deuterons into metallic titanium or palladium electrodes" and based their claim on readings from a new piece of equipment called a neutron spectrometer that the Brigham Young team had developed "over the past few years."[59] Much of the article was devoted to explaining how the neutron spectrometer worked. According to the article, the spectrometer measured neutron production by monitoring the reaction between a free neutron and ^{6}Li.[60] The Jones piece claimed far more modest fusion results than the Pons-Fleischmann *Journal of Electroanalytical Chemistry* article. The team had observed no excess heat and said that "the fusion rates observed so far are small."[61] Jones et al. suggested that the fusion reaction they had observed might be similar to muon-catalyzed fusion, a type of nuclear fusion that takes place when electrons in an atom are replaced by a larger subatomic particle with a much higher charge.[62] The article closed by suggesting that "the discovery of cold nuclear fusion in condensed matter opens the possibility, at least, of a new path to fusion energy."[63]

The Jones team's findings about cold fusion, however, were overshadowed by the more dramatic results—and more dramatic presentation—of the Pons and Fleischmann laboratory. In the months that followed the publication of the Pons-Fleischmann and Jones articles, Jones's name came up only occasionally in the debate over whether nuclear fusion at room temperature was possible. Even in *Nature*, Jones was an afterthought—an April 13 news article by associate editor Vera Rich mistakenly called Steven Jones "Robert Jones," a slip that unintentionally served to illustrate Jones's secondary status in the cold fusion discussion.[64]

The fact that *Nature* had not published the most contentious cold fusion paper, however, did not prevent the journal from carving out a major role for itself in the discussion. This time, although Maddox expressed significant skepticism of cold fusion in the news and editorial columns, he left it to *Nature*'s contributors to discredit Pons and Fleischmann. *Nature*'s columns gave a prominent voice to cold fusion critics in the months following the chemists' announcement. Contributors from a number of major research laboratories, seeking to take advantage of *Nature*'s rapid publication speed (and likely encouraged by Maddox's clear anti–cold fusion stance), made *Nature* a forum for a public evaluation of the Pons-Fleischmann findings.

In May, a group at MIT led by Dr. Richard Petrasso submitted a devastating critique of Pons and Fleischmann's gamma-ray (γ-ray) spectrum—a spectrum that had as yet not been released to scientific journals but which Pons and Fleischmann had repeatedly cited as evidence for their claims of nuclear fusion. The Petrasso team critiqued the unpublished gamma-ray spectrum based on an image that had been shown in a television broadcast about the Utah team's work.[65] In a *Nature* Letter to the Editor, Petrasso et al. noted that Pons and Fleischmann had reported observing a peak in their gamma-ray spectrum at 2.22 MeV, a peak associated with the fusion of a proton and a neutron, and had used this peak as evidence that their electrolytic cell was producing excess neutrons. Petrasso's team argued that the Pons and Fleischmann spectrum showed no evidence of a peak at 2.22 MeV. According to Petrasso, the instrumental resolution did not permit the precision Pons and Fleischmann had claimed, their gamma-ray spectrum lacked an expected Compton edge at 1.99 MeV, and they had overestimated the neutron production needed to produce such a spectrum by a factor of 50. If the Pons and Fleischmann spectrum showed anything, said Petrasso, it was that their gamma-ray line was not at 2.22 MeV at all, but at 2.5 MeV.[66] "We can offer no plausible explanation" for the reported gamma-ray spectrum, said Petrasso, "other than that it is possibly an instrumental artefact, with no relation to a γ-ray interaction."[67]

The Petrasso article was the only *Nature* piece that drew a direct response from Pons and Fleischmann. The Utah team expressed annoyance that Petrasso's critique was based on an image taken from a television broadcast of a spectrum that, according to Pons and Fleischman, was "most certainly not" made in the Utah laboratories. The Utah team's response included a complete gamma-ray spectrum. Strangely, the Pons-Fleischmann response to Petrasso contained very little discussion of the crucial 2.22 MeV peak; instead, the response focused on an "unidentified" peak at 2.496 MeV. Pons and Fleischmann admitted that "the exact interpretation of the 2.496-MeV peak is in doubt," but added, "in spite of the problems underlying the interpretation of these spectra, we consider that the measurements show the emission of γ-rays from the cell environment: removal of the cells leads to the removal of the signal peak."[68]

Nature made room for Petrasso et al. to respond directly in the same issue, the usual practice when a Letter to the Editor criticized a *Nature* paper.[69] The Petrasso team's response to Pons and Fleischmann was longer than the Utah team's own letter. The MIT group made no mention of using a spectrum derived from a television broadcast, instead devoting their article to a reiteration of their concerns about the 2.22 MeV gamma-ray line. The team dismissively noted that "In their response above, Fleischmann *et al.* fail to address our key criticisms concerning their published 2.22-MeV γ-ray line."[70]

Researchers who had attempted to replicate the Pons-Fleischmann experiments also wrote to *Nature* to report that their efforts had failed. A team of researchers from Yale University and Brookhaven National Laboratory reported that they had observed "no statistically significant deviation from the background" on either their gamma-ray or neutron detectors.[71] In November, a team from Harwell National Laboratory in the United Kingdom sent in an article announcing that their highly publicized attempts to reproduce the Pons-Fleischmann work had been a failure.[72] The nine-page article, the longest one *Nature* ran that year, carefully laid out the methods the Harwell team had used to attempt to produce cold fusion; in the end, the Harwell team said, they concluded that "our work has served to establish clear bounds for the non-observance of cold fusion in electrolysis cells."[73]

Most damningly, a team from Pons and Fleischmann's own university submitted an article to *Nature* stating that their review of Pons and Fleischmann's work had revealed no reason to believe their results could be replicated.[74] The team, led by University of Utah physicist Michael H. Salamon, had monitored the cold fusion cells in Pons and Fleischmann's own laboratory using a sodium iodine neutron detector. Salamon said that they had observed "no evidence of fusion activity."[75] The team had also found no evidence of excess heat production, although they quoted a personal communication from Pons in which he claimed that "there was a two-hour segment in which there was excessive thermal release from cell 2-1.... Unfortunately, your computer and detector were not under power at that time since they had not been reset from a power failure which had occurred in the lab."[76] Salamon et al. acknowledged that they had lost 48 hours of data because of a power outage but said that they could estimate "mean upper limits for fusion power" during that period because the material in their neutron detector would still have reacted with any neutrons present. They estimated the upper limit of the excess power to be 10^{-2} W, far less than Pons and Fleischmann had claimed.[77] Based on their measurements, Salamon's team concluded that if any excess heat had been produced at any point during their five-week monitoring of the Pons-Fleischmann experiment, "this excess did not originate from known nuclear processes."[78]

The scientific papers in *Nature* cast serious doubt on whether the Pons-Fleischmann findings could be replicated. The News section of *Nature* painted an even starker picture. In the year following Pons and Fleischmann's announcement, *Nature*'s News writers—in particular, associate editor David Lindley—told the story of a growing consensus against the cold fusion findings. As early as June 1, Lindley was writing that the claims of excess energy from cold fusion had been almost entirely discarded even among groups who thought they had reproduced some of the Pons-Fleischmann or Jones findings.[79] An article by David Swinbanks about cold fusion research in Japan reported intense interest in cold fusion in Japanese industry and government despite initially unpromising results, but it was misleadingly titled "Cold Fusion: Efforts Abandoned in Japan."[80] When Harwell National Laboratory announced that their attempts to replicate the Pons-Fleischmann experiment had failed despite Fleischmann's personal help in setting them up, Lindley described it as a "blow for cold fusion."[81] In November, Lindley reported that cold fusion had received the "official thumbs down" from the United States Department of Energy.[82] By March 1990, Lindley was prepared to declare cold fusion an "embarrassment" to the scientific community; the phenomenon, he said, was only believable if one replaced scientific evidence with "wishful thinking."[83]

Lindley's articles further portrayed Pons and Fleischmann as deviants from accepted scientific practice. He frequently implied that the pair was failing to honor promises or share data. In an article about a May cold fusion meeting at Los Alamos National Laboratory, Lindley reported that even though Pons and Fleischmann had told a congressional hearing that they would collaborate with scientists from Los Alamos, the pair was absent from the meeting, and "a spokesman for Los Alamos made it plain that no such collaboration has occurred because the University of Utah had not wished to enter into any agreement that it perceived would jeopardize its patenting and priority rights."[84]

Lindley's News articles would have been strong evidence of the *Nature* editorial board's skepticism about cold fusion on their own, but Maddox made the journal's position on the Utah claims absolutely explicit. In a series of signed editorials, *Nature*'s editor made it clear that he doubted Pons and Fleischmann's findings and did not expect them to be verified. On April 27, in the same issue that published Jones's paper, Maddox declared that "the most probable outcome" of the cold fusion controversy was that "attempts to replicate the observation of cold fusion [will] fail."[85] The closing sentences of the article made the point even more bluntly: "Robust skepticism is the only wise view. There may be something in the Brigham Young phenomenon, but that requires careful confirmation. The Utah phenomenon is literally unsupported by the evidence, could be an artefact and, given its improbability, is most likely to be one."

According to Maddox, what was at stake in the discussion was not simply the reputation of two laboratories in Utah. Maddox wrote that while the public seemed content for now to wait for confirmation of the results, failure to replicate the cold fusion experiments might damage the scientific community's reputation. He took both the Utah and Brigham Young teams to task for failing to run a control experiment in ordinary water instead of deuterated water. "How is this astonishing oversight to be explained to students repeatedly being drilled in the need that control experiments should be as conspicuous in the design of an investigation as those believed to display the phenomenon under study?" he asked. "And how should the neglect be explained to the world at large?" Notably, Maddox praised the popular press—especially the American press—for its cautious approach and for "making it plain that cold fusion was not then a proven reality, let alone a commercial source of limitless energy."[86]

One reader sarcastically suggested that perhaps Maddox ought to send an investigative team to Utah, preferably one "made up of (1) a journalist with a scientific background, preferably in a subject far removed from nuclear physics, (2) a professional conjuror, (3) an expert in scientific fraud."[87] But in contrast to the Benveniste episode, during the cold fusion controversy Maddox employed the usual channels available to *Nature*'s editor to promote skepticism. From April to July, Maddox was relatively silent on the cold fusion controversy, allowing contributors and News articles to carry on the debate—although he did take a swipe at Pons and Fleischmann in a June 29 News and Views piece about journals and scientific ethics, where he noted that referees for specialist journals "can be relied upon almost without question to draw attention to control experiments that should have been carried out." The comment was a subtle but clear reference to the Utah team's missing control experiments; Maddox was implying that waiting for referee reports would have helped Pons and Fleischmann avoid that particular criticism.[88]

Maddox returned to the fray more visibly in early July with a News and Views article titled "End of Cold Fusion in Sight." The article's subheadline declared that "there seems no doubt that cold fusion will never be a commercial source of energy," and Maddox argued that "the time has come to dismiss cold fusion as an illusion of the past four months or so."[89] As with the Benveniste episode, Maddox carefully avoided charging Pons and Fleischmann with any outright wrongdoing, writing, "none of this implies that Pons and Fleischmann have been anything but straightforward." But Maddox's piece made it clear that even if the Utah team had reported their results accurately, none of the work done since then had shown cold fusion to be a replicable phenomenon capable of solving the world's energy problems.

A leader on 29 March 1990—almost exactly a year after the original Pons and Fleischmann press conference—sought to put the final nail in cold fusion's coffin. The editorial's title bid "Farewell (Not Fond) to Cold Fusion" and opened by commenting that the "fuss" over cold fusion "will deserve a waspish footnote" in any future history of science tome. The episode, said the editorial, had been injurious to science as a whole; the scientists who claimed to have replicated the Pons-Fleischmann results had been too quick to discard established theoretical physics and too credulous when confronted with this seemingly miraculous phenomenon. Furthermore, the episode "has revealed the malign influence of extraneous considerations in modern science; Pons and Fleischmann would surely have published a full account of their work long before this if they had been concerned with the general understanding" instead of with their own potential profit from a forthcoming patent. The piece concluded with an open challenge to Pons and Fleischmann:

In short, the time has come when Pons and Fleischmann should say openly, and in as much detail as their interlocutors in Utah this week require, exactly what they had done a year ago, what they have been doing since and what reason they have in which others can still have confidence for believing that cold fusion is still to be taken seriously.[90]

But at this point the cold fusion controversy was essentially over—not just in *Nature* but in the public eye as well.

THE COLD FUSION CONTROVERSY IN THE POPULAR PRESS

Much of the literature on the cold fusion controversy has focused on the role of various forms of communication in spreading news—and skepticism—about cold fusion. In the early days after Pons and Fleischmann's press conference, scientists from across the globe tried to figure out how, exactly, the two had produced their remarkable results. In the absence of a paper describing the exact methodology, scientists began turning to their best sources of recent information: faxed copies of drafts, scientists' electronic mailing lists, and articles by science reporters in newspapers. Bruce Lewenstein has convincingly demonstrated that the media coverage of Pons and Fleischmann's work, along with details circulated by e-mail and fax, created a surfeit of information (often conflicting or of unclear value) about cold fusion that played a major role in creating uncertainty about the Pons and Fleischmann claims.[91]

In many of the existing scholarly accounts, traditional peer-reviewed journals appear almost irrelevant in the communication of ideas about cold fusion; even the weekly *Nature* seems too cumbrous to fulfill scientists' desire for the most recent information on room-temperature nuclear fusion. The mass media, not specialist scientific journals, appears to be the main source of relevant published information.[92] Thomas Gieryn has claimed an even larger role for the popular press, arguing that cold fusion proponents sought to keep cold fusion afloat by transporting "claims-making and claims-adjusting processes into places and spaces that consequentially expanded the domain of 'doing science'"—namely, the mass media and congressional hearings. Journalists became "absorbed in the process of making a scientific discovery."[93] It was the Baltimore meeting of the APS on 15 May 1989, says Gieryn, where physicists reclaimed control over the debate and returned journalists to the role of "news givers who must wait until the scientists themselves decide" the truth of cold fusion, not codiscoverers with a crucial role in judging scientific claims.[94]

But an examination of the coverage in various media outlets indicates that while it may be true that cold fusion proponents sought to enroll journalists as allies and "codiscoverers," journalists claimed no such role for themselves. Furthermore, the Baltimore meeting, while well covered, did not trigger a major change in popular cold fusion coverage. Instead, from the outset many journalists expressed cautious skepticism of Pons and Fleischmann's claims. As the controversy progressed, the lay media relied heavily on traditional sources of scientific authority, especially *Nature*, for information about cold fusion's current status. The lay media frequently cited the articles published in *Nature* as strong evidence against Pons and Fleischmann's claims and seemed to regard *Nature*'s skeptical editorial stance as proof that Pons and Fleischmann were all but finished in the eyes of the scientific community. Far from claiming a new role for themselves in the knowledge-making process, journalists for publications such as *Time*, *Scientific American*, the *Economist*, and the *Wall Street Journal* seemed to agree with Maddox that scientific journals like *Nature*—and, specifically, *Nature* itself—were the appropriate place to evaluate the truth of the Pons and Fleischmann claims.

Scientific American, a monthly popular science magazine, joined *Nature* in expressing doubt about cold fusion. The magazine's first coverage of the cold fusion claims did not appear until the May 1989 issue, when regular contributor Tim Beardsley wrote that although the results "could have far-reaching implications for the world's energy future," they faced "profound skepticism." The piece quoted Princeton's William Happer Jr. saying of Pons and Fleischmann, "I would bet my house that they're wrong."[95] A follow-up article in June twice referred to Richard Garwin's *Nature* critique of the Fleischmann-Pons work and said that Pons and Fleischmann's withdrawal of their *Nature* paper was "dismaying" to many scientists.[96] A lengthy piece in August cited Richard Petrasso's gamma-ray spectrum critique as "damning" evidence against Pons and Fleischmann's claims.[97]

Time, the widely read weekly US newsmagazine, was also cautious in its initial coverage of Pons and Fleischmann's press conference. *Time* devoted its 8 May 1989 cover story to cold fusion. Just six weeks into the controversy, the magazine's editors were confident enough to write that cold fusion "is probably an illusion. . . . It seems likely that they [Pons and Fleischmann] jumped to a hasty conclusion based on incomplete research."[98] It was *Nature*, said the *Time* reporters, that had led the way to this conclusion. The article called Maddox's April 27 editorial "damning" and implied that *Nature*'s skepticism meant Pons and Fleischmann were finished.[99] *Time* even took *Nature*'s side in their public argument with Pons and Fleischmann over the withdrawn paper, writing

> *Nature* asked for more information from Pons and Fleischmann before publishing the paper, but according to the journal the pair said they were too busy.... Says Fleischmann: "*Nature* is not the appropriate place to publish because they don't publish full papers." That peculiar sentiment might come as a surprise to James Watson and Francis Crick, whose Nobel-prizewinning discovery of the structure of DNA was first published in the British journal.[100]

The British weekly newsmagazine the *Economist* took a slightly different approach to the controversy. The magazine's initial report on the cold fusion press conference was printed on April 1 (i.e., April Fool's Day) and opened with a wry caution: "Do not be deceived by the date."[101] The *Economist*'s coverage over the next year frequently highlighted the disciplinary clash between the chemists Pons and Fleischmann and their critics in physics.[102] The April 1 piece noted the "unpredictability" of the experiment and suggested this unpredictability would be a barrier to the phenomenon's acceptance among physicists, who were used to dealing with more reliable results.[103] A May 13 article detailing the "backlash" against Pons and Fleischmann noted that "Physicists have not taken kindly to the two chemists who claim to have tamed nuclear fusion" and said that it was the physicist Jones whose work "at present holds the high ground."[104] The issue was simple: the evidence for fusion rested on neutron detection and neutrons were "the province of physicists."[105] The *Economist* was not shy about siding with the physicists and implicitly backing their claims to exclusive rights over neutron detection. By September, the *Economist* declared that cold fusion had "vanished from sight," despite a small number of researchers who continued to investigate and believe in the reality of the phenomenon Pons and Fleischmann had reported.[106] According to the *Economist*, the criticisms from Utah's Michael Salamon, soon to be printed in *Nature*, were "particularly damning" to the Pons-Fleischmann team.[107]

The most influential mass media coverage of the cold fusion controversy occurred in the *Wall Street Journal*, the famed US financial daily. It was *Wall Street Journal* science writer Jerry Bishop who coined the term "cold fusion" to describe the Pons and Fleischmann findings, and in May 1990 Bishop would win an award from the American Institute of Physics for his reporting on the controversy. Bishop and the other *Wall Street Journal* reporters who covered the controversy were particularly interested in the technical details of the experiments. The *Wall Street Journal* coverage contained several careful explanations of the experimental setups different laboratories had used to try and replicate the Pons and Fleischmann findings, including details such as the diameter of the palladium rods and the exact wattage of excess heat detected.[108]

Even the *Wall Street Journal*, however, relied heavily on traditional sources of scientific authority, especially *Nature*. Science writer Richard L. Hudson paid particularly close attention to *Nature*'s role in the developing controversy. From the beginning of the controversy, the *Wall Street Journal* portrayed the promised scientific papers and not the conversation that followed the Pons and Fleischmann press conference as the test of the two chemists' claims. Bishop's March 23 article about the Utah press conference said (not entirely accurately) that both groups had "simultaneously submitted reports on independent discoveries to the British journal *Nature*."[109] The *Wall Street Journal* regularly updated their readers on the papers' progress through the *Nature* review process, noting on March 29 that the papers had been sent out for review and that "if the papers are accepted as submitted, they could be published as early as the April 27 issue."[110] On April 20, Hudson published an article detailing the withdrawal of the Pons-Fleischmann *Nature* paper.

> *Nature*, a respected 130-year-old journal, had been expected to publish a scientific account by Messrs. Fleischmann and Pons of their experiments. But *Nature*'s editor, John Maddox, said three of the journal's scientist-reviewers raised questions about the research that the two authors declined to address. Critical peer reviews are standard procedure for scientific journals, and often end in rejection or withdrawal. But it's rare for a paper of such importance to be pulled back from the presses.[111]

Maddox used his interview with Hudson to reinforce his ideas about the advantages of standard peer review for assessing scientific truth claims. Quotes from Maddox revealed that the *Nature* reviewers criticized the absence of control experiments using plain water instead of heavy water and "inadequate data" on the heat generated by electricity passing through the palladium electrode—flaws that, he implied, should have been corrected before the paper went into print.[112] A week later, Hudson reported that *Nature*'s April 27 condemnation of Pons and Fleischmann's experimental methods was "one of the strongest attacks yet from an increasingly skeptical scientific world on cold-fusion research" and suggested that the piece would carry extra weight coming from "Europe's most prominent scientific journal ... first to report the discovery of the neutron, fission in uranium, the structure of DNA and many other scientific milestones."[113]

Unlike many other mass media publications, the *Wall Street Journal*'s coverage included several pieces on efforts to verify the Pons-Fleischmann results well into 1990, although the *Wall Street Journal* was careful to make it clear that the prevailing opinion within the scientific community ran against Pons and Fleischmann.[114] Their coverage included discussion of the conflict between the Utah chemists and *Nature* over the publication of the article by Utah physicist Michael Salamon, who had monitored Pons and Fleischmann's experiments for weeks and found no evidence of a nuclear reaction. *Nature* published the Salamon piece immediately before a cold fusion conference in Salt Lake City. According to Bishop's article, the cold fusion supporters at the conference believed the timing was deliberate.

> The Utah physicists did their radiation-detection experiments last May and submitted a paper to *Nature* last summer, Mr. Pons said. Yet, *Nature* chose not to publish the paper until this week.... "I don't know what their motives are," Mr. Pons said of *Nature*'s editors, "but they've done everything in their power to condemn this work, to trash it." He and Mr. Fleischmann accused the scientific journal of "polarizing" scientists for and against cold-fusion research early in the game.[115]

The *Wall Street Journal*'s interest in *Nature* was not limited to its interactions with cold fusion. On 15 May 1989, Hudson wrote a front-page article for the *Wall Street Journal* titled "If You Read It First in *Nature*, It's Big and (Usually) True."[116] The piece discussed *Nature*'s recent headline-grabbing interactions with cold fusion and homeopathy, citing these episodes as evidence of the "lively mix of newsiness, authority, controversy and even arrogance that *Nature* has brought to the sober-sided world of science publishing."

Hudson's piece was more than a rundown of recent controversies in *Nature*; it also tackled the question of the role *Nature* played in the world of science publishing. "Getting in the right journal helps win respect, promotion, grants and fame," Hudson explained. He then elucidated exactly why *Nature* was considered the right journal. "[*Nature*] can get important, claim-staking research into print faster than most other scholarly publications," printing articles in as little as three weeks instead of the more usual lead time of six to twelve months for other scientific publications. Although *Nature* had only 40,505 paying subscribers—compared to 150,000 subscribers for *Science*—Hudson noted that *Nature* was cited far more often than *Science* and in fact was cited more than any other scientific publication except for the *Proceedings of the National Academy of Science*.

Hudson's interview also drew out some revealing comments about how Maddox himself saw his role in the world of scientific publishing.

> Mr. Maddox, 63 years old, is a cantankerous Welshman who ... has shaped *Nature*'s newshound-in-a-laboratory personality for 17 of the past 24 years. (He quit for a while to pursue other interests.) He thrives on controversy. He says he helps create, rather than follow, fashions in science—a responsibility that he says makes him uneasy.
>
> "A journal really has to have an opinion," he says.[117]

THE ROLE AND PURPOSE OF THE SCIENTIFIC JOURNAL

The cold fusion saga might have signaled a change in the way science would be communicated. Had Pons and Fleischmann's results yielded the dreamed-of era of cheap energy, we might imagine an alternate universe in which it became acceptable, even expected, for scientists with dramatic new findings to circumvent the journal submission process and make their findings public through the popular press. Instead, the cold fusion episode became a moment of triumph for the scientific journal. Pons and Fleischmann's rapid fall from grace gave Maddox a golden opportunity to argue that it was still essential for scientists to put their work through the channels of peer review before announcing their findings to the world. Had Pons and Fleischmann followed the usual workings of the journal's peer-review process, Maddox implied, they might have corrected problems in their experiment before publication and avoided their harsh public discrediting. By using *Nature* to defend his vision of the proper workings of science, in less dramatic and controversial fashion than he had during the Benveniste episode, Maddox was also defending *Nature*'s continued relevance. In his view, the scientific journal had an essential role to play in the international scientific community even when millions of dollars in potential patents tempted scientists to take their communications elsewhere.

Nature's role in these late-twentieth-century scientific controversies also illustrates its continued role in defining science. As we saw in the early chapters of this book, *Nature* became significant in the nineteenth century because it served as a site where editors, contributors, and readers could negotiate and construct the identity of the "man of science." One hundred and twenty years later, *Nature* continued to be a site where contributors, editors, and staff writers debated the proper way to conduct scientific experiments and publicize new knowledge claims—in short, the right way to be a scientist. Davies and his staff pushed scientists to put their expertise to use in realms such as extrasensory perception; Maddox tried to police "proper" science by visiting Benveniste's laboratory in person. Notably, in both cases *Nature*'s readers and contributors rejected the editor's vision for the journal. They largely ignored Davies's suggestion that they should analyze the Targ-Puthoff paper and criticized Maddox for creating a "circus" with his visit to Paris. Davies's disappointment with the lack of discussion of the Targ-Puthoff paper and Maddox's contrasting handling of the cold fusion episode suggests that they took this sort of reaction to heart. While the editorship gave Davies and Maddox a position from which to make new claims for *Nature*'s role in the scientific community, readers and contributors had the power to reject them.

FIGURE 12 *Nature*'s cover from 18 October 2012. Reprinted by permission of the Nature Publishing Group.

Conclusion

In June 1989, in the middle of the cold fusion debate, John Maddox wrote a News and Views editorial titled "Can Journals Influence Science?" The piece was a response to a recent hearing in the US House of Representatives in which members of Congress had expressed the view that journals could do more to prevent scientific fraud.[1] Maddox took the opportunity to reflect directly on the role journals played in science. "There is a powerful school of thought, chiefly represented by the editors of journals," he said wryly,

> which holds that the scientific literature is and should be a passive means of communication—a mirror held up to the face of research in which people other than its authors can discover what is happening in laboratories the world over.
> That is, of course, an idealization which is far from the truth.

Maddox went on to point out that researchers hoping to make their careers in science had to publish papers in specialist journals, which meant that "authors will go to endless trouble to meet conditions laid down by journals and their editors. In the process, they are moulding accounts of their research in response to external demands."[2] In Maddox's view, journals—including his own—could not claim to be passive vessels when they wielded such tremendous influence over scientists' careers and over which scientific research was printed. Journals were actively involved in shaping science itself.

With the possible exceptions of Brimble and Gale, Maddox's predecessors as editor would have agreed with his sentiments. Over its long history *Nature* has not been a "passive means of communication" (if, indeed, any

CONCLUSION 229

edited journal could be considered "passive"). *Nature* has not only shaped scientific research by printing and rejecting scientific articles; *Nature* has been a site where practitioners defined the very idea of modern science.

Notably, while *Nature*'s editors had the power to decide which articles would be accepted and rejected, negotiating "science" in the pages of *Nature* has never been a matter of editors handing down judgments from their London offices. There was often tension between the editor's vision for *Nature* and the readers' and contributors' wishes for the journal. Lockyer's contributors did not write the popularizing pieces he had hoped for; contributors under Brimble and Gale insisted on competing with one another instead of collaborating as the editors suggested; few researchers responded to Davies's call to analyze the Targ-Puthoff paper; *Nature*'s readership largely rejected the idea of an editor visiting a laboratory in person to assess scientific results. Time and time again, *Nature*'s readers and contributors made their voices heard. They pushed to assert their own visions for *Nature*—and their visions for science and its practitioners.

A history of *Nature* is also, in many ways, a microcosm of the history of the print journal. *Nature* was founded exactly at the moment when the scientific journal was becoming the dominant form of communication between researchers. As we saw in chapter 1, Norman Lockyer did not intend for *Nature* to be a publication by and for specialists, but his contributors found *Nature* so useful for this purpose that their preferences remade *Nature* into a specialist journal. *Nature* became more international as its contributors increasingly found international communication central to their research, and the journal both reflected and helped build communication between international networks of scientific researchers. Sometimes *Nature* also reflected and helped *limit* communication between scientists, as the publication's uneasy interaction with Soviet contributors illustrates. Looking at *Nature*'s development also reveals some surprising things about the history of the scientific journal, such as the fact that a respectable journal could neglect systematic peer review as late as the 1970s.[3]

In the early twenty-first century, *Nature* is unquestionably at the top of the journal hierarchy, rivaled only by *Science* and *Cell*. It has gone from a journal that "might publish anything" to one that rejects about 92 percent of submissions.[4] *Nature* now has strict policies on financial disclosure, requires certain types of data from *Nature* papers (including DNA sequences) to be deposited in a database, and also requires its authors to share reagents.[5] In 2000, *Nature* adopted a new mission statement, setting aside Norman Lockyer's original 1869 declaration of *Nature*'s aims for a document that, among other things, eliminated the references to "men of science." Many scientists, however, feel that there is still work to be done to produce gender equality in *Nature*.[6] *Nature*'s editorial staff has also continued developing and experimenting with the journal's format and content—in 2007, for example, the journal began running a column of short science fiction called Futures.

Historians are often uneasy evaluating the recent past because the consequences of recent changes are still working themselves out. However, one of the major developments since Maddox's retirement deserves special attention. This book has been largely concerned with *Nature* as a print journal. But in the twenty-first century, the print journal is no longer the dominant form of scientific communication. The online journal—or, perhaps more accurately, the online *article*—has usurped the print journal's place. So what has happened to *Nature* during the online transition? And what might *Nature*'s history as a print journal tell us about the future of scientific communication?

NATURE AND THE RISE OF ELECTRONIC PUBLISHING

Two momentous events in *Nature*'s history occurred in 1995. First, John Maddox stepped down after twenty-two (nonconsecutive) years as *Nature*'s editor. Macmillan chose Philip Campbell, who had been with the journal since 1979, as its new editor in chief. As of 2015, Campbell still held *Nature*'s editorship, meaning that in nearly 150 years of publication, *Nature* has had just seven editors—an average editorial tenure of more than twenty years.

In 1995, *Nature* also opened its website, Nature.com. This was arguably a much less momentous occurrence than Maddox's departure, at least in the short term. Maddox and his fellow *Nature* staffers had been cautious about embracing an electronic version of their journal at all. As recently as 1990 Maddox had expressed skepticism about whether *Nature* could ever be electronic, noting that "storing a year's issues of this journal in a conventional computer store would be an extravagant proposition" because each issue of *Nature* contained roughly one megabyte of data, and "a 50 megabyte hard disk would cost the best part of $1,000."[7] Maddox thought it might be possible to distribute *Nature* on CD-ROM but did not see much promise in the idea that *Nature* might be published online.

Maddox had reason to wonder whether electronic journals were here to stay; in the short term, at least, most models of electronic publishing did not seem terribly viable. The early 1990s were a period of growth for electronic journals, but—much like Victorian science periodicals—many of them proved unable to sustain the costs of publication and folded rapidly. In 1990, for example, a heavyweight group of scientific publishers including Elsevier, Pergamon, Blackwell, and Springer collaborated on the *Adonis* project. This was an effort to distribute over 400 scientific journals to libraries via CD-ROM. The appeal of the plan was obvious: a single CD-ROM could contain images of several journal issues, freeing shelf space in storage-strapped libraries. But *Adonis* quickly failed because of its high cost to customers. Libraries were strapped for cash as well as storage, and most universities could not justify replacing cheaper print subscriptions with the more costly CD-ROM versions.

Adonis was an attempt to make existing journals electronic (although it is worth pointing out that *Adonis* was not, and was never intended to be, an online publication). Other groups tried to start new journals that would be electronic from the beginning. The first peer-reviewed, online-only journal was the *Online Journal of Current Clinical Trials*, the result of a partnership between the American Association for the Advancement of Science (publisher of *Nature*'s rival *Science*) and the Online Computer Library Center (OCLC), an enormous library cooperative aimed at helping member libraries share online resources. The *Online Journal of Current Clinical Trials* was, as its name suggested, a publication devoted to publishing the most recent results of clinical studies. It had no print counterpart—the entire journal was only available online. The AAAS and OCLC announced plans for the new journal (originally titled *Current Clinical Trials*) in 1990, and Dr. Edward Huth, a 19-year veteran of the *Journal of Internal Medicine*, was chosen as its editor.[8] The *Online Journal of Current Clinical Trials* launched in July 1992. However, the AAAS and OCLC had trouble attracting submissions to their new publication. The *Online Journal of Current Clinical Trials* ceased publication in 1996 and fueled some concerns about the stability of online publishing when its staff did not make provisions for the future accessibility of its articles.[9]

Not all ventures into electronic dissemination of papers folded. One online publication venue that survived and thrived was xxx.lanl.gov, the Los Alamos Preprint Server, founded in August 1991 by the Los Alamos physicist Paul Ginsparg. High-energy physics, the field that lanl.gov was designed to serve, had (and has) a strong tradition of circulating drafts of papers among colleagues before they were submitted for peer review and publication. In the 1980s many of these preprints were circulated via electronic mailing lists, but Ginsparg and his colleagues found that the number of preprints was creating an unmanageably large number of mailing list posts.

With xxx.lanl.gov (or "xxx," as it was often called), physicists could seek out their colleagues' preprints when they wanted instead of being bombarded by mailing list updates. In 1999 Ginsparg renamed the site arXiv.org, and in 2001 he moved its servers to Cornell University when he accepted a position there.

The arXiv.org site (or "the arXiv"—pronounced "archive") remains one of the major sites of publication for high-energy physics. However, the arXiv has one crucial difference from most other research publication venues: it is not peer reviewed. In their early years, both xxx.lanl.gov and arXiv.org allowed users to upload any paper they wished without review or commentary from anyone at the site. As the site expanded, however, the lack of peer review before publication on arXiv.org became a point of concern for some members of the scientific community, who worried that the arXiv's open submission policy might attract papers from fields most scientists consider unscientific. Such fears proved justified. In 2002, arXiv.org removed a paper by the creationist Robert Gentry, who had attempted to use arXiv.org to circulate his alternate Big Bang hypothesis, and revoked his right to post papers on the arXiv. Gentry sued arXiv.org and Cornell University, claiming religious discrimination.[10] Gentry filed the suit in his home state of Tennessee, where the case was dismissed because neither Cornell nor arXiv.org had sufficient presence in that state to allow for legal action.[11] As of 2004, new authors must be endorsed by either an academic institution or other arXiv.org users in order to post papers to the arXiv.[12] Despite such concerns, making a preprint available on arXiv.org is now seen as a routine part of publication in many areas of physics because it is the fastest way for physicists to make their latest results available to their colleagues. In many ways, the arXiv fulfills the same function that *Nature*'s non-peer-reviewed Letters to the Editor did during the radioactivity boom.

Faced with examples such as xxx.lanl.gov, Maddox's views on electronic publishing shifted rapidly in the early 1990s. By 1992, he was ready to declare that "Electronic journals have a future," and in 1993, following a conference of journal publishers in Frankfurt, Germany, Maddox confidently announced that "Electronic journals are already here."[13] Nature.com went online in October 1995—just two months before Maddox stepped down as editor.

When Nature.com opened, its content was limited to a table of contents for the week's issue, a copy of the News section, one of the week's opinion articles, and appendixes for scientific papers printed in *Nature*. There were no full-text scientific papers available—Maddox explained that it was not yet possible to recreate technical illustrations, Greek symbols, and mathematical equations online, and the journal's staff had not solved the problem of making sure content was available only to subscribers. In an editorial for the print version of *Nature*, Maddox candidly described Nature.com as a "learning process for *Nature*'s staff," adding that *Nature*'s readers and staff were all "in a strange state of uncertainty about the future role of the internet."[14] By 1998, most of these problems had been ironed out, and under Campbell's leadership *Nature* began making full-text articles available to its subscribers on Nature.com.

Today scientists find and access almost all of their scientific reading online—a major contrast to the days when Bertram Boltwood impatiently waited for the next issue of *Nature* to arrive in his mailbox so he could respond to Ernest Rutherford's letter. Interestingly, Charles Wenz, who was a member of *Nature*'s editorial staff when Nature.com opened, did not feel that the website changed much about *Nature*'s content or the process of selecting papers—even after the website began publishing full versions of the journal's research articles. Wenz remembered the development of Nature's web page as "organic" and said that "it happened so gradually, honestly, that . . . I don't have a sense of it having changed what we do." Early in its life the *Nature* website acted more as a digital mirror to the print *Nature* (occasionally with the addition of some online-only appendixes or news updates) than as an entity that changed life for the *Nature* staff or for *Nature*'s readers. In Wenz's words, "it was just another way of it all getting out there."[15]

In fact, although we might instinctively think of online publishing as a development that disrupted the scientific publishing landscape, *Nature* was not the only scientific journal that transitioned into the online era with relative ease. In 2007, the information scientist John Mackenzie Owen wrote a book on *The Scientific Article in the Age of Digitization*, in which he argued that scientific journals were so well established as a form of publishing that the change from physical pages to online pages was having little effect on their content or their editing. The scientific journal, Mackenzie Owen said, had simply moved online.[16]

However, since 2007 the Internet has changed dramatically with the advent of social media (such as Twitter) and especially online commenting. According to Wenz, the addition of online commenting and its near-instantaneous feedback from readers has had a more dramatic effect on *Nature* than the initial creation of the website.[17] Giving readers the ability to praise or criticize *Nature*'s content on the web page of the journal itself—

a sort of post hoc, publically available form of peer review—has been a major change in online publishing, and its effects on the scientific world are still unfolding.[18]

Today, a print issue of *Nature* shares many similarities with *Nature* under Lockyer, Gregory, Davies, and Maddox. *Nature*'s covers and pages have been printed in full color since 1997, and each front cover displays an image related to a paper or article in that week's issue.[19] Every issue opens with editorials; the editorials are followed by the "News" and "Comment" sections. "Correspondence" prints short letters from contributors that discuss recent *Nature* articles and current debates about science. The "News and Views" column still runs every week but is no longer the single venue for short news items: "Research Highlights" calls attention to interesting recent papers in other journals, and the "Seven Days" section shares news from the previous week. "News and Views" now focuses on recent research news and leads the "Research" section of the print version of *Nature*. The "Research" section still prints two types of research articles: longer "Articles" and shorter "Letters."

However, most readers will now access *Nature* by visiting Nature.com. The Nature.com home page displays headlines from the "Latest News" and the "Latest Research," as well as an image of the current week's cover. Visitors can browse recent articles by discipline or visit the Nature Publishing Group's regional websites, which are published in English, Chinese, Japanese, Korean, Spanish, and Portuguese. Subscribers can access all of *Nature*'s print content online in both text and pdf format. Nature.com also hosts a complete, searchable archive of *Nature*'s past issues. Nonsubscribers can generally access abstracts or first paragraphs, but not full articles. Nature.com also maintains a lively collection of publically accessible blogs that post online-only news articles. This blog network includes *Nature*'s *Newsblog*, a community blog for guest writers called *Soapbox Science*, and discipline-specific blogs such as *The Skeptical Chymist* and *Spoonful of Medicine*. As of 2015, *Nature* allowed online comments on its blogs, news articles, and some letters to the editor, but not on its research articles.

In 1990, Maddox worried that giving readers the ability to print individual articles would lead to a loss of journal identity and that "browsing may die off" when the electronic journal took over.[20] Of course, to echo Oliver Lodge's observation from the 1880s, even in the nineteenth century few scientists read any journal, even *Nature*, in its entirety. But the ability to search for articles by topics and keywords and access them online has changed the way most scientists interact with the literature in their field—it is far more common for scientists to read a handful of individual articles from different journals than to browse through a full issue of a single journal. One important feature of print-journal identity has carried over to the online world, however: prestige. Some journals are still considered more impressive or more prestigious publication venues than others. A paper downloaded from Nature.com still carries extra cachet—that is, if the reader has access to the article.

OPEN-ACCESS VERSUS FOR-PROFIT PUBLISHERS

The opening of Nature.com was not the only significant operational change for *Nature* in the past twenty years. Under Maddox and Campbell, *Nature* has expanded its publishing empire to include a number of sister publications. During his last years at *Nature* Maddox oversaw the creation of *Nature Genetics*, which published its first issue in 1992, and *Nature Structural Biology* (now *Nature Structural and Molecular Biology*), which followed in 1994. Four more sister publications were formed before the end of the century: *Nature Medicine* in 1995, *Nature Biotechnology* in 1996, *Nature Neuroscience* in 1998, and *Nature Cell Biology* in 1999. The expansion has continued in the twenty-first century. At the time of writing there were nearly forty publications bearing the *Nature* name, including *Nature Climate Change*, *Nature Geoscience*, *Nature Methods*, *Nature Plants*, and several disciplinary versions of *Nature Reviews*, which publish review articles about recent scientific research. In addition, there has been a major institutional change in the way Macmillan manages *Nature* and its namesakes. In 1999, Macmillan collaborated with another publisher of academic journals, the Stockton Press, to form the Nature Publishing Group (NPG). NPG is a division of Macmillan Publishers that oversees all of the publications that bear the *Nature* name as well as other specialist publications such as the *Journal of Human Genetics* and prominent lay publications such as *Scientific American*.

NPG's lifetime has coincided with a period of growing discontent with the current publishing model in the sciences, especially as online access becomes more and more central to scientists' use of the literature. In the 1990s, many university libraries began expressing concerns about the high cost of subscribing to journals. Some saw electronic publishing as a potential solution to the cost of print journals and even offered incentives for faculty members and departments to choose electronic journals over print ones.[21] However, electronic journals have not solved the problem of high journal costs. Rising site license fees (the price for full access to a journal's website) quickly became a bone of contention between publishers and academic libraries—and between journals and their readers. Between 1986 and 2006, the average journal subscription budget for a North American research library increased by 321 percent; the US Consumer Price Index, in comparison, rose by 84 percent.[22]

Those sorts of numbers have given powerful ammunition to the open-access movement, a growing group of scholars and readers who argue that scholarly papers should be freely available to anyone on the Internet.[23] Some open-access pressure comes from governments whose taxpayers fund research through organizations such as the US National Science Foundation or the UK Medical Research Council; if taxpayers funded the work, many argue, they should be able to read the resulting research. In 2000, the US National Institutes of Heath (NIH) opened PubMed Central, an open-access repository where researchers could post papers resulting from government-funded research. Many of these papers were originally published in subscriber-only journals. Other governments have followed suit. In 2007, the Wellcome Trust and the British Library partnered to create a UK equivalent, UK PubMed Central (now Europe PubMed Central), and in 2009 Canada created PubMed Central Canada.

The open-access movement has not limited itself to repositories. Open-access advocates have also ventured into publishing original research. The most influential open-access publisher has arguably been the Public Library of Science, or PLOS. In 2000, three scientists—Patrick Brown of Stanford University, Michael Eisen of the University of California, Berkeley, and Harold Varmus, the director of the National Cancer Institute—founded PLOS as an advocacy organization aimed at convincing researchers to make their published work freely available in repositories such as PubMed Central. In 2003, PLOS ventured into the publishing world with its first open-access journal, *PLOS Biology*. This was followed quickly by *PLOS Medicine* in 2004 and *PLOS Computational Biology*, *PLOS Genetics*, and *PLOS Pathogens* in 2005. The organization's most innovative publication, however, is arguably *PLOS ONE*. This open-access, peer-reviewed journal pledges to publish "all papers that are judged to be technically sound" and instructs its referees to ignore the perceived importance or "impact" of the article when evaluating submissions.[24] Instead, *PLOS ONE* keeps track of each paper's number of citations, views, downloads, comments, and mentions on social media and blogs in order to present "article-level metrics" that will help researchers decide which of *PLOS ONE*'s published articles have had the most impact. Importance, in other words, is decided by the entire research community after publication, not by a handful of editors and referees before publication.

The debate over open access has found its way into *Nature* several times over the past decade and a half. Some contributors have insisted that scientific articles should be available at no cost to all interested readers. As early as 2001, computer scientist Steven Harnad wrote that online journals and archives had the potential to "free the literature" and enable scientists to share their work with the entire research community, not just those scientists whose institutions subscribe to a particular journal.[25] The two most commonly described forms of open-access journals are "gold" journals, which make all papers open access immediately, and "green" journals, which permit authors to deposit their papers in open-access repositories such as PubMed Central after an embargo period. Open-access advocates have also written to *Nature* to suggest different funding models for gold and green open-access journals, such as charging authors a fee to publish their work or requiring readers to pay to read any journal content that is still under an embargo.[26] Other contributors, however, have pointed to the costs of paying editors and hosting web pages as inevitable barriers. Several observers have also argued that the pay-to-publish model of open access places an untenable financial burden on contributors from smaller universities or developing countries; in 2014, Philip Campbell estimated that a single *Nature* article costs around $31,000–$47,000 to bring to press.[27] Despite these potential obstacles, it appears that the pressure for open access is having some effect on NPG and Macmillan. In December 2014, Macmillan announced the launch of a pilot program that will provide limited open access to forty-eight NPG journals, including *Nature*. Macmillan plans to test a system that will allow NPG subscribers to generate and share links to NPG papers via Macmillan's proprietary ReadCube software. Nonsubscribers will be able to view papers in ReadCube, but not download or print them.[28]

In addition to the open-access movement, NPG has also had to face an academic climate that is increasingly skeptical of the large profits commercial journals generate for their publishers. Many scientists are questioning whether for-profit publishers add enough value to the world of scientific publishing to justify their prices when the marginal cost of online access is seemingly so low, publishers do not pay scientific authors for their work, and peer review at most journals relies on volunteer labor from the referees. Criticism of for-profit journal prices is on the rise and has, in some cases, been accompanied by threats of canceled subscriptions or boycotts. In 2003, for instance, researchers at the University of California, San Francisco, wrote an open letter to their scientific colleagues urging them to boycott Elsevier's Cell Press to protest their high site license fees.[29]

Although Elsevier and the University of California eventually reached a compromise, in January 2012 Elsevier came under more intense fire for its support of the Research Works Act, a bill proposed in the US Congress. Some publically funded organizations, such as the National Science Foundation and the NIH, had begun requiring certain grant recipients to submit copies of their grant-funded publications to open-access repositories such as PubMed Central. The Research Works Act sought to forbid this sort of requirement on the grounds that it was infringing on the publishers' distribution and copyright privileges. Scholars protested that Elsevier was supporting a money-grubbing measure that would require US citizens to pay to read the work that their taxes had funded.[30] Elsevier eventually withdrew its support for the bill, but its detractors were not mollified. Later that month University of Cambridge mathematician and Fields Medal winner Timothy Gowers announced that he would boycott Elsevier journals by refusing to act as a referee, be a member of an editorial board, or submit an article to any Elsevier journal.[31] Gowers then created a website, TheCostOfKnowledge.com, where scholars could pledge to follow his example. At the time of writing, TheCostOfKnowledge.com had over 14,000 pledges.

Nature has also faced this sort of controversy. In June 2010, the University of California library system threatened to boycott NPG's journals after NPG attempted to impose a substantial increase in digital subscription prices for the university's libraries.[32] The librarians said they would suspend both physical and digital subscriptions to *Nature* and its sister publications if the licensing fee was not reduced; they also urged University of California faculty members to send their research to other journals and to refuse to referee articles for NPG journals. NPG responded by saying that the University of California's statements about its pricing system were "entirely untrue" and argued that the California library system had been paying an exceptionally low rate for many years.[33] By August, NPG and the University of California agreed to meet to renegotiate the site licenses, and a boycott was averted.[34]

NATURE AND THE FUTURE OF THE SCIENTIFIC JOURNAL

The growth of online publishing has placed scientific communication in a moment of transition not unlike the moment in the nineteenth century when the scientific journal rose as the dominant form of communication. It is increasingly clear that the print scientific journal has *not* simply moved online. Social media and commenting functions are changing the way journal staffs interact with their readers and are also changing the way that researchers communicate with each other about the latest papers in their fields. Furthermore, editors and readers have seen potential for online journals to accept and publish far more articles than their print counterparts. Print journals have to reject articles because of space constraints (only so many pages can be bound in a single issue), and many journals have settled on "importance" as the essential criterion for deciding between a large number of solid submissions. But few space restrictions apply online. *PLOS ONE* represents a new model in which peer review filters out obviously problematic papers, but no articles are rejected because of "lack of space" or "insufficient importance"; the task of determining "importance" is reassigned to the research community.

There is even some thought that peer review might be opened and democratized in the online era. Why restrict the peer-review process to a handful of chosen referees when dozens or even hundreds of researchers might be willing to give their opinions about a paper online? In June 2006, *Nature* conducted a trial of "open peer review" in which prospective *Nature* authors were given the opportunity to have their papers posted online for public comment. The authors of 71 potential *Nature* papers agreed to participate in the experiment. Although *Nature*'s traditional peer-review procedure is anonymous, online commenters were required to provide their full names.

The experiment had mixed results. Most authors who received open-peer-review comments rated them as either "somewhat useful" or "very useful." However, of the 71 papers, nearly half (33 papers) received no comments at all. Furthermore, authors in highly competitive fields where priority or patent rights might be at stake generally declined to participate and expressed discomfort with the idea of making their papers available on the Internet before they were accepted for publication.[35] In December 2006, *Nature*'s editorial team discussed the experiment and said,

> [The open peer-review trial] was not a controlled experiment, so in no sense does it disprove the hypothesis that open peer review could one day become accepted practice. But this experience, along with informal discussions with researchers, suggests that most of them are too busy, and lack sufficient career incentive, to venture onto a venue such as *Nature*'s website and post public, critical assessments of their peers' work.[36]

As a companion piece to the open peer-review trial, Nature.com published an online-only debate about the current state of peer review and also opened a blog called *Peer-to-Peer* explicitly devoted to peer-review issues.[37]

Although *Peer-to-Peer* stopped updating in 2010, peer review in the online era has remained a source of discussion and frustration for many scientists, some of whom question whether editorial peer review assures the quality of the scientific literature, or simply slows the publication of scientific papers and places too much power in the hands of a small number of elite scientists and editors.[38]

The twenty-first century is likely to be one of tremendous change for scientific publishing. As this history of *Nature* has shown, during the era of the print journal, scientists consistently adopted the publication venues that fulfilled their perceived needs. *Nature* rose to prominence because it was able to satisfy so many researchers' desires. In the 1880s, British scientists wanted a publication where they could debate scientific questions before a readership of fellow specialists. During the radioactivity boom, researchers wanted to publish their results quickly in order to secure priority. And in the twentieth century, even when the editors did little to attract contributions from outside Britain, researchers all over the world continued submitting their articles to *Nature* because it was a convenient place to share their latest theories and findings.

Nature's speed of publication has been at the root of much of its success over the past 146 years. In the nineteenth century, a weekly such as *Nature* was the fastest way to make findings known to scientific colleagues—certainly faster than writing a monograph or waiting for a scientific society to approve a new issue of its *Transactions*. But in the online era, a weekly journal with a demanding peer-review process is no longer the most rapid form of dissemination; online publication has become the key to sharing results quickly, and print journals are respectable but slow by comparison. The online era has also made access into a perceived need. Researchers want their work to be read and cited. When many scientists already feel buried under a deluge of literature, an article that requires a subscription or a payment to read runs the risk of being ignored.[39]

Nature does not offer instantaneous publication and has only begun to experiment with open access, but it does offer something that many researchers find even more important: prestige. Because *Nature* rejects so many submissions, many scientists and observers consider an article in *Nature* to be a more impressive accomplishment than an article in almost any other publication (the obvious rivals are *Science* and *Cell*). A *Nature* acceptance can still make a scientist's career, can help secure tenure and funding and laboratory space in a way few other accomplishments can. So long as this is the case, scientists will continue to submit their papers to *Nature*, give hiring and tenure preference to colleagues whose work has appeared in *Nature*'s pages (both physical and web), and ask their institutions to pay for access to Nature.com's archives—meaning that *Nature* articles will be easily accessible to anyone with an institutional affiliation. *Nature* is therefore better positioned than most journals to weather a potential open-access storm or a backlash against high costs.

But *Nature*'s continued importance is not preordained. *Nature*'s current reputation is, as we have seen, partly the result of a conscious effort by several *Nature* editors to attract interesting papers and make *Nature* a desirable place to publish new research. Given *Nature*'s current selectivity, it is tempting to view *Nature*'s editorial staff as all-powerful gatekeepers of scientific success. But the scientists who submit articles to *Nature* are not passive nodes in the network of science publishing. Since 1869, researchers have chosen *Nature* as a publication venue not because an anonymous authority decreed that *Nature* would be important but because they found that journal particularly useful. The qualities researchers desire in a specialist publication, however, have changed over time and will continue to change in the future. *Nature* has been a steady part of the print-journal landscape because its editors and contributors have continually reshaped it to meet their community's publishing needs. What scientists want from their journals in the twenty-first century, and *Nature*'s ability to fulfill those desires, will be the factors that shape *Nature*'s future.

Acknowledgments

No scholarly book, especially a first book, reaches print without a lot of advice, support, and encouragement. I owe special thanks to five people in particular. Anthony Grafton has always had new insight on the intellectual context of my actors. Janet Browne, whom I am privileged to have as a colleague at Harvard University, has been an unfailing source of enthusiasm and support as I have worked to put the finishing touches on this manuscript. Angela Creager is both an extraordinary scholar and a dedicated mentor, and I hope it is not too presumptuous to say that I aspire to one day live up to her example. Bernard Lightman not only wrote some of the books and articles that most influenced my work, he also read this book in its entirety and gave me some of my most helpful comments and suggestions. Finally, Michael Gordin has always gone above and beyond the call of duty as both a mentor and a colleague. Every conversation with Michael has made this book better.

I also want to thank Ruth Barton, Geoffrey Belknap, Ann Blair, Dan Bouk, Keynyn Brysse, Graham Burnett, Henry Cowles, Alex Csiszar, Gowan Dawson, James Elwick, Jim Endersby, Peter Galison, Benjamin Gross, Katja Guenther, Katrina Gulliver, David Kaiser, John Lynch, the late Michael Mahoney, Everett Mendelsohn, Erika Milam, Dael Norwood, Eileen Reeves, Michael Reidy, Alistair Sponsel, Matt Stanley, Jenna Tonn, Brigid Vance, Iain Watts, Paul White, the Princeton Program Seminar, the Modern Sciences Working Group at Harvard, and everyone else who has read part of this book or attended one of my talks and improved my work with their thoughtful comments. The University of Chicago Press and my editor, Karen Darling, have been a delight to work with. I also owe thanks to the anonymous readers who reviewed my manuscript for Chicago, who provided thorough and extremely helpful feedback.

Several organizations provided me with financial support as I worked on this book. Princeton University supported my research during its early years, and the National Science Foundation funded my archival research in England. York University and the Situating Sciences Cluster of the Social Sciences and Humanities Research Council of Canada provided me with a postdoctoral fellowship that enabled me to research and draft the concluding chapters of this book.

This project assumed its final form while I was the Sarton Visiting Scholar at the American Academy of Arts and Sciences, and all of my colleagues there asked incisive questions and challenged me to think about my project in new ways. Many thanks to the Academy (especially John Tessitore) and the Sarton Fellowship and to Francesca Ammon, Hillary Chute, Matt Karp, Chris Loss, Nikki Skillman, Peter Wirzbicki, Patricia Spacks, and especially Mary Dunn.

The staff members at the libraries and archives where I conducted my research have been unfailingly professional and helpful. Thank you to the librarians and archivists at University of Exeter Special Collections, Imperial College London Records and Archives, University of Reading Special Collections, University of Sussex Special Collections, and the British Library, who pulled hundreds of files and boxes on my behalf, and to *Nature*'s Head Librarian Caroline McLean, who obtained scans of past covers of *Nature* for me. Thank you as well to the people who agreed to be interviewed for this book: David Davies, Walter Gratzer, Brenda Maddox, Mary Sheehan, and Charles Wenz. Brenda Maddox also shared her late husband's personal papers with me, for which I am exceedingly grateful. Thanks also go to Walter Stewart, who kindly granted permission to have one of his unpublished letters quoted in this book.

Portions of chapters 1 and 2 were previously published as "The Shifting Ground of *Nature*: Establishing an Organ of Scientific Communication in Britain, 1869–1900," *History of Science* 50 (2012): 125–154. Parts of chapter 2 appear in altered form as "The Successors to the X Club? Late Victorian Naturalists and *Nature*, 1869–1900," in Gowan Dawson and Bernard Lightman, eds., *Victorian Scientific Naturalism: Community, Identity, Continuity* (Chicago: University of Chicago Press, 2014). Parts of chapters 4 and 5 appear in different form in "'Keeping in the Race': Physics, Publication Speed and National Publishing Strategies in *Nature*, 1895–1939," *British Journal for the History of Science* 47 (2014): 257–279.

Finally, I owe so much to the support and love of my friends and family—especially my mother Linda, my father Robert, my brother Ross, and my husband Scott. This book is dedicated to you.

Notes

INTRODUCTION

1. Norman R. Campbell, "The Word 'Scientist' or Its Substitute," *Nature* 114 (29 November 1924): 788.
2. Editor, "The Word 'Scientist' or Its Substitute," *Nature* 114 (29 November 1924): 788.
3. R. Ruggles Gates, "Cell-Wall Formation," *Nature* 114 (29 November 1924): 788–789; Paul D. Foote, "Nitrogen and Uranium," *Nature* 114 (29 November 1924): 789; David Hooper, "Edible Earth from Travancore," *Nature* 114 (29 November 1924): 789.
4. For a useful essay on the history of the word *scientist* and on the debates surrounding its acceptance, see Ross, "Scientist: The Story of a Word." A discussion similar to the one in *Nature* took place in 1894 in the pages of the popular science magazine *Science-Gossip*, but in that instance, the editor, J. T. Carrington, brought up the issue and solicited opinions from eight prominent men on the suitability of the word. Seven of them responded. Alfred Russel Wallace and Lord Rayleigh wrote that the word was likely to endure and ought to be tolerated. The other five—Sir John Lubbock, Thomas H. Huxley, the Duke of Argyll, the popular writer Grant Allen, and Albert Günther of the zoological department at the British Museum—all wrote that the word was illegitimate and urged Carrington not to use it. Huxley wrote that the term *scientist* "must be about as pleasing a word as 'Electrocution'" (see Ross, "Scientist: The Story of a Word," 76–78). Notably, Wallace, Rayleigh, Huxley, and Argyll were all frequent nineteenth-century *Nature* contributors. On the history of the word *scientist*, see also Snyder, *Philosophical Breakfast Club*; Lucier, "Professional and the Scientist."
5. [Richard Gregory], "Words, Meanings, and Styles, I," *Nature* 115 (21 February 1925): 253–255.
6. Clifford Allbutt, "The Word 'Scientist' or Its Substitute," *Nature* 114 (6 December 1924): 823.
7. Oliver Lodge, "The Word 'Scientist' or Its Substitute," *Nature* 114 (6 December 1924): 823.
8. D'Arcy W. Thompson, "The Word 'Scientist' or Its Substitute," *Nature* 114 (6 December 1924): 824. Thompson was the son of a renowned classicist who was also named D'Arcy Wentworth Thompson. The elder Thompson died in 1902.
9. E. Ray Lankester, "The Word 'Scientist' or Its Substitute," *Nature* 114 (6 December 1924): 823.
10. Herbert Dingle, "The Word 'Scientist' or Its Substitute," *Nature* 114 (20 December 1924): 897.
11. A Chemist, "The Word 'Scientist' or Its Substitute," *Nature* 114 (20 December 1924): 898.
12. Henry E. Armstrong, "The Word 'Scientist' or Its Substitute," *Nature* 115 (17 January 1925): 85.

13. [Richard Gregory], "Words, Meanings, and Styles, I," *Nature* 115 (21 February 1925): 253–255.

14. [Richard Gregory], "Words, Meanings, and Styles, II," *Nature* 115 (28 February 1925): 289–291.

15. On the Republic of Letters and pre-1800 scientific communication more generally, see Broman, "Habermasian Public Sphere"; Daston, "Ideal and Reality"; Goldgar, *Impolite Learning*; Goodman, *Republic of Letters*; Habermas, *Structural Transformation*; Johns, *Nature of the Book*; Lux and Cook, "Closed Circles or Open Networks?"; Mayhew, "Mapping Science's Imagined Community"; Shapin, *Social History of Truth*, chap. 5; Sibum, "Experimentalists in the Republic of Letters."

16. Mayhew, "Mapping Science's Imagined Community," 76.

17. For more on *Nature*'s role as a forum for discussion and a venue for scientific controversies, see Kjærgaard, "'Within the Bounds of Science.'"

18. Alysoun Sanders, archivist for the Macmillan Publishing Group, personal communication, 8 May 2007.

19. See Lightman, *Victorian Popularizers of Science*, chap. 6.

20. On national styles of science, see Crosland, "History of Science"; Daston and Otte, "Styles of Science"; Duhem, *Aim and Structure*; Harwood, "National Styles," *Styles of Scientific Thought*; Reingold, "National Style."

21. On professionalization, see Desmond, "Redefining the X Axis"; Meadows, *Victorian Scientist*; Morrell, "Professionalisation"; Porter, "Gentlemen and Geology," "Scientific Naturalism"; Secord, "Geological Survey"; Turner, "Victorian Conflict." For challenges to the "professionalization" narrative in nineteenth-century British science, see Barton, "'Men of Science'"; Endersby, *Imperial Nature*.

22. On the growth of English as an international language, see Ammon and McConnell, *English as an Academic Language*; Bailey, *Nineteenth-Century English*; Crystal, *English as a Global Language*.

23. On distrust in science and demarcating science from nonscience, see, e.g., Oreskes and Conway, *Merchants of Doubt*, and Gordin, *Pseudoscience Wars*.

24. On the *Philosophical Transactions*, see Atkinson, *Scientific Discourse*; Johns, *Nature of the Book*, chap. 3; "Miscellaneous Methods." On the *Journal des Sçavans*, see Costabel, "'L'à-peu-près n'est pas leur fait'"; Morgan, *Histoire du* Journal des Sçavans. On the *Acta Eruditorum*, see A. H. Laeven, *Acta Eruditorum*.

25. For a thorough discussion of this assumption, see Csiszar, "Broken Pieces of Fact," 13–22.

26. Meadows, *Communication in Science*, 66–68.

27. Csiszar, "Broken Pieces of Fact," 7. On nineteenth-century changes in scientific publishing, see also Watts, "'We Want No Authors.'"

28. Csiszar, "Broken Pieces of Fact," 6.

29. Secord, "Science, Technology and Mathematics," 443–444.

30. On popular science writing in Victorian Britain, see Barton, "Just before *Nature*"; Cantor, *Science in the Nineteenth-Century Periodical*; Cantor and Shuttleworth, *Science Serialized*; Dawson, *Darwin, Literature and Victorian Respectability*; Fyfe, *Science and Salvation*; Henson et al., *Culture and Science*; Lightman, *Victorian Popularizers of Science*; Lightman, "Visual Theology"; Lightman et al., *Victorian Science in Context*; Mussell, *Science, Time and Space*; Secord, *Victorian Sensation*. On the twentieth century, see Bowler, *Science for All*.

31. Pieces that analyze the history of the scientific journal include Brock, "Development of Commercial Science Journals"; Broman, "Periodical Literature"; Csiszar, "Broken Pieces of Fact"; Johns, *Nature of the Book*, chap. 3; "Miscellaneous Methods," 159–186; Meadows, *Communication in Science*, chap. 3; Secord, "Science, Technology and Mathematics."

32. E.g., Atkinson, *Scientific Discourse*; Gross et al., *Communicating Science*.

33. Krementsov, *International Science*, 4.

34. R. M. Macleod, various articles, *Nature* 224 (1 November 1969): 417-461; Gary Werskey, "*Nature* and Politics between the Wars," *Nature* 224 (1 November 1969): 462-472.
35. Maddox, "Introduction."
36. Barton, "Scientific Authority"; Kjærgaard, "'Within the Bounds of Science.'"
37. Roos, "The 'Aims and Intentions' of *Nature*."
38. James Secord wrote perhaps the richest history of reading to date in his 2001 book *Victorian Sensation*, in which he analyzed the way readers interacted with the popular evolutionary tract *Vestiges of the Natural History of Creation*. See Secord, *Victorian Sensation*. See also Secord, *Visions of Science*, which discusses the readers of several early nineteenth century scientific books.
39. Walter Gratzer, interview by author, Toronto, Canada, 31 May 2012.
40. Mary Sheehan, interview by author, London, United Kingdom, 12 April 2012.

CHAPTER ONE

1. On Lockyer, see Lockyer et al., *Life and Work*; Meadows, *Science and Controversy*.
2. Joseph Hooker to Alexander Macmillan, 27 July 1869, Norman Lockyer Papers, Special Collections, University of Exeter Library, Exeter, United Kingdom (hereafter NLP), MS 236.
3. Meadows, *Science and Controversy*, 6.
4. Lockyer's ease at being accepted as a colleague by more experienced and highly trained astronomers is a striking illustration of the permeable boundary between "professional" and "amateur" scientists in the mid-Victorian era. Experienced astronomers such as the Italian physicist Angelo Secchi read Lockyer's articles and debated his findings; when Secchi and Lockyer clashed over their differing maps of Mars's surface in 1862, Lockyer was widely regarded as the winner despite his comparatively paltry experience as an observer. Meadows, *Science and Controversy*, 45. A satisfactory study of Lockyer's astronomical career has yet to be written, although Meadows's biography contains much useful information on the subject.
5. For more on solar spectra, see Meadows, *Science and Controversy*, 48-49.
6. For the paper's abstract, see Norman Lockyer, "Spectroscopic Observations of the Sun," *Proceedings of the Royal Society of London* 15 (1866-1867): 256-258.
7. Thomas A. Hirst to Norman Lockyer, 2 October 1864, NLP, MS 110.
8. G. J. Allman to Norman Lockyer, 2 December 1863, NLP, MS 110.
9. On Huxley, see Barr, *Thomas Henry Huxley's Place*; Desmond, *Huxley*; Di Gregorio, *T. H. Huxley's Place*; Huxley, *Life*; Paradis, *T. H. Huxley*; White, *Thomas Huxley*.
10. On the X Club, see Barton, "X Club"; "'Huxley, Lubbock, and Half a Dozen Others'"; Jenson, "X Club"; Macleod, "X-Club."
11. John Fiske, quoted in Barton, "X Club," 117.
12. For more on the X Club and the *Reader*, see Barton, "X Club," 223-226.
13. On the Macmillan family and Macmillan Publishing, see van Arsdel, "Macmillan Family"; Graves, *Life and Letters of Alexander Macmillan*; James, *Macmillan*; Morgan, *House of Macmillan*.
14. Alexander Macmillan, quoted in Graves, *Life and Letters of Alexander Macmillan*, 262.
15. Lockyer was incensed over this turn of events and wrote several letters to his superiors at the War Office attempting to have his old title and salary reinstated. See Norman Lockyer, "Private and confidential," 16 November 1868, NLP, MS 110 ZP.
16. On Victorian popular science publishing, see Introduction, note 30.
17. See Broman, "Periodical Literature"; Feltes, *Modes of Production*; Fyfe, *Steam-Powered Knowledge*; Hughes and Lund, *Victorian Serial*; Jordan and Patten, *Literature in the Marketplace*.
18. Brock, "Development of Commercial Science Journals."
19. Dawson, "*Cornhill Magazine*."
20. See Lightman, *Victorian Popularizers of Science*, 356-369.

21. The gendered language here is not accidental. While *Nature* came out in support of increased science education for women in its very first issue ("Notes," *Nature* 1 [4 November 1869]: 25–26), many of *Nature*'s contributors did not consider women capable of producing original scientific work. Huxley in particular was well known for his belief that women were too susceptible to religion and superstition to make good researchers. See Lightman, "Marketing Knowledge," 102. Lockyer was an exception. In 1902, when the Royal Society considered nominating the astronomer Hertha Ayrton, he was a vocal supporter of her candidacy. However, the Fellows eventually decided that Ayrton's status as a married woman made her ineligible. See Mason, "Hertha Ayrton." More will be said about *Nature* and female contributors in chapter 3.

22. Desmond, *Huxley*, 372.

23. Crookes, like Lockyer, had previous editorial experience before founding his journal; Crookes had been the editor of several London and Liverpool photography journals in the 1850s. Crookes also edited the *Quarterly Journal of Science*, a popular science periodical, from 1864 to 1879. On the *Chemical News*, Crookes, and his publishing interests, see Brock, *"Chemical News"*; *William Crookes*; Knight, "Science and Culture."

24. E.g., John E. Gray, "The Culture of Salmon," *Athenaeum* 2103 (15 February 1868): 243.

25. The 10 percent figure is based on an analysis of the number of pages devoted to scientific articles in *British Quarterly Review*, *Fortnightly Review*, and *Nineteenth Century* in the 1860s and 1870s. For more information on these publications, see Houghton, *Wellesley Index*.

26. "Advertisement: *Nature*, an Illustrated Journal of Science," *Athenaeum* 2191 (23 October 1869): 538.

27. "Forthcoming Publications: Nature," *Journal of the Society of Arts* 17 (1869): 860; "Advertisement: 'Nature,' a Weekly Illustrated Journal of Science," *Cambridge University Gazette* (10 November 1869): 231; "Advertisement: Nature: an Illustrated Journal of Science," *Academy* 1 (1869): 11.

28. See James, "Reporting Royal Institution Lectures." James suggests that newspapers such as the *Times* might have surrendered coverage of the sciences to the popular shilling monthlies and that the decline of science coverage might not actually have reflected a loss of interest as men such as Lockyer claimed.

29. On the history and significance of the *Academy*, see Beer, *"Academy"*; Roll-Hansen, *Academy*.

30. On Foster, see Geison, *Michael Foster*.

31. Michael Foster to Norman Lockyer, 4 August 1869, NLP, MS 110.

32. Beer, *"Academy,"* 192.

33. Roll-Hansen, *Academy*, 159.

34. Advertisements for *Nature* contained a list of "eminent Scientific Men" who "have already promised to contribute Articles, or otherwise aid in the work." See, e.g., "Advertisement: *Nature*, an Illustrated Journal of Science," *Athenaeum* 2191 (23 October 1869): 538. The list included such luminaries as Huxley, Hooker, Michael Foster, Alfred Russel Wallace, John Tyndall, Charles Darwin, Peter Guthrie Tait, and William Thomson.

35. F. A. Abel to Norman Lockyer, 26 July 1869, NLP, MS 110.

36. Sir George Airy to Norman Lockyer, 24 July 1869, NLP, MS 110.

37. William Pengelly to Norman Lockyer, 29 July 1869, NLP, MS 236.

38. Charles Pritchard to Norman Lockyer, 8 July 1869, NLP, MS 110.

39. Michael Foster to Norman Lockyer, 4 August 1869, NLP, MS 110.

40. We know Hooker put Lockyer in contact with Bennett from several letters he wrote to Huxley in the 1870s, e.g., Joseph Hooker to Thomas Huxley, 13 November 1872, Thomas Henry Huxley Collection, Records and Archives, Imperial College London Library, London (hereafter THHC), 3.181.

41. T. H. Huxley, quoted in Gowan Dawson, *"Review of Reviews,"* 177.

42. Alfred Russel Wallace to George Lillie Craik, 2 February 1876, Macmillan Papers, British Library, London (hereafter MP:BL), MS 55221.5.

43. Archibald Geikie to "Jack," 29 December 1878, MP:BL, MS 55212.43.

44. Tennyson, "Sacramental Imagination," 371. On the British Association for the Advancement of Science's efforts to employ "Nature" to suggest a normative moral agenda to their scientific work, see Morrell and Thackray, *Gentlemen of Science*, 29-34.

45. On Romanticism's influence on nineteenth-century science, see Cunningham and Jardine, *Romanticism and the Sciences*; Richards, *Romantic Conception of Life*.

46. Meadows, *Science and Controversy*, 28.

47. T. H. Huxley, "Nature: Aphorisms by Goethe," *Nature* 1 (4 November 1869): 9-11.

48. Alfred Bennett, "On the Fertilisation of Winter-Flowering Plants," *Nature* 1 (4 November 1869): 11-13; "Protoplasm at the Antipodes," *Nature* 1 (4 November 1869); Norman Lockyer, "The Recent Total Eclipse of the Sun," *Nature* 1 (4 November 1869): 14-15.

49. "Societies and Academies," *Nature* 1 (4 November 1869): 29-30.

50. Gross, Harmon, and Reidy, *Communicating Science*, 117-124.

51. J. Norman Lockyer, "Spectroscopic Observations of the Sun. No. II," *Philosophical Transactions of the Royal Society of London* 159 (1869): 425-444.

52. Michael Foster, "The Retardation of the Beat of the Heart: Review of *Das Hemmungsnervensystem des Herzens*," *Nature* 1 (4 November 1869): 17.

53. R. A. Proctor, "Tables of Pomona: Review of *Tafeln der Pomona, mit Berucksichtigung der Storungen durch Jupiter, Saturn, und Mars* by Dr. Otto Lesser," *Nature* 1 (4 November 1869): 18.

54. "The Dulness of Science," *Nature* 1 (11 November 1869): 43-44. The title was meant to be ironic.

55. H. Woodward, "Geology and Agriculture," *Nature* 1 (11 November 1869): 46-48.

56. H [Thomas H. Huxley], "Darwinism and National Life," *Nature* 1 (16 December 1869): 183-184.

57. The literary historian David Roos interprets Lockyer's January notice as a confident statement of a cohesive editorial policy. See Roos, "'Aims and Intentions' of *Nature*," 165. Roos draws this interpretation from a 1919 retrospective that Lockyer wrote for the journal's fiftieth anniversary issue. However, it seems likely that Lockyer's 1919 essay retroactively imposes cohesion and stability on the early years of the journal rather than being an accurate description of the actual situation in 1869.

58. No title, *Nature* 1 (20 January 1870): 323.

59. For *Nature*'s current mission statement, see "About the Journal," *Nature.com*, accessed 9 June 2014, http://www.nature.com/nature/about/index.html.

60. G. S. Brady, "The Microscopic Fauna of the English Fen District," *Nature* 1 (10 March 1870): 483-484.

61. J. Ericsson, "The Source of Solar Energy," *Nature* 6 (31 October 1872): 539-540.

62. G. C. Foster, "M. Fizeau's Experiments on 'Newton's Rings'," *Nature* 2 (9 June 1870): 105.

63. James Clerk Maxwell, "The Dynamical Evidence of the Molecular Constitution of Bodies (Part I)," *Nature* 11 (4 March 1875): 357-359; "The Dynamical Evidence of the Molecular Constitution of Bodies (Part II)," *Nature* 11 (11 March 1875): 374-377.

64. There are a few, though not many, exceptions to this division; e.g., a chatty first-person piece about an eclipse expedition to Siam appeared after the Letters to the Editor in the 22 July 1875 issue of *Nature*. Arthur Schuster, "Science in Siam," *Nature* 12 (22 July 1875): 233-234.

65. For more on Kingsley, see Chitty, *Beast and the Monk*; Manlove, "Charles Kingsley"; Straley, "Of Beasts and Boys." Kingsley carried on an intense and extremely interesting correspondence with Huxley centering on religious issues and theological reform; see White, *Thomas Huxley*, 114-121.

66. Charles Kingsley to Norman Lockyer, 8 November 1869, NLP, MS 110.

67. Charles Kingsley, "The World of the Sea: Review of *The World of the Sea* by Moquin Tandon, trans. H. Martyn Hunt," *Nature* 1 (11 November 1869): 78-80.

68. Charles Kingsley to Norman Lockyer, 8 November 1872, NLP, MS 110. Kingsley's letter stands in contrast to David Roos's argument that during Lockyer's tenure, "*Nature* was neither written for nor by 'scientists'" and that the journal continued to be accessible "to all interested readers" well into the 1880s. Roos, " 'Aims and Intentions' of *Nature*," 171–173. Both the journal's contents and the correspondence of the journal's readers clearly indicate otherwise. Roos cites the participation of "amateurs" in an 1884 debate over the significance of solar phenomena as evidence that *Nature* was accessible to a wide cross section of British society; however, more recent scholarship on Victorian science has challenged older ideas about "amateurs" and "professionals" in the nineteenth century. See Barton, " 'Huxley, Lubbock, and Half a Dozen Others' "; " 'Men of Science' "; Endersby, *Imperial Nature*.

69. For more on the social status of nineteenth-century science journalists, see Fyfe, "Conscientious Workmen."

70. For a more detailed account of the background of the Ayrton-Hooker affair, see Endersby, *Imperial Nature*, 282–286.

71. "Mr. Ayrton and Dr. Hooker, " *Nature* 6 (11 July 1872): 211–216.

72. Joseph Hooker, "Dr. Hooker's Reply to Prof. Owen," *Nature* 6 (4 October 1872): 516–517.

73. "Mr. Ayrton and Dr. Hooker," *Nature* 6 (8 August 1982): 280–281.

74. Richard Owen, "The National Herbarium," *Nature* 7 (7 November 1872): 6.

75. Jim Endersby observes that Hooker's high emotions and intemperate language during the Ayrton controversy likely cost him an outright victory over Ayrton's plans for Kew. See Endersby, *Imperial Nature*, 293.

76. Joseph Hooker to Thomas Huxley, 13 November 1872, THHC, 3.181.

77. Joseph Hooker to Thomas Huxley, 16 November 1872, THHC, 3.183. Hooker sent Huxley a transcript of the conversation between Huxley and Lockyer for verification before he showed it to Bennett, who was evidently also disavowing responsibility for the letter.

78. Joseph Hooker, "Kew Gardens and the National Herbarium," *Nature* 7 (21 November 1872): 46. Alfred Bennett (1833–1902) was a lecturer on botany at Bedford College for Women. He had wide-ranging publishing experience, including a ten-year stint as the proprietor and editor of the Quaker journal *Friends*, and Hooker himself had recommended Bennett to Lockyer when Lockyer was seeking someone with knowledge of botany to handle the life sciences aspects of *Nature*. However, Bennett was not a Fellow of the Royal Society and did not travel in the same scientific circles as men like Lockyer, Owen, Huxley, and Hooker. It likely would not have occurred to him to reject a contribution from a scientific man as famous as Owen. For more on Bennett, see Cleevely, "Bennett, Alfred William." Hooker's annoyance with Bennett is clear in a letter to Huxley: "I cannot understand Bennett being so utterly false & foolish (I must underline).... I got him the subeditorship.... He again to-day before the officers of the Harbour (but in my presence), gave his assurance, that he did not even see the article till it was corrected by Owen for print, & that it was sent to him in that state marked urgent. I cannot conceive such folly as this, if this is false." Joseph Hooker to Thomas Huxley, 13 November 1872, THHC, 3.181.

79. Joseph Hooker to Thomas Huxley, 11 November 1872, THHC, 3.178.

80. Thomas Huxley to Norman Lockyer, 24 November 1872, THHC, 21.255.

81. For samples of Tait's work for *Nature*, see Peter Guthrie Tait, "Energy, and Prof. Bain's Logic," *Nature* 3 (1 December 1870): 89–90; "True and Spurious Metaphysics," *Nature* 5 (30 November 1871): 81. For Tyndall's contributions, see John Tyndall, "Atmospheric Effect," *Nature* 6 (1 August 1872): 260; "Effects of Resistance in Modifying Spectra," *Nature* 7 (20 March 1873): 384. For more on Tyndall, see Kim, *John Tyndall's Transcendental Materialism*; de Young, *Vision of Modern Science*. On Tait, see Wilson, "P.G. Tait."

82. See Wilson, "P.G. Tait," 276.

83. Peter Guthrie Tait, "Tyndall and Forbes," *Nature* 8 (11 September 1873): 381–382.

84. John Tyndall, "Tyndall and Tait," *Nature* 8 (18 September 1873): 399.

85. Norman Lockyer, "Tyndall and Tait," *Nature* 8 (18 September 1873): 399.

86. "Tait and Tyndall," *Nature* 8 (25 September 1873): 431.

87. Peter Guthrie Tait to Norman Lockyer, 26 September 1873, NLP, MS 110.

88. John Tyndall to Rudolf Clausius, 26 March 1874, John Tyndall Papers, Royal Institution of Great Britain, London, T206.

89. On Spencer, see Carneiro and Perrin, "Herbert Spencer's Principles of Sociology"; Elliott, "Erasmus Darwin"; Francis, *Herbert Spencer*; Jones and Peel, *Herbert Spencer*; Ridley, "Coadaptation"; Smith, "Evolution and the Problem of Mind."

90. Barton, "Scientific Authority," 235.

91. Anonymous, "Herbert Spencer," *British Quarterly Review* 58 (October 1873): 253.

92. Herbert Spencer, "Replies to Criticism [I]," *Fortnightly Review* 54 (1 November 1873): 581–595; "Replies to Criticism [II]," *Fortnightly Review* 54 (1 December 1873): 715–739; Anonymous, "Note to the Article on Herbert Spencer," *British Quarterly Review* 49 (January 1874): 111–113.

93. Herbert Spencer, *Mr. Herbert Spencer and the British Quarterly Review* (London: H. Spencer, 1874).

94. It is unclear whether Tait was deliberately caricaturing Spencer's argument or whether his interpretation was a genuine misunderstanding. If it was the latter, he was not the only reader puzzled by Spencer's verbose writing style. Spencer's X Club nickname was "The Xhaustive Mr. Spencer," a nickname that, as Roy Macleod observes, was accurate but not exactly flattering. See Macleod, "X-Club," 310–311.

95. Peter Guthrie Tait, "Herbert Spencer versus Thomson and Tait," *Nature* 9 (26 March 1874): 402–403.

96. Herbert Spencer, "Prof. Tait and Mr. Spencer," *Nature* 9 (2 April 1874): 420–421.

97. The Author of the Article in the British Quarterly Review, "Herbert Spencer versus Sir I. Newton," *Nature* 9 (2 April 1874): 421.

98. Robert B. Hayward, "Herbert Spencer and *à priori* Axioms," *Nature* 9 (30 April 1874): 499–500; "Mr. Spencer and *à priori* Axioms," *Nature* 10 (14 May 1874): 25; "Physical Axioms," *Nature* 10 (28 May 1874): 61–62; "Proportionality of Cause and Effect," *Nature* 10 (11 June 1874): 104; "Mr. Herbert Spencer and Physical Axioms," *Nature* 10 (27 August 1874): 335.

99. A Senior Wrangler, "Herbert Spencer and *à priori* Axioms," *Nature* 9 (30 April 1874): 500.

100. Not a Metaphysician, "Mr. Spencer and *à priori* Axioms," *Nature* 10 (14 May 1874): 25.

101. Frederick Guthrie, "Mr. Herbert Spencer and Physical Axioms," *Nature* 10 (20 August 1874): 306.

102. James Collier, "Quantitative Relations of Cause and Effect," *Nature* 10 (21 May 1874): 44.

103. Robert Hayward, "Physical Axioms," *Nature* 10 (28 May 1874): 62.

104. A Senior Wrangler, "Physical Axioms," *Nature* 10 (28 May 1874): 62.

105. Charles Root, "The Germans and Physical Axioms," *Nature* 10 (18 June 1874): 123.

106. Herbert Spencer to Norman Lockyer, 19 May 1874, NLP, MS 110.

107. P. G. Tait, quoted in R. M. Macleod, "Is It Safe to Look Back?," *Nature* 224 (1 November 1969): 449.

108. Alexander Macmillan to Norman Lockyer, 10 November 1871, NLP, MS 236.

CHAPTER TWO

1. C. William Siemens to Norman Lockyer, 24 February 1879, NLP, MS 110. Siemens was a Hanoverian by birth and had moved to England at the age of 20, where he distinguished himself as an electrical engineer and, along with his brother Werner, the inventor of the regenerative furnace. On Siemens, see Pole, *Life of Sir William Siemens*; Scott, *Siemens Brothers*.

2. For more on Lockyer and his fondness for scientific controversies, see his aptly titled biography: Meadows, *Science and Controversy*, 314–316.

3. In addition, between 1869 and 1881, Hooker, Huxley, Tyndall, Spottiswoode, and Lubbock

all served as president of the British Association for the Advancement of Science (BA), and many other X Club members served as BA trustees, council members, and section presidents. See Barton, "'An Influential Set of Chaps'"; "X Club," 116-191.

4. Turner, "Victorian Conflict," 362; *Contesting Cultural Authority*, 180. The word *professionalize* is placed in quotation marks because it is a problematic and not entirely appropriate term in the context of Victorian science. This issue will be discussed further in chap. 3.

5. On the decline of the X Club, see Macleod, "X-Club," 314-316.

6. On Haeckel, see Richards, *Tragic Sense of Life*.

7. E. Ray Lankester to T. H. Huxley, 18 December 1872, THHC, 21.39.

8. On Lankester, see Lester, *E. Ray Lankester*; Milner, "Huxley's Bulldog."

9. On Romanes, see England, *Design after Darwin*; Forsdyke, *Origin of Species, Revisited*; Lesch, "Isolation in Evolution"; Schwartz, "George John Romanes's Defense of Darwinism"; "Out from Darwin's Shadow"; Turner, *Between Science and Religion*, 134-163.

10. There is very little scholarly literature on Thiselton-Dyer. See Thomason, "Dyer, Sir William Turner Thiselton."

11. On Meldola, see Eyre and Rodd, "Raphael Meldola"; Gay, "Chemist, Entomologist, Darwinian"; Marchant, *Raphael Meldola*; Webb, "Raphael Meldola."

12. See Hunt, *The Maxwellians*.

13. On Lodge, see Clow, "Laboratory of Victorian Culture"; Hunt, "Experimenting on the Ether"; Jolly, *Sir Oliver Lodge*; Rowlands, *Oliver Lodge and the Liverpool Physical Society*; Rowlands and Wilson, *Oliver Lodge and the Invention of Radio*; Wilson, "Thought of the Late Victorian Physicists." Much of the literature on Lodge relates to his interest in spiritualism, which will be discussed further in chap. 3.

14. On Perry, see Burchfield, *Lord Kelvin*; England, Moinar, and Richter, "Kelvin, Perry and the Age of the Earth"; Gooday, "The Morals of Energy Metering."

15. There is an enormous amount of secondary literature on these six men. Helpful biographies include Browne, *Charles Darwin: The Power of Place*; *Charles Darwin: Voyaging*; Desmond, *Huxley*; Endersby, *Imperial Nature*; Raby, *Alfred Russel Wallace*; Rylance, *Victorian Psychology*; Slotten, *Heretic in Darwin's Court*; Taylor, *Men Versus the State*; White, *Thomas Huxley*; Wilson, "P.G. Tait."

16. Kjærgaard, "'Within the Bounds of Science,'" 211-221.

17. George J. Romanes, "Permanent Variation of Colour in Fish," *Nature* 8 (5 June 1873): 101.

18. For a complete account of Romanes's correspondence with Darwin, see Schwartz, "George John Romanes's Defense of Darwinism."

19. George J. Romanes, "Physiological Selection: An Additional Suggestion on the Origin of Species [I]," *Nature* 34 (5 August 1886): 314-316; "Physiological Selection: An Additional Suggestion on the Origin of Species [II]," *Nature* 34 (12 August 1886): 336-340; "Physiological Selection: An Additional Suggestion on the Origin of Species [III]," *Nature* 34 (19 August 1886): 362-365. For the full paper, see George J. Romanes, "Physiological Selection: An Additional Suggestion on the Origin of Species," *Journal of the Linnean Society* 19, no. 115 (1886): 337-411.

20. George J. Romanes, "Physiological Selection: An Additional Suggestion on the Origin of Species [I]," *Nature* 34 (5 August 1886): 316.

21. On Weismann, see Churchill, "August Weismann"; Churchill and Risler, *August Weismann*.

22. George J. Romanes, "Panmixia," *Nature* 41 (13 March 1890): 438. Romanes's ideas on the inheritance of disuse and the shrinking of useless organs follow quite closely Darwin's views in chap. 13 of *On the Origin of Species*, where Darwin wrote, "On my view of descent with modification, the origin of rudimentary organs is simple.... I believe that disuse has been the main agency; that it has led in successive generations to the gradual reduction of various organs, until they have become rudimentary.... An organ, when rendered useless, may well be variable, for its variations cannot be checked by natural selection.... If each step of the process of reduction were to be inherited, not at the corresponding age, but at an extremely early period of life (as we have

good reason to believe to be possible) the rudimentary part would tend to be wholly lost." Charles Darwin, *On the Origin of Species* (1859; facsimile ed., Cambridge, MA: Harvard University Press, 1964), 454-455.

23. A. R. Wallace, "Romanes *versus* Darwin: An Episode in the History of Evolution Theory," *Fortnightly Review* 60 (1 September 1886): 300-316.

24. Raphael Meldola, "Physiological Selection and the Origin of Species," *Nature* 34 (26 August 1886): 384.

25. George J. Romanes, "Co-adaptation," *Nature* 43 (26 March 1891): 489-90; "Co-adaptation and Free Intercrossing," *Nature* 43 (23 April 1891): 582-583; "Co-adaptation," *Nature* 44 (14 May 1891): 28; "Co-adaptation," *Nature* 44 (21 May 1891): 55; Raphael Meldola, "Co-adaptation," *Nature* 43 (16 April 1891): 557-558; "Co-adaptation," *Nature* 44 (7 May 1891): 7; "Co-adaptation," *Nature* 44 (14 May 1891): 28-29.

26. George J. Romanes, "Mr. Wallace on Physiological Selection," *Nature* 43 (11 December 1890): 127-128.

27. George J. Romanes, "Co-adaptation and Free Intercrossing," *Nature* 43 (23 April 1891): 582.

28. Alfred R. Wallace, "Dr. Romanes on Physiological Selection," *Nature* 43 (18 December 1890): 150.

29. Alfred Russel Wallace to W. T. Thiselton-Dyer, 26 September 1893, Alfred Russel Wallace Papers, British Library, London (hereafter ARWP:BL), MSS 46435.300. The copyright of literary works by Alfred Russel Wallace that were unpublished at the time of his death and that are published in this book belongs to the A. R. Wallace Literary Estate. These works are licensed under Creative Commons Attribution-NonCommercial-ShareAlike 3.0 Unported (http://creativecommons.org/licenses/by-nc-sa/3.0/legalcode).

30. W. T. Thiselton-Dyer to Alfred Russel Wallace, 29 October 1889, ARWP:BL, MSS 46435.213. See George J. Romanes, "Mr. Dyer on Physiological Selection," *Nature* 39 (29 November 1888): 103-104; "Natural Selection and the Origin of Species," *Nature* 39 (20 December 1888): 173-175; W. T. Thiselton-Dyer, "Mr. Romanes's Paradox," *Nature* 39 (1 November 1888): 7-9; "Mr. Romanes on the Origin of Species," *Nature* 39 (6 December 1888): 126-127.

31. Raphael Meldola, "Co-adaptation," *Nature* 44 (14 May 1891): 29.

32. E. Ray Lankester to J. Norman Lockyer, 25 September [1886], NLP, MSS 110. Unfortunately, Lankester's rejected letter to *Nature* has not survived.

33. Following the death of Lockyer's wife Winifred in 1879, Lockyer became quite close to George and Ethel Romanes and carried on a warm personal correspondence with them. See George J. Romanes to J. Norman Lockyer, various letters, NLP, MSS 110.

34. George J. Romanes to J. Norman Lockyer, 30 October 1886, NLP, MSS 110.

35. For more information about the *Fortnightly Review*, see "The Fortnightly Review" in Houghton, *Wellesley Index*.

36. Alfred Russel Wallace, "Physiological Selection and the Origin of Species," *Nature* 34 (16 September 1886): 467-468.

37. E.g., Francis Darwin, "Physiological Selection and the Origin of Species," *Nature* 34 (2 September 1886): 407; E. Ray Lankester, "Darwinism," *Nature* 41 (7 November 1889): 9; R. Meldola, "Physiological Selection and the Origin of Species," *Nature* 34 (26 August 1886): 384-385.

38. George J. Romanes, "Physiological Selection and the Origin of Species," *Nature* 34 (9 September 1886): 439.

39. Thomson was made Baron Kelvin in 1892; for consistency's sake, we shall continue to refer to him as Thomson in this passage.

40. William Thomson, "On Compass Adjustment in Iron Ships," *Nature* 17 (28 February 1878): 352-354; "Approximate Photometric Measurements of Sun, Moon, Cloudy Sky, and Electric and Other Artificial Lights," *Nature* 27 (18 January 1883): 277-279; William Thomson and Osborne Reynolds, "Storage of Electric Energy," *Nature* 24 (16 June 1881): 156-157.

41. See Smith and Wise, *Energy and Empire*, 552–611. See also Burchfield, *Lord Kelvin*.
42. John Perry, "The Age of the Earth," *Nature* 51 (3 January 1895): 224–227.
43. John Perry, "On the Age of the Earth," *Nature* 51 (7 February 1895): 341–342.
44. John Perry, "The Age of the Earth," *Nature* 51 (18 April 1895): 582–585.
45. William Thomson, "The Age of the Earth," *Nature* 51 (7 March 1895): 438–440.
46. On Stokes, see Wilson, *Kelvin and Stokes*; "A Physicist's Alternative."
47. "'M.P., P.R.S.,'" *Nature* 37 (17 November 1887): 49–50.
48. W. T. Thiselton-Dyer to T. H. Huxley, 7 December 1887, THHC, 27.214. For the letter to *Nature* referred to, see W. T. Thiselton-Dyer, "Politics and the Presidency of the Royal Society," *Nature* 37 (1 December 1887): 103–104.
49. Balfour Stewart, "Politics and the Presidency of the Royal Society," *Nature* 37 (24 November 1887): 76.
50. Alex. W. Williamson, "Politics and the Presidency of the Royal Society," *Nature* 37 (24 November 1887): 76–77.
51. On the X Club's mentorship of the younger generation, see Stanley, "Where Naturalism and Theism Met."
52. On Romanes's writings in literary publications, see Schwartz, "Out from Darwin's Shadow," 133–159.
53. Maddox, "Introduction," 3.
54. As James Secord has observed, scientific conversations and lectures were another important form of scientific communication in the early nineteenth century, and like monographs, lectures, and conversations, they had become far less central by the end of the nineteenth century. Secord, "How Scientific Conversation Became Shop Talk."
55. I would like to thank Gowan Dawson for suggesting this point.
56. Henry Draper to Norman Lockyer, 8 November 1873, NLP, MS 110.
57. Brake and Demoor, *Dictionary of Nineteenth-Century Journalism*, s.v. "*Nature* (1869–)."
58. Kjærgaard, "'Within the Bounds of Science,'" 212–217.
59. William Crookes to Norman Lockyer, 20 August 1895, NLP, MS 110. Crookes's helium paper was published shortly after he wrote the letter. William Crookes, "The Spectrum of Helium," *Nature* 52 (29 August 1895): 428–430. A note on the article indicates that Crookes's paper also appeared in that week's edition of *Chemical News*.
60. On the relationship between priority and publication, see Csiszar, "Broken Pieces of Fact," chap. 2.
61. I thank Michael Taylor and James Elwick for drawing my attention to this point. On the importance of publication in general periodicals to Huxley's career and finances, see White, *Thomas Huxley*, chap. 3.
62. Brock, "Advancing Science," 116.
63. The *Wellesley Index to Victorian Periodicals* indexes the contents and authors of forty-five Victorian publications, including *British Quarterly Review*, *Contemporary Review*, *Cornhill Magazine*, *Edinburgh Review*, *Fortnightly Review*, *Macmillan's Magazine*, *Nineteenth Century*, and *Westminster Review*. The *Index* includes two popular science publications, *Macmillan's Magazine* and the *Cornhill Magazine*, but no scientific periodicals aimed at an audience of men of science.
64. "Index of Authors," in Houghton, *Wellesley Index*.
65. Schwartz, "Out from Darwin's Shadow," 133–159.
66. See Kjærgaard, "'Within the Bounds of Science,'" 211–221.
67. See, e.g., Charles Darwin, "Pangenesis," *Nature* 3 (27 April 1871): 502–503; "Perception in the Lower Animals," *Nature* 7 (13 March 1873): 360; "Recent Researches on Termites and Honey-Bees," *Nature* 9 (19 February 1874): 308–309; "Sexual Selection in Relation to Monkeys," *Nature* 15 (2 November 1876): 18–19; "Fertility of Hybrids from the Common and Chinese Goose," *Nature* 21 (1 January 1880): 207; "Inheritance," *Nature* 24 (21 July 1881): 257. Darwin's fondness for

using *Nature* to publicize his research and discuss current theories may be related to his distaste for traveling to meetings. See Browne, "I Could Have Retched All Night."

68. I owe this observation to Dr. Michele Aldrich, and I am grateful for her willingness to share her research on Darwin's publishing patterns with me.

69. George J. Romanes, "Physiological Selection: An Additional Suggestion on the Origin of Species [II]," *Nature* 34 (12 August 1886): 340.

70. A. R. Wallace, "Romanes *versus* Darwin: An Episode in the History of Evolution Theory," *Fortnightly Review* 60 (1 September 1886): 300–316.

71. Raphael Meldola, "Physiological Selection and the Origin of Species," *Nature* 34 (26 August 1886): 384.

72. E. Ray Lankester, "Darwinism," *Nature* 41 (7 November 1889): 9.

73. Francis Darwin, "Physiological Selection and the Origin of Species," *Nature* 34 (2 September 1886): 407.

74. George J. Romanes, "Physiological Selection," *Nature* 36 (11 August 1887): 341.

75. Herbert Spencer, "The Inheritance of Acquired Characters," *Nature* 41 (6 March 1890): 415.

76. See, e.g., George J. Romanes, "Panmixia," *Nature* 41 (3 April 1890): 511–512.

77. E. Ray Lankester, "The Transmission of Acquired Characters, and Panmixia," *Nature* 41 (27 March 1890): 487–488.

78. Bowler, *Eclipse of Darwinism*, 43.

79. John Tyndall, quoted in Barton, "Scientific Authority," 223.

80. H. J. S. Smith, quoted in Macleod, "Is It Safe to Look Back?," 449.

81. Lockyer, Lockyer, and Dingle, *Life and Work*, 114.

82. See Meadows, *Science and Controversy*, 210.

83. Secord, "Science, Technology and Mathematics," 458.

84. Frederick Macmillan to George Macmillan, 25 December 1886, MP:BL, MSS 54788.66.

85. William Thomson, "Sir William Thomson's Baltimore Lectures," *Nature* 31 (5 March 1885): 407.

86. John Michels, "Salutatory," *Science* 1 (3 July 1880): 6. *Science*'s early road proved more difficult than *Nature*'s. Edison withdrew his financial support in early 1882. The journal briefly ceased publication before returning in 1883 with a new editor, Samuel H. Scudder, and a new financial backer, Alexander Graham Bell. In 1894, under the leadership of its third editor, N. D. C. Hodges, *Science* secured the financial and institutional support of the American Association for the Advancement of Science. That same year, Hodges turned over *Science*'s editorship and its publishing rights to the psychologist James McKeen Cattell, who would be its editor for the next fifty-one years. For more on *Science*, see Kohlstedt, "*Science*"; Sokal, "*Science* and James McKeen Cattell, 1894 to 1945"; Vandome, "Advancement of *Science*."

87. For more on Proctor and *Knowledge*, see Lightman, "*Knowledge* Confronts *Nature*." See also Lightman, *Victorian Popularizers of Science*, chap. 6.

88. Oliver J. Lodge, "Thoughts on the Bifurcation of the Sciences Suggested by the Nottingham Meeting of the British Association," *Nature* 48 (12 October 1893): 565.

CHAPTER THREE

1. Matthew, "Campbell, George Douglas." See also Gillespie, "Duke of Argyll."

2. Duke of Argyll, "On Tertiary Leaf-Beds in the Isle of Mull," *Quarterly Journal of the Geological Society* 7 (1851): 89–103.

3. John Murray, "On the Structure and Origin of Coral Reefs and Islands," *Proceedings of the Royal Society of Edinburgh* 10 (1880): 505–518.

4. Charles Darwin, *The Structure and Distribution of Coral Reefs, Being the First Part of the Geol-*

ogy of the Voyage of the Beagle, under the Command of Capt. Fitzroy, R.N. during the Years 1832 to 1836 (London: Smith Elder and Company, 1842).

5. Alfred R. Wallace, "A Critic Criticised," *Nature* 49 (8 February 1894): 336.

6. Alfred R. Wallace, "Another Substitute for Darwinism," *Nature* 50 (4 October 1894): 541.

7. Alfred R. Wallace, "Another Darwinian Critic," *Nature* 43 (9 April 1891): 530.

8. Raphael Meldola, "An Evolutionary Castigation," *Nature* 44 (10 September 1891): 441.

9. Raphael Meldola, "An Anti-Darwinian Contribution," *Nature* 43 (5 March 1891): 412.

10. See, e.g., Turner, "Victorian Conflict"; "Public Science in Britain"; White, *Thomas Huxley*.

11. Thomas Carlyle, quoted in Gross, *Rise and Fall*, xiii. Granted, Carlyle himself was a man of letters, and his description was intended to be self-promoting.

12. An excellent example of this is the critical furor surrounding the anonymous 1845 evolutionary work *Vestiges of the Natural History of Creation* (which, it was later revealed, had been written by the Scottish publisher Robert Chambers). See Secord, *Victorian Sensation*.

13. Alfred R. Wallace, "Another Darwinian Critic," *Nature* 43 (9 April 1891): 529.

14. In 1903, Lockyer married Thomazine Mary Brodhurst, who had been instrumental in the campaign to create a women's hall of residence for the University of London. See Meadows, *Science and Controversy*, 280-283.

15. On Ayrton's attempted election, see Mason, "Hertha Ayrton."

16. "Women and the Fellowship of the Chemical Society," *Nature* 78 (9 July 1908): 226-228.

17. "Women and the Fellowship of the Chemical Society," *Nature* 79 (11 February 1909): 429-431.

18. T., "Women and the Chemical Society," *Nature* 79 (24 December 1908): 221.

19. On the Geological Society, see W. J. Atkinson, "Scientific Societies and the Admission of Women Fellows," *Nature* 79 (25 February 1909): 488; T. E. Thorpe, "Scientific Societies and the Admission of Women Fellows," *Nature* (28 March 1909): 67-68. On the Paris Academy of Sciences, see "The Admission of Women to the French Academies," *Nature* 85 (12 January 1911): 342; "The Admission of Women to the Paris Academy of Sciences," *Nature* 85 (19 January 1911): 372-373. On the Royal Society, see "Notes," *Nature* 94 (28 January 1915): 594.

20. The Chemical Society first admitted women in 1920; the Geological Society voted to admit women in 1919; the Paris Academy of Sciences admitted their first female member in 1979; the Royal Society inducted its first female Fellows in 1945.

21. On the Murray-Dana debates, see Sponsel, "Coral Reef Formation," 302-311. On Darwin's theory of coral reef formation and debates surrounding the theory more generally, see ibid., chap. 2-4.

22. See George Granville Campbell, *Log Letters from "The Challenger"* (London: Macmillan, 1877).

23. Argyll, "A Great Lesson," *Nineteenth Century* 22 (September 1887): 305.

24. T. H. Huxley, "Science and the Bishops," *Nineteenth Century* 22 (November 1887): 625-641.

25. T. G. Bonney, "A Conspiracy of Silence," *Nature* 37 (10 November 1887): 26.

26. Argyll, "'A Conspiracy of Silence,'" *Nature* 37 (17 November 1887): 53-54.

27. Samuel F. Clarke, "'The Conspiracy of Silence,'" *Nature* 37 (29 December 1887): 200; An Old Pupil of Wyville Thomson's, "'The Conspiracy of Silence,'" *Nature* 37 (29 December 1887): 200; John W. Judd, "'A Conspiracy of Silence,'" *Nature* 37 (19 January 1888): 272; "The Duke of Argyll's Charges against Men of Science," *Nature* 37 (2 February 1888): 317-318; T. H. Huxley, "The Duke of Argyll's Charges against Men of Science," *Nature* 37 (9 February 1888): 342.

28. T. Mellard Reade, "The Theories of the Origin of Coral Reefs and Islands," *Nature* 37 (17 November 1887): 54.

29. Argyll, "'The Conspiracy of Silence,'" *Nature* 37 (12 January 1888): 246.

30. Argyll, "Acquired Characters and Congenital Variation," *Nature* 41 (26 December 1889): 173-174. Argyll's charges were not entirely accurate, as *Nature* had printed several pieces by evo-

lutionary theorists who argued that acquired characters could be inherited. For more on *Nature* debates over the inheritance of acquired characters, see chap. 2.

31. W. T. Thiselton-Dyer, "The Duke of Argyll and the Neo-Darwinians," *Nature* 41 (16 January 1890): 247.

32. Ibid., 248.

33. On spiritualism in Britain and its connection to men of science, see Noakes, "'Bridge Which Is Between Physical and Psychical Research'"; "Ethers, Religion and Politics"; Oppenheim, *Other World*; Owen, *Place of Enchantment*; Winter, *Mesmerized*.

34. See, e.g., William Crookes, "The Radiometer and Its Lessons," *Nature* 17 (15 November 1877): 43; Alfred R. Wallace, "The Radiometer and Its Lessons," *Nature* 17 (15 November 1877): 44; William Crookes, "Mr. Crookes and Eva Fay," *Nature* 17 (10 January 1878): 200.

35. William Crookes, "On Radiant Matter II," *Nature* 20 (4 September 1879): 439-440.

36. William Crookes, "The Radiometer and Its Lessons," *Nature* 17 (15 November 1877): 43-44.

37. Alfred R. Wallace, "The Radiometer and Its Lessons," *Nature* 17 (15 November 1877): 44. For Dr. Carpenter's letter, see W. B. Carpenter, "The Radiometer and Its Lessons," *Nature* 17 (8 November 1877): 26-27.

38. Oliver J. Lodge, "Peculiarities of Psychical Research," *Nature* 51 (10 January 1895): 247.

39. Edward T. Dixon, "Peculiarities of Psychical Research," *Nature* 51 (27 December 1894): 200.

40. "Our Book Shelf: *An Essay on Spiritual Evolution Considered in Its Bearing upon Modern Spiritualism, Science, and Religion* by J. P. B. (Trübner and Co., 1879)," *Nature* 20 (23 October 1879): 602.

41. On Pearson, see Porter, *Karl Pearson*. More will be said about Pearson in chap. 4.

42. Karl Pearson, "Peculiarities of Psychical Research," *Nature* 51 (17 January 1895): 274.

43. See Buchanan, "Science and Engineering." On late nineteenth century engineering and its relationship to academic physics, see Gooday, *Morals of Measurement*; Harwood, "Engineering Education"; Hunt, *The Maxwellians*; Marsden and Smith, *Engineering Empires*; Smith and Wise, *Energy and Empire*.

44. On the "scientific engineering" movement in Great Britain, see Buchanan, "Rise of Scientific Engineering."

45. "A 'Practical Man' on Electrical Units," *Nature* 39 (4 April 1889): 529.

46. Ibid.

47. P. G. T., "The *Engineer* on the Dimensions of Physical Quantities," *Nature* 35 (17 March 1887): 462.

48. A. M. Worthington, "On the Use of the Words 'Mass' and 'Inertia': A Suggestion," *Nature* 39 (10 January 1889): 248.

49. Oliver J. Lodge, "Mass and Inertia," *Nature* 39 (17 January 1889): 270-271.

50. Andrew Gray, "Mass and Inertia," *Nature* 39 (7 February 1889): 342.

51. A. G. Greenhill, "Weight and Mass," *Nature* 39 (21 February 1889): 390.

52. On the "professionalization" of nineteenth-century British science, see Desmond, "Redefining the X Axis"; Meadows, *Victorian Scientist*; Morrell, "Professionalisation"; Turner, "Victorian Conflict." For a more recent reassessment of the professionalization argument, see Porter, "Scientific Naturalism."

53. Endersby, *Imperial Nature*, 7, 21-27.

54. Barton, "'Men of Science.'"

55. For more on Gregory, see Armytage, *Sir Richard Gregory*.

56. Wells, of course, is best known as the author of such classic novels as *The Time Machine* and *The War of the Worlds*. For more on Wells, see Mackenzie and Mackenzie, *Time Traveller*; Smith, *H. G. Wells*. See also Wells, *Experiment in Autobiography*.

57. Richard Gregory, *Elementary Physical and Astronomical Geography* (London: Joseph Hughes,

1891); *A Description of the Laws and Wonders of Nature* (London: Joseph Hughes, 1892); *The Vault of Heaven: An Elementary Textbook of Modern Physical Astronomy* (London: Methuen, 1893); Richard Gregory and J. C. Christie, *Advanced Physiography* (London: Joseph Hughes, 1893); Richard Gregory and H. G. Wells, *Honours Physiography* (London: Joseph Hughes, 1893).

58. R. A. Gregory, "On the Determination of Masses in Astronomy," *Nature* 40 (24 May 1889): 80-82; "Professor Loomis on Rainfall," *Nature* 40 (1 August 1889): 330-332; "Comets of Short Period," *Nature* 42 (8 May 1890): 31-32; "Lunar Photography," *Nature* 42 (9 October 1890): 568-571; "Electrical Storms on Pike's Peak," *Nature* 42 (16 October 1890): 595-596.

59. Sections of Gregory's extensive correspondence with Macmillan about various manuscripts he was reviewing for the publishing house are available in the Macmillan Papers, Special Collections, University of Reading Library, Reading (hereafter MP:UR).

60. See, e.g., "The Scientific Education of Women," *Nature* 2 (16 June 1870): 117-118; J. L. Prereton, "Degrees to Women," *Nature* 23 (24 February 1881): 394; "Science at the Women's International Congress," *Nature* 60 (6 July 1899): 228.

61. Richard A. Gregory, "The Spectroscope in Astronomy," *Nature* 68 (13 August 1903): 339.

62. Richard A. Gregory, "Stars and Nebulae," *Nature* 73 (29 March 1906): 505.

63. Henry E. Armstrong, "Obituary: Mrs. Hertha Ayrton," *Nature* 112 (1 December 1923): 800-801.

64. The Editor of *Nature* [Sir Richard Gregory] to Barbara Ayrton Gould, 10 December 1923, MP:UR, 11/17. Gould was not the only reader who objected to the tone of the obituary. For responses, see H. H. Mills, "Mrs. Hertha Ayrton," *Nature* 112 (15 December 1923): 865; T. Mather, "Mrs. Hertha Ayrton," *Nature* 112 (29 December 1923): 939. A mention of the obituary in the Current Topics and Events section strongly suggests that there was a large amount of additional correspondence on the Ayrton obituary that was not published. See "Current Topics and Events," *Nature* 112 (22 December 1923): 910-911.

65. Henry E. Armstrong, "*The World of William Clissold*: A Novel at a New Angle," *Nature* 118 (20 November 1926): 723-728.

66. F. W. H., "The Future of the Human Race," *Nature* 71 (29 December 1904): 193. F. W. H. was very likely Captain F. W. Hutton, a biologist who worked at New Zealand's Canterbury Museum. For another *Nature* contribution by Hutton, see F. W. Hutton, "Hering's Theory of Heredity, and Its Consequences," *Nature* 70 (18 February 1904): 366-369.

67. Richard Gregory to H. G. Wells, 4 January 1905, Sir Richard Gregory Papers, Addition from the University of Sheffield, Special Collections, University of Sussex Library, Falmer (hereafter SRGP:SA), MS 14. When Wells's letter was finally printed in February, it opened with a complaint about his editorial treatment: "I addressed a letter to the editor of *Nature* replying to what I allege to be misrepresentations and misstatements in a review.... After a delay of some weeks due to the absence of 'F.W.H.' abroad, the editor of *Nature* has written to ask me to modify and shorten my protest." H. G. Wells, "Fact in Sociology," *Nature* 71 (2 February 1905): 319. For the reviewer's response, see F. W. H., "Fact in Sociology," *Nature* 71 (16 February 1905): 366.

68. Norman Lockyer to Frederick Macmillan, 12 October 1918, MP:BL, MSS 55218.167. Unfortunately, the previous letter from Macmillan to Lockyer has not survived.

69. Norman Lockyer to Frederick Macmillan, 7 November 1918, MP:BL, MSS 55218.169.

70. On Armstrong, see Brock, ed., *H.E. Armstrong*.

71. On the British Science Guild, see MacLeod, "Science for Imperial Efficiency"; Meadows, *Science and Controversy*, 273; Turner, "Public Science in Britain." British Science Guild meetings were occasionally written up in *Nature*; see, e.g., "The British Science Guild," *Nature* 70 (11 August 1904): 343-344; "The British Science Guild," *Nature* 73 (2 November 1905): 10-13.

72. On the "neglect of science" debate, see Mayer, "Reluctant Technocrats."

73. Richard Gregory, *Discovery; or, The Spirit and Service of Science* (London: Macmillan, 1926 [1916]), 12-13.

74. Gregory, *Discovery*, 13-14.

75. R. A. Gregory to Maurice Macmillan, 28 September 1918, MP:UR, 75/126.

76. Richard Gregory, "The Promotion of Research," *Nature* 104 (6 November 1919): 259-261.

77. Gregory, like Lockyer, did not write all of the editorials that were printed during his editorial tenure. See Gary Werskey, "*Nature* and Politics between the Wars," *Nature* 224 (1 November 1969): 462-472.

78. The wartime years are omitted from this analysis because the skyrocketing price of paper forced *Nature* to cut its number of pages per issue.

79. E.g., "The 'Death Ray': Sir Richard Gregory's Criticism," *Times*, 29 May 1924; "No New Rays: Sir Richard Gregory on the Possibilities," *Manchester Guardian*, 1 June 1924; "Isaac Newton: Sir Richard Gregory's Tribute," *Times*, 23 March 1927.

80. E.g., "Discoveries that Cost Jobs: Why Labour Is Suspicious of Science; Abuse of Knowledge; Sir Richard Gregory's Defense," *Manchester Guardian*, 8 September 1930; "The Riches of Science: Their Use and Abuse, Sir Richard Gregory's Address," *Times*, 1 January 1931; "Long Editorship of '*Nature*': Dinner to Sir Richard Gregory," *Times*, 17 November 1938.

81. See Gerald W. Wollaston to Sir Richard Gregory, 12 February 1931, Sir Richard Gregory Papers, Addition from the University of Sheffield, Special Collections, University of Sussex Library, Falmer (hereafter SRGP), 2/5.

82. "Statutes of the Royal Society," SRGP, 2/6.

83. R. Ruggles Gates to R. A. Gregory, 19 May 1933, SRGP:SA, 8/2.

CHAPTER FOUR

1. On Rutherford's correspondence leading up to the Brussels conference, see Wilson, *Rutherford: Simple Genius*, 252-254.

2. Ultimately, despite his efforts to cultivate his colleagues, Rutherford's proposals were not adopted in full. Madame Curie argued that the curie should not be based on such an infinitesimal amount of radium. The congress decided that a curie would be the number of radioactive decays per second in one gram of radium, or 3.7×10^{10} decays per second. One curie is a very large amount of radiation; most scientists dealt with quantities more on the lines of a nanocurie (1×10^{-9} curie). In 1975, the International System of Units (Système international d'unités [SI]) changed the standard unit of radiation from the curie to the becquerel. One becquerel is equal to one decay per second, and one curie is equal to 3.7×10^{10} becquerels.

3. For more on networks of communication and research in the field of radioactivity, see Hughes, "The Radioactivists."

4. There are a few examples of non-British scientists writing pieces specifically for *Nature* as opposed to pieces from foreign journals translated by *Nature* staff or contributors. Many of them were cowritten with or communicated by a British contributor. See H. Helmholtz, "Rayleigh's 'Theory of Sound,'" *Nature* 17 (24 January 1878): 237-239; Hermann L. F. Helmholtz, "Mathematical and Physical Papers," *Nature* 32 (14 May 1885): 25-28;Wilhelm Ostwald, "Scientific Education in Germany and England," *Nature* 54 (27 August 1896): 405-406; John Tyndall and Louis Pasteur, "Prof. Tyndall on Germs," *Nature* 13 (17 February 1876): 305-306.

5. On x-rays, see Dibner, *Wilhelm Conrad Röntgen*; Glasser, *Wilhelm Conrad Röntgen*.

6. On the Curies, see Brian, *Curies: A Biography*; Boudia, *Marie Curie et son laboratoire*; Ogilvie, *Marie Curie: A Biography*; Quinn, *Marie Curie: A Life*.

7. On Soddy, see Kauffman, *Frederick Soddy*; Merricks, *World Made New*; Morrison, *Modern Alchemy*.

8. Ernest Rutherford and Frederick Soddy, "Radioactive Change," *Philosophical Magazine* 5 (1903): 576-591.

9. On Mendeleev's reaction to the "alchemy" of atomic transmutation, see Gordin, *Well-Ordered Thing*, chap. 8.

10. See, e.g., J. J. Thomson, "Radium," *Nature* 67 (30 April 1903): 601–602.

11. "Notes," *Nature* 53 (16 January 1896): 253.

12. W. C. Röntgen, "On a New Kind of Rays," trans. Arthur Stanton, *Nature* 53 (23 January 1896): 274–276.

13. A. A. C. Swinton, "Professor Röntgen's Discovery," *Nature* 53 (23 January 1896): 276–277.

14. For examples of *Chemical News* articles on Röntgen rays, see "Professor Röntgen's New Discovery," *Chemical News* 73 (31 January 1896): 49; W. Ackroyd, "Action of the Metals and Their Salts on the Ordinary and on the Röntgen Rays," *Chemical News* 74 (20 November 1896): 257; Jean Perrin, "Certain Properties of Röntgen's Rays," *Chemical News* 73 (7 February 1896): 61; John Waddell, "The Permeability of Various Elements to the Röntgen Rays," *Chemical News* 74 (18 December 1896): 298–299.

15. On the nature of the rays, see, e.g., Oliver Lodge, "On the Present Hypotheses Concerning the Nature of Röntgen's Rays," *Electrician* 36 (7 February 1896): 471–473. For an original research article, see, e.g., John Burke, "Some Experiments with Röntgen Rays," *Electrician* 37 (17 July 1896): 373–375.

16. E.g., Henry A. Rowland, N. R. Charmichael, and L. J. Briggs, "Notes of Observations on the Röntgen Rays," *Philosophical Magazine*, 5th ser., 41 (April 1896): 381–382; Franz Streinitz, "On an Electrochemical Action of the Röntgen Rays on Silver Bromide," *Philosophical Magazine*, 5th ser., 41 (May 1896): 462–463; R. W. Wood, "Note on 'Focus Tubes' for Producing X-rays," *Philosophical Magazine*, 5th ser., 41 (April 1896): 382–383.

17. For mentions of Becquerel's work, see, e.g., "Societies and Academies," *Nature* 53 (5 March 1896): 430–432; "Recent Work with Röntgen Rays," *Nature* 53 (2 April 1896): 522–524; "The Röntgen Rays," *Nature* 54 (30 July 1896): 302–306. For Thomson's article, see J. J. Thomson, "The Röntgen Rays," *Nature* 53 (23 April 1896): 581–583.

18. "Notes: Röntgeniana," *Electrician* 38 (4 December 1896): 173–174.

19. Henri Becquerel, "On the Invisible Radiations Emitted by the Salts of Uranium," *Chemical News* 73 (10 April 1896): 167–168; "On the Different Properties of the Invisible Radiations Emitted by Uranium Salts and the Radiation of the Antikathodic Wall of a Crookes Tube," *Chemical News* 73 (24 April 1896): 189–190; "Emission of New Radiations by Metallic Uranium," *Chemical News* 73 (26 June 1896): 295.

20. Badash, "Radium, Radioactivity," 145.

21. Nye, *Before Big Science*, 151.

22. On radioactivity and radium in the popular press, see Badash, *Radioactivity in America*, 19. See also Lavine, "Cultural History of Radiation," 29–90.

23. For mentions of the Curies and their work on radioactivity in *Nature* before 1900, see, e.g., "Societies and Academies," *Nature* 57 (21 April 1898): 599–600; "The British Association," *Nature* 58 (8 September 1898): 436–460; "Notes," *Nature* 60 (25 May 1899): 84–88; "Societies and Academies," *Nature* 60 (26 October 1899): 635–636.

24. For mentions of Marie Curie's work on the magnetic properties of steel, see, e.g., "Notes," *Nature* 51 (21 February 1895): 392–395; "Notes," *Nature* 55 (17 December 1896): 159–162; "Societies and Academies," *Nature* 56 (24 June 1897): 189–192; David K. Morris, "The Magnetic Properties and Electrical Resistance of Iron at High Temperatures," *Nature* 57 (6 January 1898): 232–234; W. C. Roberts-Austen, "Micrographic Analysis," *Nature* 52 (15 August 1895): 367–369.

25. Pierre Curie and Marie Curie, "Chemical Effects Produced by Becquerel's Rays," *Philosophical Magazine*, 5th ser., 49 (February 1900): 242–244.

26. P. Curie and Mdme. P. Curie, "Radio-Activity Due to Becquerel Rays," *Chemical News* 80 (8 December 1899): 269; Mdme. Sklowdowska Curie, "Radio-Active Substances," *Chemical News* 88 (1903): 85–86, 97–99, 134–135, 145–147, 159–160, 169–171, 175–177, 187–188, 199–201, 211–212, 223–224, 235–236, 247–249, 259–261, 271–272.

27. "Notes," *Electrician* 48 (14 March 1902): 803.

28. On Marie Curie's publishing patterns after Pierre's death, see Davis, "Research School of Marie Curie."

29. For more on the French academic community, see Nye, "N-rays"; Pestre, "Moral and Political Economy"; von Gizycki, "Centre and Periphery."

30. On Rutherford's New Zealand background, see Badash, "Influence of New Zealand."

31. On Rutherford, see Badash, *Ernest Rutherford and Theoretical Physics*; Bunge and Shea, *Rutherford and Physics*; Wilson, *Rutherford: Simple Genius*.

32. Ernest Rutherford to Mary Newton, 22 April 1898, as printed in Eve, *Rutherford*, 50.

33. On the Macdonald Physics Laboratory, see Pyenson, "Incomplete Transmission." See also Badash, "Origins of Big Science"; Heilbron, "Physics at McGill."

34. M. et Mme. P. Curie, "Sur la radioactivité provoquée par les rayons de Becquerel," *Comptes Rendus* 129 (20 November 1899): 714-716.

35. Ernest Rutherford, "On Radioactivity Produced in Substances by the Action of Thorium Compounds," *Philosophical Magazine*, 5th ser., 49 (February 1900): 161-192.

36. Wilson, *Rutherford: Simple Genius*, 142.

37. Rutherford, quoted in Wilson, *Rutherford: Simple Genius*, 147.

38. On Rutherford's favored publications, see Badash, "Origins of Big Science," 32-33.

39. E.g., E. Rutherford, "Emanations from Radio-Active Substances," *Nature* 64 (13 June 1901): 157-158; "The Amount of Emanation and Helium from Radium," *Nature* 68 (20 August 1903): 366-367; "Slow Transformation Products of Radium," *Nature* 71 (9 February 1905): 341-342; "Production of Radium from Actinium," *Nature* 75 (17 January 1907): 270-271.

40. E. Rutherford and F. Soddy, "Radio-Activity of Thorium Compounds," *Chemical News* 85 (1902): 261, 271-272, 282-285, 293-295, 304-308; 86 (1902): 97-101, 132-135, 169-170. Rutherford did coauthor a *Chemical News* article with Boltwood in 1905 announcing Boltwood's forthcoming paper in the *American Journal of Science*: E. Rutherford and B. B. Boltwood, "Relative Proportion of Radium and Uranium in Radio-Active Minerals," *Chemical News* 92 (28 July 1905): 38-39.

41. In 1897, J. J. Thomson sent the *Electrician* a shorter version of his protégé's forthcoming paper in the *Philosophical Magazine*. E. Rutherford, "On the Electrification of Gases Exposed to Röntgen Rays, and the Absorption of Röntgen Radiation by Gases and Vapours," *Electrician* 38 (23 April 1897): 865-868.

42. Kjærgaard, "'Within the Bounds of Science,'" 212-217.

43. Ernest Rutherford to J. J. Thomson, 26 March 1901, printed in Eve, *Rutherford*, 77.

44. Ernest Rutherford to Martha Thompson Rutherford, 17 December 1906, printed in Eve, *Rutherford*, 148.

45. See William Dawson to Norman Lockyer, 10 May 1898, NLP, MSS 110; Ernest Rutherford to J. Norman Lockyer, 31 October 1906, NLP, MSS 110.

46. Wilson, *Rutherford: Simple Genius*, 240.

47. Badash, "Origins of Big Science," 33.

48. E.g., M. L. Oliphant, P. Harteck, and E. Rutherford, "Transmutation Effects Observed with Heavy Hydrogen," *Nature* 133 (17 March 1934): 413; E. Rutherford, "The Boiling Point of the Radium Emanation," *Nature* 79 (18 February 1909): 457-458; E. Rutherford and J. Chadwick, "The Bombardment of Elements by α-Particles," *Nature* 113 (29 March 1924): 457.

49. For the full correspondence, see Badash, *Rutherford and Boltwood*.

50. On Boltwood and American radioactivity research more generally, see Badash, *Radioactivity in America*.

51. Bertram Boltwood to Ernest Rutherford, 11 April 1905, printed in Badash, *Rutherford and Boltwood*, 60. The *Nature* article to which Boltwood refers is "Societies and Academies," *Nature* 71 (13 April 1905): 574, which includes an abstract of Otto Hahn's preliminary communication to the Royal Society, "A New Radio-Active Element, Which Evolves Thorium Emanation."

52. Bertram Boltwood to Ernest Rutherford, 7 November 1906, printed in Badash, *Rutherford and Boltwood*, 142-143. The communications Boltwood mentions in his letter were both printed; see Bertram Boltwood, "The Production of Radium from Actinium," *Nature* 75 (15 November 1906): 54; "Note on the Production of Radium by Actinium," *American Journal of Science* 22 (December 1906): 537-538.

53. Ernest Rutherford to Bertram Boltwood, 20 June 1904, printed in Badash, *Rutherford and Boltwood*, 32. The *Nature* letter Rutherford mentions is Frederick Soddy, "The Life-History of Radium," *Nature* 70 (12 May 1904): 30.

54. Ernest Rutherford to Bertram Boltwood, 14 October 1906, as printed in Badash, *Rutherford and Boltwood*, 139. For the *Nature* letter, see Ernest Rutherford, "Absorption of the Radio-Active Emanations by Charcoal," *Nature* 74 (25 October 1906): 634.

55. For post-1908 references to *Nature* in the Boltwood-Rutherford correspondence, see Badash, *Rutherford and Boltwood*, 182, 212-213, 224, 227-228, 257, 264-265, 282, 311-312, 343, 347-348.

56. A radioactive element, such as thorium, will slowly transmute into other elements as it loses alpha particles. Rutherford and Boltwood were trying to determine which elements scientists should expect to encounter when a radioactive element decayed and in which sequence those new elements would appear during the decay process.

57. Ernest Rutherford to Bertram Boltwood, 28 July 1907, as printed in Badash, *Rutherford and Boltwood*, 159-160. For the exchange in *Nature* between Boltwood and Soddy, see Frederick Soddy, "The Origin of Radium," *Nature* 76 (13 June 1907): 150; Bertram Boltwood, "The Origin of Radium," *Nature* 76 (25 July 1907): 293.

58. Bertram Boltwood to Ernest Rutherford, 2 August 1907, as printed in Badash, *Rutherford and Boltwood*, 162-164.

59. Otto Hahn, "A New Product of Actinium," *Nature* 73 (12 April 1906): 559-560; "The Origin of Radium," *Nature* 77 (14 November 1907): 30-31; Otto Hahn and Otto von Baeyer, "Magnetic Deflection of β Rays," *Nature* 83 (26 May 1910): 369.

60. Hahn, quoted in Wilson, *Rutherford: Simple Genius*, 242.

61. M. Berthelot to Norman Lockyer, 13 October 1879, NLP, MSS 110. Notably, Berthelot was writing to Lockyer to ask for aid in helping his recent paper reach the Royal Society, not *Nature*.

62. Norman Lockyer, "The Advancement of Natural Knowledge," *Nature* 65 (30 January 1902): 289-291; William Abney, "Science and the State," *Nature* 71 (24 November 1904): 90-91.

63. Editor, "The Death of the Queen," *Nature* 63 (24 January 1901): 293; "The Death of the King," *Nature* 83 (12 May 1910): 301.

64. E.g., see "Professor Ostwald on English and German Science," *Nature* 54 (27 August 1896): 385-386; Henry Dyer, "Education and National Efficiency in Japan," *Nature* 71 (15 December 1904): 150-151; H. R. Reichel, "Some Characteristics of American Universities," *Nature* 73 (9 November 1905): 44-46; W. A. C., "Some German Public Laboratories," *Nature* 70 (26 May 1904): 83.

65. When Rutherford moved to Manchester in 1908, Eve took over as the head of the Macdonald Physics Laboratory.

66. A. S. Eve, "Some Scientific Centres. VIII. The Macdonald Physics Building, McGill University, Montreal," *Nature* 74 (19 July 1906): 272.

67. Otto Hahn, quoted in Eve, *Rutherford*, 143.

68. Carl Correns, "G. Mendel's Regel über das Verhalten der Nachkommenschaft der Rassenbastarde," *Berichte der deutschen botanischen Gesellschaft* 18 (1900): 158-168; Hugo de Vries, "Das Spaltungsgesetz der Bastarde," *Berichte der deutschen botanischen Gesellschaft* 18 (1900): 83-90; Eric von Tschermak, "Über künstliche Kreuzung bei Pisum sativum," *Berichte der deutschen botanischen Gesellschaft* 18 (1900): 232-239. For Mendel's original papers, see Mendel, *Experiments in Plant Hybridisation*.

69. On Bateson, see Bateson, *William Bateson*; Darden, "William Bateson"; Olby, "William Bateson's Introduction of Mendelism"; Richmond, "Women in the Early History of Genetics."

70. On Weldon, see Magnello, "Karl Pearson's Gresham Lectures"; "Karl Pearson's Mathematization of Inheritance."

71. See W. T. Thiselton-Dyer, "Origin of the Cultivated Cineraria," *Nature* 52 (2 May 1895): 3-4; W. Bateson, "The Origin of the Cultivated Cineraria," *Nature* 52 (9 May 1895): 29; W. F. R. Weldon, "The Origin of the Cultivated Cineraria," *Nature* 52 (16 May 1895): 54.

72. On the debate between Bateson and the biometricians, see Farrall, "Controversy and Conflict"; Kevles, "Genetics in the United States and Great Britain"; Robert Olby, "Dimensions of Scientific Controversy."

73. Kevles, "Genetics in the United States and Great Britain," 448.

74. Ibid. Historian Jan Sapp has similarly described the rise of "genetics," based on the Mendelian theory of heredity, as the outcome of "a struggle for power and authority" between competing theories of heredity. See Sapp, "Struggle for Authority."

75. W. Bateson, "The Alleged 'Aggressive Mimicry' of *Vollucellae*," *Nature* 46 (20 October 1892): 585; "The Alleged 'Aggressive Mimicry' of *Vollucellae*," *Nature* 47 (24 November 1892): 77-78. For the original review and Poulton's response, see Edward B. Poulton, "Natural Selection and Alternative Hypotheses," *Nature* 46 (6 October 1892): 533-537; Poulton, "The *Vollucellae* as Examples of Aggressive Mimicry," *Nature* 47 (10 November 1892): 28-30; "The *Vollucellae* as Alleged Examples of Variation 'Almost Unique among Animals,'" *Nature* 47 (8 December 1892): 126-127.

76. W. Bateson, "Mendel's Principles of Heredity in Mice," *Nature* 67 (19 March 1903): 462-463.

77. W. F. R. Weldon, "Mendel's Principles of Heredity in Mice," *Nature* 67 (2 April 1903): 512.

78. W. Bateson, "Mendel's Principles of Heredity in Mice," *Nature* 67 (23 April 1903): 585-586; W. F. R. Weldon, "Mendel's Principles of Heredity in Mice," *Nature* 67 (30 April 1903): 610.

79. G. Archdall Reid, "The Interpretation of Mendelian Phenomena," *Nature* 76 (3 October 1907): 566.

80. Articles in this discussion include R. H. Lock, "The Interpretation of Mendelian Phenomena," *Nature* 76 (17 October 1907): 616; G. Archdall Reid, "The Interpretation of Mendelian Phenomena," *Nature* 76 (31 October 1907): 662-663; J. T. Cunningham, "The Interpretation of Mendelian Phenomena," *Nature* 77 (21 November 1907): 54; W. T. Thiselton-Dyer, "Specific Stability and Mutation," *Nature* 77 (28 November 1907): 77-79.

81. There are many excellent scholarly works on genetics in different national contexts in the early twentieth century, e.g., Adams, *Wellborn Science*; Harwood, *Styles of Scientific Thought*; Kevles, *In the Name of Eugenics*.

82. On the growth in international congresses, see Crawford, *Nationalism and Internationalism in Science*, 35-41; Everett-Lane, "International Scientific Congresses." On scientific internationalism more generally, see Crawford, Shinn, and Sörlin, *Denationalizing Science*.

83. Everett-Lane, "International Scientific Congresses," 5-11.

84. See Jungnickel and McCormmach, *Intellectual Mastery of Nature*, 1:37.

85. Johann Christian Poggendorff, quoted in ibid., 1:38.

86. Karl Scheel, quoted in ibid., 1:34.

87. See "Jacques Danne (1882-1919)," *Le Radium* 11 (May 1919): 193-194.

88. Examples of articles by foreign physicists in *Le Radium* include Bertram Boltwood, "Sur les quantités relatives de radium et d'uranium contenus dans quelques minéraux," *Le Radium* 1 (15 August 1904): 45-48; E. Rutherford and O. Hahn, "Masse et vitesse des particules α émises par le radium et l'actinium," *Le Radium* 3 (November 1906): 321-326; F. Soddy, "La table périodique des elements," *Le Radium* 11 (January 1914): 6-8.

89. These articles contain a brief postscript note indicating the name of the person who translated it into French. Frequent translators include Léon Bloch, A. Laborde, P. Razet, and Gaston Danne (Jacques Danne's younger brother).

90. The members of *Le Radium*'s Comité de Direction were Jacques-Arsène d'Arsonval, Henri Becquerel, Antoine Béclère, René Blondlot, Charles Bouchard, Pierre Curie, Jean Danysz, Andre

Debierne, Charles Féry, Niels Finsen, Charles Edouard Guillaume, Paul Marie Oudin, Heinrich Rubens, and Ernest Rutherford. Notably, Marie Curie's name was not on this list, even after Pierre's death in 1906.

91. See, e.g., Crawford, "Universe of International Science"; Schroeder-Gudehus, "Nationalism and Internationalism."

92. Crawford, *Nationalism and Internationalism in Science*, 44.

CHAPTER FIVE

1. Bernhard Rust, untitled decree of 12 November 1937, Assets of the Zoologist Victor Frenz, Archives of the Ernst Haeckel House, Jena. Available online at http://www.nature.com/nature/history/pdf/essays/greg-1.pdf.

2. Henry E. Roscoe, "Science Education in Germany I," *Nature* 1 (9 December 1869): 157-159; "Science Education in Germany II," *Nature* 1 (10 March 1870): 475-477. Note that this is before German unification; Roscoe's "Germany" is the German-speaking lands, including Austria.

3. S. Stricker, "The Medical Schools of England and Germany I," *Nature* 2 (1 September 1870): 349-350; "The Medical Schools of England and Germany II," *Nature* 2 (8 September 1870): 369; "Medical Schools in England and Germany III," *Nature* 2 (1 December 1870): 81-82.

4. E.g., "The Dilution of Dog Poison," *Nature* 30 (29 May 1884): 97-98; "Pasteur's Researches," *Nature* 34 (17 June 1886): 144-145. For more on Pasteur in *Nature*, see "Hydrophobia," *Nature* 33 (5 November 1885): 3; "The Opening of the Pasteur Institute," *Nature* 39 (22 November 1888): 74. On Pasteur, see Geison, *Private Science*.

5. William Ramsay, "Scientific Education in Germany and England," *Times*, 25 August 1896.

6. "Professor Ostwald on English and German Science," *Nature* 54 (27 August 1896): 385-386; W. Ostwald, "Scientific Education in Germany and England," *Nature* 54 (27 August 1896): 405-406.

7. On British-German relations during the reign of Kaiser Wilhelm II, see McLean, "Kaiser Wilhelm II."

8. R. A. Gregory, "Primary Education and Beyond," *Nature* 93 (16 April 1914): 173-175.

9. [Norman Lockyer], "The War—And After," *Nature* 94 (10 September 1914): 29-30. The editorial is unsigned, but Roy M. MacLeod identifies Lockyer as its author. See MacLeod, "Is It Safe to Look Back?," 459.

10. On British feelings toward Germany during the war, see Badash, "British and American Views."

11. W. Ramsay, "Germany's Aims and Ambitions," *Nature* 94 (8 October 1914): 137-139.

12. On Anglo-German scientific relations during and after the First World War, see Kevles, "'Into Hostile Political Camps'"; Schroeder-Gudehus, "Challenge to Transnational Loyalties."

13. Badash, "British and American Views," 96.

14. Ernest Rutherford to Bertram Boltwood, 14 September 1915, in Badash, *Rutherford and Boltwood*, 311-315. Rutherford, who was among Schuster's supporters, told Boltwood that the proposed boycott had failed to have any effect on attendance at that year's BA meeting but that Schuster's voice was a little weak owing to his emotions over his son's injury.

15. Badash, "British and American Views," 96-98.

16. Ernest Rutherford, quoted in Crawford, "Internationalism in Science," 174.

17. On Moseley, see Heilbron, *H.G.J. Moseley*.

18. Moseley's missing elements were those with atomic numbers 43, 61, 72, 75, 85, 87, and 91. All of these elements would be discovered by 1945.

19. Rutherford also blamed the British army's stubbornness in assigning men like Moseley to the front lines instead of special scientific research units. See E. Rutherford, "Henry Gwyn Jeffreys Moseley," *Nature* 96 (9 September 1915): 33-34.

20. On the Rutherford-Boltwood relationship, see Badash, *Rutherford and Boltwood*, 1–23.

21. "The War," *Nature* 94 (14 January 1915): 527–528.

22. Lord Walsingham, "German Naturalists and Nomenclature," *Nature* 102 (5 September 1918): 4.

23. See "Notes," *Nature* 102 (17 October 1918): 129.

24. H. H. Godwin-Austen, "Future Treatment of German Scientific Men," *Nature* 102 (26 September 1918): 64–65.

25. See, e.g., "International Organisation in Science," *Nature* 104 (23 October 1919): 154–155; E. Ray Lankester, "International Relations in Science," *Nature* 104 (30 October 1919): 172. *Nature* contributors' lingering anti-German feelings following the war were not unusual in Britain. On Britain in the interwar period, see Hynes, *War Imagined*; Kent, *Aftershocks*; Winter, *Great War*.

26. See, e.g., D'Arcy W. Thompson, "International Relations in Science," *Nature* 104 (23 October 1919): 154; "International Science," *Nature* 116 (4 July 1925): 1–3.

27. For examples of material about relativity in *Nature*, see E. Cunningham, "Einstein's Relativity Theory of Gravitation I," *Nature* 104 (4 December 1919): 354–356; A. S. Eddington, "Gravitation and the Principle of Relativity," *Nature* 98 (28 December 1916): 328–330; Joseph Larmor, "Gravitation and Light," *Nature* 104 (25 December 1919): 412; Oliver J. Lodge, "Gravitation and Light," *Nature* 104 (11 December 1919): 372; Arthur Schuster, "The Deflection of Light during a Solar Eclipse," *Nature* 104 (8 January 1920): 468. On the acceptance of relativity theory in Great Britain, see Sponsel, "Constructing a 'Revolution in Science'"; Stanley, "'An Expedition to Heal the Wounds of War'"; Warwick, *Masters of Theory*, chap. 9.

28. A. Einstein, "A Brief Outline of the Development of the Theory of Relativity," trans. Robert W. Lawson, *Nature* 106 (17 February 1921): 782–784; H. A. Lorentz, "The Michelson-Morley Experiment and the Dimensions of Moving Bodies," *Nature* 106 (17 February 1921): 793–795; H. Weyl, "Electricity and Gravitation," trans. Robert W. Lawson, *Nature* 106 (17 February 1921): 800–802.

29. Bohr's model of the atom built off Rutherford's nuclear model of the atom. Using Johann Jakob Balmer's work on the spectral lines in hydrogen, Bohr suggested that atoms contained a positive nucleus surrounded by electrons orbiting the nucleus like planets orbiting the sun. These orbitals occurred at specific discrete distances from the nucleus, and the angular momentum of each electron was equal to an integer multiple of the reduced Planck constant ($h/2\pi$). Bohr suggested that Balmer's spectral lines corresponded to electrons moving between orbitals, releasing energy in the process. The Bohr atom is still taught to students as the basic model of atomic structure, although it is now considered more an approximation than a perfectly accurate model.

30. Niels Bohr, "The Spectra of Hydrogen and Helium," *Nature* 95 (4 March 1915): 6–7. The signature on the letter indicates that it was submitted from Manchester.

31. Rutherford died unexpectedly following surgery for a minor hernia at the age of sixty-six. For a helpful summary of the Bohr-Rutherford correspondence, see Peierls, "Rutherford and Bohr," 229–241.

32. For examples of letters to the editor from Copenhagen, see O. R. Frisch and E. T. Sørensen, "Velocity of Slow Neutrons," *Nature* 136 (17 August 1935): 258; O. R. Frisch, H. von Halban Jr., and Jørgen Koch, "A Method of Measuring the Magnetic Moment of Free Neutrons," *Nature* 139 (1 May 1937): 756–757; "Capture of Slow Neutrons in Light Elements," *Nature* 140 (20 November 1937): 895; G. Gamow, "The Quantum Theory of Nuclear Disintegration," *Nature* 122 (24 November 1928): 805–806; S. Rosseland, "Origin of Radioactive Disintegration," *Nature* 111 (17 March 1923): 357.

33. D. Coster and G. Hevesy, "On the Missing Element of Atomic Number 72," *Nature* 111 (20 January 1923): 79. The French scientists Georges Urbain and Alexandre Dauvillier briefly challenged the Coster-Hevesy priority claim; Urbain and Dauvillier claimed they had discovered element 72 first and had named it "celtium." See Kragh, "Anatomy of a Priority Conflict."

34. Irène Curie and Frederic Joliot, "Effet d'absorption de rayons γ très haute fréquence par projection de noyaux légers," *Comptes Rendus* 194 (1932): 708–711.

35. J. Chadwick, "Possible Existence of a Neutron," *Nature* 129 (27 February 1932): 312.

36. There were a vast number of letters about neutron research published in *Nature* during the 1930s. For examples of letters involving research on neutrons and neutron-induced radioactivity, see Irène Curie and F. Joliot, "New Evidence for the Neutron," *Nature* 130 (9 July 1932): 57; W. Ehrenberg and Hu Chien Shan, "Absorption of Slow Neutrons," *Nature* 135 (15 June 1935): 993-994; D. Iwanenko, "The Neutron Hypothesis," *Nature* 129 (28 May 1932): 798; Leo Szilard and T. A. Chalmers, "Radioactivity Induced by Neutrons," *Nature* 135 (19 January 1935): 98-99.

37. E.g., E. Fermi, "Quantum Mechanics and the Magnetic Moment of Atoms," *Nature* 118 (18 December 1926): 876; "Magnetic Moments of Atomic Nuclei," *Nature* 125 (4 January 1930): 16; "Radioactivity Induced by Neutron Bombardment," *Nature* 133 (19 May 1933): 757.

38. See, e.g., Segrè, *Enrico Fermi, Physicist*, 71-72. The paper was eventually published in Italian as Enrico Fermi, "Tentativo di una teoria dei raggi β," *La Ricerca Scientifica* 4 (1933): 491-495.

39. E.g., Hiroo Aoki, "Gamma Ray Excitation by Fast Neutrons," *Nature* 139 (27 February 1937): 372-373; Seishi Kikuchi, Hiroo Aoki, and Kodi Husimi, "Emission of Beta-Rays from Substances bombarded with Neutrons," *Nature* 138 (14 November 1936): 841; Itaru Nonaka, "Resonance Capture of Slow Neutrons and Emission of Gamma-Rays," *Nature* 144 (11 November 1939): 831-832.

40. For examples of international letters on genetics, see M. José Capinpin, "Chromosome Behaviour of Triploid Œnothera," *Nature* 126 (27 September 1930): 469-470; Dontcho Kostoff, "Effect of the Fungicide 'Granosan' on Atypical Growth and Chromosome Doubling in Plants," *Nature* 144 (19 August 1939): 334; B. N. Sidorov, N. N. Sokolov, and I. E. Trofimov, "Forces of Attraction of Homologous Loci and Chromosome Conjugation," *Nature* 136 (20 July 1935): 108-109.

41. News and Views, *Nature* 133 (10 February 1934): 203.

42. For more information on science in Asia in the early twentieth century, see Bartholomew, *Formation of Science in Japan*; "Science in Twentieth-Century Japan"; Fan, "Redrawing the Map"; Needham, *Grand Titration*.

43. News and Views, *Nature* 133 (14 April 1934): 558; 134 (19 January 1935): 94; 137 (22 February 1936): 306. The journal also began printing fifty-word summaries of the week's letters at the end of the column.

44. News and Views, *Nature* 137 (22 February 1936): 306.

45. Otto Hahn and Fritz Strassmann, "Über den Nachweis und das Verhalten der bei der Bestrahlung des Urans mittels Neutronen entstehenden Erdalkalimetalle," *Die Naturwissenschaften* 27 (1939): 11-15.

46. Lise Meitner and Otto Frisch, "Disintegration of Uranium by Neutrons: A New Type of Nuclear Reaction," *Nature* 143 (11 February 1939): 239; italics in the original.

47. Ibid.

48. Meitner's job in Stockholm proved disappointing. The salary was poor and her scientific equipment was out-of-date; furthermore, she and Manne Siegbahn, the director of the Nobel Institute, took an almost immediate dislike to each other, making an improvement in her professional conditions unlikely. See Sime, *Lise Meitner*, 210-230; Stueland Yruma, "How Experiments Are Remembered," 70-78, 195-196.

49. News and Views, *Nature* 142 (12 November 1938): 865.

50. Hahn and Strassman published their experimental findings in *Die Naturwissenschaften* on January 6, 1939; see chap. 5, n. 45. On the Hahn-Meitner correspondence about Hahn and Strassman's results, see Stueland Yruma, "How Experiments Are Remembered," 78-85.

51. E.g., O. R. Frisch and E. T. Sørensen, "Velocity of Slow Neutrons," *Nature* 136 (17 August 1935): 258; O. R. Frisch, H. von Halban Jr., and Jørgen Koch, "A Method of Measuring the Magnetic Moment of Free Neutrons," *Nature* 139 (1 May 1937): 756-757; "Capture of Slow Neutrons in Light Elements," *Nature* 140 (20 November 1937): 895.

52. Stueland Yruma, "How Experiments Are Remembered," 120-129.
53. N. Feather, "Fission of Heavy Nuclei: A New Type of Nuclear Disintegration," *Nature* 143 (27 May 1939): 877-879.
54. For Feather's communication of Hahn's letter, see N. Feather, "Fission of Heavy Nuclei," *Nature* 143 (17 June 1939): 1027.
55. "Tribute of Science to the Royal Jubilee," *Nature* 135 (4 May 1935): 669-670; "The Death of His Majesty King George V," *Nature* 137 (25 January 1936): 123-125.
56. On Richard Gregory and *Discovery*, see chap. 3.
57. Boris Sokoloff, "The 'Proletarisation of Science' in Russia," *Nature* 108 (1 September 1921): 20-22. Sokoloff's letter drew a reply from the British Marxist H. Lyster Jameson, who accused Sokoloff of being "more concerned with spreading propaganda against 'Bolshevism' than with giving an outline of all the features, good or bad, of the attitude of Bolshevism to science," and he argued that the Soviet government had made great strides in improving scientific education in Russia. H. Lyster Jameson, "The 'Proletarisation of Science' in Russia," *Nature* 108 (29 September 1921): 147.
58. J. W. Mellor, "Communism and Science," *Nature* 108 (22 September 1921): 113-114.
59. Bohuslav Brauner, "The 'Proletarisation of Science' in Russia," *Nature* 108 (17 November 1921): 367-368.
60. V. Korenchevsky, "Scientific Workers in Russia," *Nature* 108 (8 December 1921): 469.
61. R. A. Gregory and C. Hagberg Wright, "Scientific Literature for Russia," *Nature* 109 (16 February 1922): 208.
62. On Kapitza, see Badash, *Kapitza, Rutherford, and the Kremlin*; Boag, Rubinin, and Schoenberg, *Kapitza in Cambridge and Moscow*; Kozhevnikov, "Piotr Kapitza."
63. News and Views, *Nature* 135 (4 May 1935): 755. For further discussion of Kapitza's detainment, see News and Views, *Nature* 136 (23 November 1935): 825.
64. A great deal of scholarship has been written on Lysenkoism. See Rossianov, "Editing nature"; Soifer, Gruliow, and Gruliow, *Lysenko and the Tragedy of Soviet Science*; Weiner, "The Roots of 'Michurinism.'" On the response to Lysenko outside Russia, see Harman, "C. D. Darlington"; Krementsov, "'Second Front' in Soviet Genetics"; Teich, "Haldane and Lysenko Revisited"; Wolfe, "What Does It Mean?"
65. V. H. Blackman, "Light and Temperature and the Reproduction of Plants," *Nature* 137 (13 June 1936): 971-973.
66. News and Views, *Nature* 139 (23 January 1937): 142.
67. Although the 1937 reports of Vavilov's arrest were incorrect, Vavilov would be arrested in 1940. He died in a Soviet prison in January 1943.
68. News and Views, *Nature* 139 (30 January 1937): 185.
69. News and Views, *Nature* 139 (30 January 1937): 185.
70. "Genetics and Plant Breeding in the U.S.S.R," *Nature* 140 (21 August 1937): 296-297.
71. On Bryan, see Cherny, *Righteous Cause*; Koenig, *Bryan*.
72. "The Proscription of Darwinism," *Nature* 115 (4 April 1925): 485-486.
73. Useful sources on the Scopes trial include Gieryn, Bevins, and Zehr, "Professionalization of American Scientists"; Larson, *Summer for the Gods*; Lienesch, *In the Beginning*; Moran, *Scopes Trial*; Ryan, *Darwinism and Theology*.
74. "Evolution and Intellectual Freedom: Supplement to *Nature*," *Nature* 116 (11 July 1925): 70.
75. Rev. J. Scott Lidgett, "Evolution and Intellectual Freedom: Supplement to *Nature*," *Nature* 116 (11 July 1925): 82.
76. "Evolution and Intellectual Freedom," *Nature* 116 (18 July 1925): 102-105; Rev. Charles F. Armagh, "Science and Intellectual Freedom," *Nature* 116 (1 August 1925): 172; Henry E. Armstrong, "Science and Intellectual Freedom," *Nature* 116 (1 August 1925): 172.
77. J. McKeen Cattell, "Science and Intellectual Freedom," *Nature* 116 (5 September 1925): 358.

78. Beyerchen, *Scientists under Hitler*, 40. For more scholarship on science in Germany under National Socialism, see Harwood, "German Science"; Renneberg and Walker, *Science, Technology, and National Socialism*; Szöllösi-Janze, *Science in the Third Reich*; Walker, *German National Socialism and the Quest for Nuclear Power*.

79. News and Views, *Nature* 133 (20 January 1934): 96-97.

80. News and Views, *Nature* 136 (28 September 1935): 506.

81. News and Views, *Nature* 132 (5 August 1933): 198.

82. "Peace and War," *Nature* 133 (17 February 1934): 229-231.

83. "The Aryan Doctrine," *Nature* 134 (18 August 1934): 229-231.

84. News and Views, *Nature* 137 (4 April 1936): 570-571. Huxley suggested using the term *ethnic group* or *ethnos* to describe cultural and linguistic groups. See also the follow-up note in News and Views, *Nature* 137 (18 April 1936): 649, and a later editorial, "The Delusion of Race," *Nature* 137 (18 April 1936): 635-637.

85. "Intellectual Freedom," *Nature* 133 (24 February 1934): 269-272. The editorial also indicated that the "devotion to a political theory" in the USSR and the "loyalty to a leader who exercises a dictatorial power" in Italy were causes for concern.

86. News and Views, *Nature* 133 (24 February 1934): 284. A translation of the circular can be found in "The Teaching of History and Prehistory in Germany," *Nature* 133 (24 February 1934): 298-300.

87. See, e.g., News and Views, *Nature* 133 (23 June 1934): 941; 134 (13 October 1934): 564-565.

88. "Intellectual Freedom and the Progress of Science," *Nature* 137 (8 February 1936): 203-204.

89. "The Protection of Scientific Freedom," *Nature* 137 (13 June 1936): 963-964.

90. P. F. F., "The Ceremonial Dedication of the Philipp-Lenard-Institut at Heidelberg," *Nature* 137 (8 February 1936): 233. For the initial *Nature* coverage of the speeches at the opening of the Philipp Lenard Institute, see "Philipp-Lenard-Institut at Heidelberg," *Nature* 137 (18 January 1936): 93-94.

91. "Freedom of the Mind," *Nature* 139 (5 June 1937): 941-942.

92. Stone, *Responses to Nazism in Britain*.

93. See Werskey, *Visible College*.

94. On Stark, see Beyerchen, *Scientists under Hitler*, chap. 6; Walker, *Nazi Science*, chap. 2.

95. J. Stark, "International Status and Obligations of Science," *Nature* 133 (24 February 1934): 290.

96. J. Stark, "The Attitude of the German Government towards Science," *Nature* 133 (21 April 1934): 614.

97. R. Woltereck, "Science and Intellectual Liberty," *Nature* 134 (7 July 1934): 27-28.

98. A. V. Hill, "International Status and Obligations of Science," *Nature* 133 (24 February 1934): 290.

99. J. B. S. Haldane, "The Attitude of the German Government towards Science," *Nature* 133 (12 May 1933): 726.

100. Editor of *Nature*, "Science and Intellectual Liberty," *Nature* 134 (7 July 1934): 28.

101. The Sturmabteilung ("storm detachment"), also called the SA or "brownshirts," was a paramilitary organization that had played a crucial role in aiding Hitler's rise to power. By 1934, Hitler feared the group had become too powerful in its own right, and the Führer ordered a purge of the SA leadership. At least 85 men were executed in secret on 30 June 1934, and over a thousand were imprisoned for allegedly conspiring against Hitler and the Nazi government. The purge is known as the "Night of the Long Knives." See Evans, *Third Reich*, 39-41.

102. "Heidelberg, Spinoza and Academic Freedom," *Nature* 137 (22 February 1936): 303-304.

103. News and Views, *Nature* 137 (29 February 1936): 352.

104. News and Views, *Nature* 137 (7 March 1936): 394.
105. "University of Heidelberg and New Conceptions of Science," *Nature* 139 (16 January 1937): 98-100.
106. "Centenary of the University of Göttingen," *Nature* 139 (24 April 1937): 701-703. A British lecturer from Trinity College, Oxford, named J. D. Lambert wrote a response in the Letters to the Editor a month later. Lambert, a scientific visitor working in a Göttingen laboratory, emphasized that much good scientific work was still being done at Göttingen and urged his fellow Britons not to abandon contact with German scientists. J. D. Lambert, "The University of Göttingen," *Nature* 139 (29 May 1937): 930. The writer of the article responded in the same issue; see The Writer of the Article, "The University of Göttingen," *Nature* 139 (29 May 1937): 930.
107. Uwe Hossfeld and Lennart Olsson, "*Nature* and Hitler," *Nature*, accessed 1 January 2014, http://www.nature.com/nature/history/full/nature06242.html.
108. H. Rügemer, "Die 'Nature' ein Greuelzeitschrift," *Zeitschrift für die gesamte Naturwissenschaft* 3 (1938): 475-479.
109. In the absence of an editorial archive, we can only speculate on why the Nazi restrictions on *Nature* received so little coverage in the journal. It is possible that Gregory did not wish to fan the flames of the controversy; however, this seems unlikely given the unquestionably inflammatory nature of the articles by *Nature*'s anonymous correspondent in Germany. It seems more likely that Gregory and the *Nature* staff sought to create the impression that they were indifferent to or even contemptuous of Rust's decree.
110. News and Views, *Nature* 141 (22 January 1938): 151.
111. "Reich Ban of Paper Shocks the British," *New York Times* (30 January 1938): 28.

CHAPTER SIX

1. Richard Gregory to Daniel Macmillan, May 1938 (no day given), SRGP, 3/1. The letter appears to be a draft; it has no salutation and no signature. If it is a draft, it is not clear how close it is to the text of the actual letter Gregory sent Macmillan.
2. Richard Gregory to Daniel Macmillan, May 1938 (no day given), SRGP, 3/1.
3. There are several letters regarding Gale's appointment in the Macmillan Papers at Reading; see MP:UR, 75/30. In 1921, the appointment was increased to £300 per year for four days' work per week. Richard Gregory to Maurice Macmillan, 8 October 1921, MP:UR, 75/126. See also A. J. V. Gale to Sir Richard Gregory, 10 October 1921, MP:UR, 75/30.
4. For Brimble's letter accepting the position, see Lionel J. F. Brimble to Daniel Macmillan, 12 July 1930, MP:UR, 127/130.
5. A. J. V. Gale to Sir Richard Gregory, 18 November 1938, SRGP, 3/1.
6. [Rainald Brightman], "The Future of Civilization," *Nature* 144 (16 September 1939): 491.
7. E.g., "Reconstruction in Europe," *Nature* 145 (6 January 1940): 1-4; "Fundamentals in Post-War Reconstruction," *Nature* 146 (19 October 1940): 500-501; "Educational Reconstruction in Europe," *Nature* 152 (14 August 1943): 169-172.
8. E.g., "Colonial Development and Reconstruction," *Nature* 149 (4 April 1942): 365-368; "Development of Colonial Dependencies," *Nature* 151 (10 April 1943): 399-402; "Research and Colonial Development," *Nature* 154 (15 July 1944): 63-65.
9. On the dating and mailing of *Nature*, see News and Views, *Nature* 146 (14 September 1940): 361.
10. News and Views, *Nature* 145 (8 June 1940): 887.
11. News and Views, *Nature* 147 (21 June 1941): 771.
12. News and Views, "Announcements," *Nature* 145 (25 May 1940): 817.
13. *Nature* 147 (1 March 1941): 266. A notice at the top of the Letters to the Editor column

reads, "In the present circumstances, proofs of 'Letters' will not be submitted to correspondents outside Great Britain."

14. News and Views, *Nature* 150 (26 December 1942): 763.

15. See "Problems of Scientific Publication," *Nature* 157 (23 February 1946): 205-206; News and Views, *Nature* 158 (2 November 1946): 613.

16. See News and Views, *Nature* 159 (15 February 1947): 224; 159 (1 March 1947): 297.

17. News and Views, "Soviet Genetics: The Real Issue," *Nature* 165 (6 May 1950): 711.

18. On Brightman's work during the Gregory regime, see Gary Werskey, "*Nature* and Politics between the Wars," *Nature* 224 (1 November 1969): 462-472.

19. See, e.g., "Expansion of Scientific and Technical Education," *Nature* 183 (27 June 1959): 1763-1765; "Quality in Technical Education," *Nature* 181 (31 May 1958): 1491-1493; "Apprenticeship and Technical Education," *Nature* 191 (19 August 1961): 737-740.

20. [Rainald Brightman], "University Expansion in Britain," *Nature* 179 (22 June 1957): 1259-1261.

21. David Davies, interview by author, London, United Kingdom, 11 April 2012.

22. Mary Sheehan, interview by author, London, United Kingdom, 12 April 2012.

23. Davies, interview.

24. [L. J. F. Brimble], "'Letters to the Editor,'" *Nature* 165 (1 April 1950): 497-498.

25. Virginia E. Davis et al., "Riboflavin of Sow's Milk," *Nature* 165 (1 April 1950): 522-523.

26. For secondary sources on the research that led to the publication of the structure of DNA, see de Chadarevian, *Designs for Life*; Maddox, *Rosalind Franklin*; McElheny, *Watson and DNA*; Olby, *Path to the Double Helix*; Sayre, *Rosalind Franklin and DNA*.

27. Crick, *What Mad Pursuit*, 58-59.

28. Wilkins, *Third Man*, 197-198. Raymond Gosling's own recollection largely supports Wilkins's version of the story. Raymond Gosling, interview on *Nova*, "The Secret of Photo 51."

29. Watson, *Double Helix*, 98.

30. On the importance of the MRC report, see McElheny, *Watson and DNA*, 54-55. MRC biophysics committee member Max Perutz was responsible for giving Watson and Crick the report; however, as he later pointed out, it contained no data that Watson would not have obtained had he chosen to take notes at earlier colloquia given by Franklin at King's College London. See Perutz, "DNA Helix," 1537-1539. Perutz's article includes a partial reprint of the MRC report.

31. On Franklin's career after 1953, see Creager and Morgan, "After the Double Helix."

32. Watson, *Double Helix*, 124-128; Wilkins, *Third Man*, 216.

33. Maddox, *Rosalind Franklin*, 208-209.

34. Watson, *Double Helix*, 16.

35. See, e.g., J. A. V. Butler, L. A. Gilbert, and K. A. Smith, "Radiometric Action of Sulphur and Nitrogen 'Mustards' on Deoxyribonucleic Acid," *Nature* 165 (6 May 1950): 714-716; J. A. V. Butler and K. A. Smith, "Degradation of Deoxyribonucleic Acid by Free Radicals," *Nature* 165 (27 May 1950): 847-848; "Human Desoxypentose Nucleic Acid," *Nature* 165 (13 May 1950): 746-757; Waldo E. Cohn and Elliot Volkin, "Nucleoside-5'-Phosphates from Ribonucleic Acid," *Nature* 167 (24 March 1951): 483-484; E. Klein and G. Klein, "Nucleic Acid Content of Tumor Cells," *Nature* 166 (11 November 1950): 832-833; J. Pasteels and L. Lison, "Deoxyribonucleic Acid Content of the Egg of *Sabellaria* during Maturation and Fertilization," *Nature* 167 (9 June 1951): 948-949; B. R. Seshachar, "Deoxyribonucleic Acid Content of the Ciliate Micronucleus," *Nature* 165 (13 May 1950): 848.

36. F. H. C. Crick, "Is α-Keratin a Coiled Coil?" *Nature* 170 (22 November 1952): 882-883; W. Cochran and F. H. C. Crick, "Evidence for the Pauling-Corey α-Helix in Synthetic Polypeptides," *Nature* 169 (9 February 1952): 234-235; M. H. F. Wilkins, R. G. Gosling, W. E. Seeds, "Physical Studies of Nucleic Acid," *Nature* 167 (12 May 1951): 759-760.

37. On British molecular biology, see de Chadarevian, *Designs for Life*.

38. See, e.g., Lawrence Bragg, "Organisation and Work of the Cavendish Laboratory," *Nature* 161 (24 April 1948): 627-628; "Famous Experimental Apparatus in the Cavendish Laboratory, Cambridge," *Nature* 166 (1 July 1950): 7-9; "Physicists after the War," *Nature* 150 (18 July 1952): 75-80; "Budgets of the Scientific Departments of the University of Cambridge," *Nature* 171 (11 April 1953): 642-643; "X-Ray Studies of Biological Molecules," *Nature* 174 (10 July 1954): 55-59.

39. Watson, *Double Helix*, 129.

40. Maddox, *Rosalind Franklin*, 209. See also Wilkins, *Third Man*, 216.

41. Unfortunately, *Nature* purged its collection of papers in 1963 when they moved offices, so any internal correspondence about the DNA papers has not survived. See Maddox, *Rosalind Franklin*, 210-211.

42. Wilkins discusses his disappointment over not being the first to solve DNA's structure quite frankly in his autobiography; see Wilkins, *Third Man*, 204-245.

43. In a 2003 article for *Nature*, historian Robert Olby argued that the double helical model had a "quiet debut" and did not gain much attention until other scientists suggested models that explained how the double helical DNA molecule might be replicated and take part in protein synthesis. See Robert Olby, "Quiet Debut for the Double Helix." However, historian Yves Gingras has used bibliographic data to demonstrate that the Watson-Crick paper was the most frequently cited 1953 *Nature* paper in every year between 1953 and 1970, indicating that even though citations of the double helix paper experienced a sharp increase in the 1960s, the paper received far more citations than the average *Nature* paper throughout the mid-1950s. Gingras, "Revisiting the 'Quiet Debut.'" For another perspective on the reception of the Watson-Crick paper, see Strasser, "Who Cares about the Double Helix."

44. Gingras compares the Watson-Crick citation history with the citation history for another major paper in the history of molecular biology: F. Jacob and J. Monod, "Genetic Regulatory Mechanisms in the Synthesis of Proteins," *Journal of Molecular Biology* 3 (1961): 318-356. However, as Gingras acknowledges, this comparison requires a great deal of correction for the expansion in the number of biologists and biology journals between 1953 and 1961. I have chosen the 1952 Hershey-Chase paper as my point of comparison to obtain a clearer picture of journals in the early 1950s.

45. Impact factors, the usual (though contested) method of evaluating a journal's significance in its field, were not calculated until 1971, although Eugene Garfield began collecting citation data on articles as early as 1955. See Eugene Garfield, "Citation Indexes for Science," *Science* 122 (15 July 1955): 108-111; "Citation Analysis as a Tool in Journal Evaluation," *Science* 178 (3 November 1972): 471-479.

46. Alfred Hershey and Martha Chase, "Independent Functions of Viral Protein and Nucleic Acid in Growth of Bacteriophage," *Journal of General Physiology* 36 (1952): 39-56.

47. On the Hershey-Chase experiment and its significance, see Morange, *A History of Molecular Biology*, chap. 4. On Watson's response to the Hershey-Chase findings, see Watson, *Double Helix*, chap. 17.

48. Chase was a twenty-five-year-old laboratory assistant at the time of the Hershey-Chase experiments and left science in the 1960s; she does not appear to have been seriously considered for the prize. A satisfactory account of Chase's scientific career has yet to be written.

49. All numbers obtained from the Thomson-Reuters Web of Science Cited Reference Search tool. The comparison is a bit skewed in Watson and Crick's favor because Watson and Crick published in April and were cited 15 times by the end of 1953; Hershey and Chase published in September and were cited only once in 1952.

50. In 1915, Wegener published a book titled *The Origin of Continents and Oceans* in which he elaborated on his theory of continental drift. On Wegener's theory, see Oreskes, "The Rejection of Continental Drift." American geologists in particular were dismissive of Wegener's ideas, arguing

that Wegener was simply seeking evidence to support his pet theory rather than undertaking a truly "scientific" investigation of the structure of the earth.

51. LeGrand, *Drifting Continents and Shifting Theories*, 121.
52. Frank Vine, quoted in Frankel "Development, Reception, and Acceptance," 12.
53. Ibid.
54. Tharp, "Connect the Dots," 33.
55. Robert S. Dietz, "Continent and Ocean Basin Evolution by Spreading of the Sea Floor," *Nature* 190 (3 June 1961): 854-857. Dietz would later acknowledge that Hess had been the first to formulate the theory and ceded any claim to priority for the theory of seafloor spreading.
56. A. Cox and R. R. Doell, "Review of Palaeomagnetism," *Bulletin of the Geological Society of America* 71 (1960): 645-768.
57. Richard R. Doell and Allan Cox, "Palaeomagnetism of Hawaiian Lava Flows," *Nature* 192 (18 November 1961): 645-646; "Reversals of the Earth's Magnetic Field," *Science* 144 (1964): 1537-1538; Allan Cox, Richard R. Doell, and G. Brent Dalrymple, "Geomagnetic Polarity Epochs and Pleistocene Geochronometry," *Nature* 198 (15 June 1963): 1049-1051.
58. Vine, "Reversals of Fortune," 57.
59. Vine, "Reversals of Fortune," 57-58.
60. Morley, "The Zebra Pattern," 83.
61. Lawrence Morley, quoted in Frankel, "Development, Reception, and Acceptance," 17.
62. Frankel, "Development, Reception, and Acceptance," 17; Vine, "Reversals of Fortune," 58.
63. Vine, "Reversals of Fortune," 58.
64. Morley, "The Zebra Pattern," 84.
65. Bruce Heezen, quoted in Stewart, *Drifting Continents*, 193.
66. See Hamblin, "Science in Isolation," 297; Stewart, *Drifting Continents*, 188.
67. See, e.g., Lear, "Canada's Unappreciated Role."
68. Morley, quoted in Frankel, "Development, Reception, and Acceptance," 17-18, n33.
69. Morley, "The Zebra Pattern," 84.
70. Walter Gratzer, interview by author, Toronto, Canada, 31 May 2012.
71. Vine, quoted in Frankel, "Development, Reception, and Acceptance," 21-22.
72. F. J. Vine and D. H. Matthews, "Magnetic Anomalies over Oceanic Ridges," *Nature* 199 (7 September 1963): 947-949.
73. Drummond Matthews, quoted in Stewart, *Drifting Continents*, 79.
74. Vine, quoted in Stewart, *Drifting Continents*, 182.
75. J. Tuzo Wilson, "Transform Faults, Oceanic Ridges, and Magnetic Anomalies Southwest of Vancouver Island," *Science* 150 (22 October 1965): 482-485. In a 1965 letter to the editor in *Nature*, Wilson argued that transform faults create a "continuous network of mobile belts about the earth which divide the surface into several large rigid plates"; J. T. Wilson, "A New Class of Faults and Their Bearing on Continental Drift," *Nature* 207 (24 July 1965): 343-347.
76. F. J. Vine and J. Tuzo Wilson, "Magnetic Anomalies over a Young Oceanic Ridge off Vancouver Island," *Science* 150 (1965): 485-489.
77. W. C. Pittman and J. R. Heirzler, "Magnetic Anomalies over the Pacific-Antarctic Ridge," *Science* 154 (2 December 1966): 1164-1171.
78. F. J. Vine, "Spreading of the Ocean Floor: New Evidence," *Science* 154 (16 December 1966): 1405-1415.
79. See Stewart, *Drifting Continents*, 134.
80. Allan Cox, *Plate Tectonics*.
81. See Stewart, *Drifting Continents*, chap. 4.
82. They are Hugo Benioff, "Orogenesis and Deep Crystal Structure," *Bulletin of the Geological Society of America* 65 (1954): 385-400; A. Cox and R. R. Doell, "Review of Palaeomagnetism," *Bulletin of the Geological Society of America* 71 (1960): 645-768; "Reversals of the Earth's Magnetic

Field," *Science* 144 (1964): 1537-1538; Allan Cox, Richard R. Doell, and G. Brent Dalrymple, "Geomagnetic Polarity Epochs and Pleistocene Geochronometry," *Nature* 198 (15 June 1963): 1049-1051; Robert S. Dietz, "Continent and Ocean Basin Evolution by Spreading of the Sea Floor," *Nature* 190 (3 June 1961): 854-857; Richard R. Doell and Allan Cox, "Palaeomagnetism of Hawaiian Lava Flows," *Nature* 192 (18 November 1961): 645-646; W. C. Pitmann and J. R. Heirzler, "Magnetic Anomalies over the Pacific-Antarctic Ridge," *Science* 154 (2 December 1966): 1164-1171; F. J. Vine, "Spreading of the Ocean Floor: New Evidence," *Science* 154 (16 December 1966): 1405-1415; F. J. Vine and D. H. Matthews, "Magnetic Anomalies over Oceanic Ridges," *Nature* 199 (7 September 1963): 947-949; F. J. Vine and J. Tuzo Wilson, "Magnetic Anomalies over a Young Oceanic Ridge off Vancouver Island," *Science* 150 (1965): 485-489; J. T. Wilson, "A New Class of Faults and Their Bearing on Continental Drift," *Nature* 207 (24 July 1965): 343-347; "Transform Faults, Oceanic Ridges, and Magnetic Anomalies Southwest of Vancouver Island," *Science* 150 (22 October 1965): 482-485.

83. J. McKeen Cattell to Richard Gregory, 15 May 1936, SRGP, 2/7.

84. Gregory would ultimately accept the opportunity to deliver an Elihu Root lecture in December 1938 followed by a university lecture tour in the United States.

85. "Information Overload," unpublished manuscript chapter draft from Kaiser, *American Physics and the Cold War Bubble*.

86. Sheehan, interview.

87. Davies, interview.

88. Ziman, *Public Knowledge*, 111.

89. On the "heroic narrative" of the scientific journal and on the historical problems with this narrative, see Csiszar, "Broken Pieces of Fact," 15-22.

90. Interestingly, little scholarly work has been done on the history of scientific or scholarly peer review. John Burnham has given an extremely useful overview of peer review's history in two pieces: Burnham, "Editorial Peer Review"; "Journal Editors." Csiszar, "Broken Pieces of Fact," presents a further challenge to the idea that peer review has been the essential determinant of a journal's credibility. See also Baldwin, "Tyndall and Stokes."

91. Burnham, "Journal Editors," 56-57.

92. John Maddox, interview on *Nova*, "The Secret of Photo 51."

93. Interestingly, this is extremely similar to the process by which Albert Einstein's famous 1905 paper "On the Electrodynamics of Moving Bodies" reached the pages of the *Annalen der Physik*: editorial board member Max Planck found it interesting and endorsed its publication without further vetting. See Fölsing, *Albert Einstein*, 146-148; Isaacson, *Einstein*, 140.

94. John Maddox, "Valediction from an Old Hand," *Nature* 378 (7 December 1995): 521-522.

CHAPTER SEVEN

1. "*Nature* in North America," *Nature* 224 (13 December 1969): 1041.

2. Brenda Maddox, interview by author, London, United Kingdom, 10 April 2012.

3. E.g., John Maddox, "Abnormal Chromosomes Give Clue to Leukemia: Important Result of Hospital Tests," *Manchester Guardian*, 3 January 1961; "Civil Airlines Want to Use Blind Landing System: Electronics Industry Solving Problems," *Manchester Guardian*, 3 February 1961; "Russians May Be Just Testing New Rocket," *Manchester Guardian*, 6 February 1961; "The Rivalry in Outer Space," *Manchester Guardian*, 28 February 1961; "The Prospects for Atomic Energy," *Manchester Guardian*, 4 April 1961; "US Difficulties in Putting Man into Full Orbit," *Manchester Guardian*, 6 May 1961; "Cigarettes and Self-Destruction," *Guardian*, 6 February 1962; "The Nation's Teeth: John Maddox Examines the State of Dental Health in Britain," *Guardian*, 10 December 1963.

4. Maurice Macmillan, the son of Prime Minister Harold Macmillan, was elected to Parliament in 1955 and served until his death in 1984.

5. John Maddox, "Valediction from an Old Hand," *Nature* 378 (7 December 1995): 521-522.

6. "Editor of *Nature*: Mr. J. Maddox," *Nature* 209 (22 January 1966): 349-350.

7. Walter Grazer, *"Nature:* The Maddox Years," *Nature.com*, accessed 9 November 2012, http://www.nature.com/nature/history/full/nature06241.html.

8. Maddox, "Valediction," 521.

9. Walter Gratzer, "John Maddox (1925-2009)," *Nature* 458 (23 April 2009): 983.

10. Mary Sheehan, interview by author, London, United Kingdom, 12 April 2012.

11. Grazer, *"Nature:* The Maddox Years."

12. Sheehan, interview.

13. Walter Gratzer, interview by author, Toronto, Canada, 31 May 2012.

14. Gratzer, interview.

15. See Stewart, *Drifting Continents*, 188.

16. Under Lockyer and Gregory, *Nature* had printed the date of submission of each letter to the editor, but Gale and Brimble discontinued the practice in the late 1950s, probably because the information would not have reflected favorably on *Nature*'s publication speed.

17. Maddox, interview.

18. Sheehan, interview. See also Maddox, "Valediction," 522.

19. "University Crisis (continued)," *Nature* 224 (4 October 1969): 1-2; "University Unrest Persists," *Nature* 228 (3 October 1970): 3-4; "Can and Should the Universities Survive?" *Nature* 230 (26 March 1971): 197-198; "Mrs Thatcher Comes Down from Sinai," *Nature* 240 (15 December 1972): 369-370.

20. "Governor and President," *Nature* 213 (28 January 1967): 323; "Are Graduate Students Worth Keeping?" *Nature* 224 (14 March 1970): 985-986; "Science after the Cultural Revolution," *Nature* 217 (30 March 1968): 1196-1197. The last editorial suggested that the Cultural Revolution would likely be good for science in China, which looks somewhat strange in retrospect.

21. "The Case against Hysteria," *Nature* 235 (14 January 1972): 63-64; "DDT May Be Good for People," *Nature* 233 (15 October 1971): 437-438. DDT, or dichloro-diphenyl-trichloroethane, was the pesticide whose use inspired Rachel Carson's *Silent Spring*.

22. See "Euratom Research: Budget Trouble Again," *Nature* 221 (22 March 1969): 1087; "Another Echo of Received Doctrine," *Nature* 226 (16 May 1970): 587-588.

23. Gratzer, interview.

24. David Davies, interview by author, London, United Kingdom, 11 April 2012.

25. Sheehan, interview.

26. Davies, interview.

27. Philip Campbell, "Maddox by His Successor," *Nature* 458 (17 April 2009): 985-986.

28. "Keeping Up to Date," *Nature* 229 (1 January 1971): 9.

29. Davies, interview.

30. Gratzer, interview; Davies, interview.

31. "Keeping Up to Date," *Nature* 229 (1 January 1971): 9.

32. Davies, interview.

33. See Nicholas Byam Shaw, "A Publisher's Perspective," *Nature* 458 (17 April 2009): 984-985.

34. Sheehan, interview.

35. Mary Sheehan and David Davies both indicated that Hughes and Maddox did not get along. Sheehan, interview; Davies, interview.

36. Maddox, interview.

37. Davies, interview.

38. Davies, interview.

39. [David Davies], *"Nature* in the Future," *Nature* 244 (24 August 1973): 475.

40. [David Davies], "What Is a *Nature* Paper?" *Nature* 246 (30 November 1973): 237-238.
41. [David Davies], "It's Your Journal," *Nature* 252 (29 November 1974): 337.
42. Davies, interview.
43. [David Davies], "Fearless, Blunt Message to Our Readers," *Nature* 258 (11 December 1975): 465.
44. [David Davies], "Dear sir ..." *Nature* 257 (2 October 1975): 345.
45. [David Davies], untitled , *Nature* 254 (10 April 1975): 469.
46. On postwar American dominance in world scientific research, see Krige, *American Hegemony*. See also Creager, "Tracing the Politics"; Zeitlin and Herrigel, *Americanization and Its Limits*."
47. See, e.g., Kaiser, "Cold War Requisitions."
48. Sheehan, interview.
49. Ibid.
50. Davies, interview.
51. Ibid.
52. [David Davies], "*Nature* in the Future," *Nature* 244 (24 August 1973): 475.
53. On Abelson's editorship, see Abelson, "Scientific Communication"; Wolfle, "*Science*."
54. Philip Abelson, quoted in Wolfle, "*Science*," 58.
55. Figure for *Science* from Wolfle, "*Science*," 60; figure for *Nature* from Davies, interview. The huge difference is likely due to, at least in part, *Nature*'s substantially higher subscription price.
56. Some of the common objections to the use of impact factors to determine a journal's influence include the difficulty of comparing impact factors across scientific subfields and the tendency for a few highly cited papers to skew a journal's impact factor. *Nature*'s current editor, Philip Campbell, articulated some of the concerns about impact factors in a 1998 article for *Ethics in Science and Environmental Politics*; see Campbell, "Escape from the Impact Factor."
57. See Krementsov, *International Science*; *Stalinist Science*.
58. Krementsov, "In the Shadow of the Bomb."
59. Ibid., 59.
60. Ibid., 60.
61. A. I. Alichanian and A. I. Alichanow, "Concerning New Elementary Particles in Cosmic Rays," *Nature* 163 (14 May 1949): 761.
62. C. M. G. Lattes, G. P. S. Occhialini, and C. F. Powell, "Observations on the Tracks of Slow Mesons in Photographic Emulsions," *Nature* 160 (11 October 1947): 486-492.
63. A. Alichanian, A. Alichanow, and A. Weissenberg, *C.R. (Doklady) Armenian S.S.R.* 5 (1946): 129; *Journal of Physics* 11 (1947): 97.
64. This was not Alichanow's first article in *Nature*. See A. I. Alichanow, A. I. Alichanian, and B. S. Delepow, "β-Spectra of Some Radioactive Elements," *Nature* 135 (9 March 1935): 393.
65. C. F. Powell, "Concerning New Elementary Particles in Cosmic Rays," *Nature* 163 (14 May 1949): 762.
66. On changes in Soviet science following the death of Stalin, including a crucial reorganization of the USSR Academy of Sciences, see Ivanov, "Science after Stalin."
67. Jiri Zemlicka and Stanislav Chladek, "Scientific Censorship," *Nature* 228 (7 November 1970): 589.
68. Some Western scientists accepted invitations to visit the USSR, but programs that attempted to encourage Western scientists to go to the USSR often found themselves with a dearth of applicants. For example, a reciprocal exchange program between the USSR Academy of Sciences and the Royal Society of London regularly filled and even exceeded the number of slots available to scientists from the USSR who wished to spend time in the United Kingdom, but British scientists were comparatively reluctant to visit the USSR. See Cox, "Royal Society."
69. See "Two Views of *Nature*," *Nature* 240 (15 December 1972): 372, which shows two images: one of the official British-printed *Nature*, and one of the Soviet photocopy, which has replaced a scientific article by Soviet dissident Zhores Medvedev with an advertisement.

70. See Zhores A. Medvedev, "Medvedev's Complaint," *Nature* 228 (19 December 1970): 1236.

71. [Rainald Brightman], "Rehabilitation of European Culture," *Nature* 155 (10 March 1945): 283–286.

72. [Rainald Brightman], "Conditions of Survival: The Moral Basis of Civilization," *Nature* 158 (28 September 1946): 425–427.

73. J. D. Bernal, "'Conditions of Survival,'" *Nature* 158 (26 October 1946): 590–591.

74. John R. Baker and A. G. Tansley, "The Course of the Controversy on Freedom in Science," *Nature* 158 (26 October 1946): 574–576.

75. [L. J. F. Brimble], "Freedom in Science," *Nature* 158 (26 October 1946): 565–567.

76. See, e.g., "Economic Development in the U.S.S.R.," *Nature* 167 (13 January 1951): 63–64; News and Views, "Science in Communist States," *Nature* 170 (20 December 1952): 1049; "Science, Technology and Man-Power," *Nature* 177 (14 January 1956): 51–52.

77. See, e.g., Eric Ashby, "Genetics in the Soviet Union," *Nature* 162 (11 December 1948): 912–913; "Lysenko in Perspective," *Nature* 174 (24 July 1954): 148–149; Julian Huxley, "Soviet Genetics: The Real Issue," *Nature* 163 (18 June 1949): 935–942; P. Maheshwari, "Lysenko's Latest Discovery: The Conversion of Wheat into Rye, Barley and Oats," *Nature* 170 (12 July 1952): 66–68; A. Quintanilha, "Social Implications of Mendelism versus Michurinism," *Nature* 183 (2 May 1959): 1222–1224.

78. [John Maddox], "A Voice from the East," *Nature* 227 (19 September 1970): 1177–1178; Zhores Medvedev, "The Closed Circuit: A Record of Soviet Scientific Life," *Nature* 227 (19 September 1970): 1197–1202.

79. [John Maddox], "Zhores Medvedev and the Reputation of Russian Science," *Nature* 238 (14 July 1972): 61–62.

80. [John Maddox], "Sad Case of Dr Zhores Medvedev," *Nature* 244 (17 August 1973): 379.

81. See, e.g., [David Davies], "Ban the Bomb—Properly," *Nature* 249 (10 May 1974): 97; "What Can Scientists Contribute to Arms Control?" *Nature* 256 (21 August 1975): 607; "Time to Speak Out against Cosy Bilateralism," *Nature* 259 (15 January 1976): 71; "Shadowy Threshold," *Nature* 262 (19 August 1976): 635; "Nuclear Secrets: No Clear Frontiers," *Nature* 281 (18 October 1979): 511; "Nuclear Power: The Critics Must Be Heard," *Nature* 283 (3 January 1980): 1.

82. [David Davies], "Nuclear Defence: The Need for Debate," *Nature* 244 (31 August 1973): 531–532.

83. [David Davies], "And Now . . . Nuclear Hypocrisy in Britain," *Nature* 249 (28 June 1974): 787.

84. Vera Rich, "USSR," *Nature* 261 (20 May 1976): 185.

85. Vera Rich, "Russian Plate Tectonics: Drift of Change," *Nature* 286 (14 August 1980): 652.

86. Vera Rich, "Happy Birthday, Soviet Style," *Nature* 249 (17 May 1974): 204–205.

87. Vera Rich, "Soviet Abuse of Psychiatry," *Nature* 252 (22 November 1974): 266.

88. See, e.g., Vera Rich, "Dissident Voices," *Nature* 253 (6 February 1975): 390; "The Clampdown Continues," *Nature* 261 (3 June 1976): 363; "Messages from Without," *Nature* 268 (21 July 1977): 192; "Black Year for Soviet Refusniks," *Nature* 227 (18 January 1979): 163.

89. [David Davies], "Scientists Don't Move," *Nature* 250 (26 July 1974): 275.

90. Following Maddox's recent death, David Davies wrote a letter to the editor reflecting on how they came to "swap" positions in 1980. David Davies, "Leading the Tributes to Editor John Maddox," *Nature* 459 (14 May 2009): 163.

91. Gratzer, interview.

92. Maddox, interview.

93. Davies, interview.

94. John Maddox to Michael Swann, 3 October 1980, personal papers of Brenda Maddox.

CHAPTER EIGHT

1. "Academy Elections in a Muddle," *Nature* 338 (30 March 1989): 361.
2. "Cold (Con)fusion," *Nature* 338 (30 March 1989): 361.
3. Ibid., 361–362.
4. Ibid., 362. On superconductivity research in the 1980s, see Nowotny and Felt, *After the Breakthrough*; Matricon and Waysand, *Cold Wars*, chap. 21.
5. "Farewell (Not Fond) to Cold Fusion," *Nature* 344 (29 March 1990): 365.
6. R. W. Wood, "The N-Rays," *Nature* 70 (29 September 1904): 530–531.
7. On the N-rays controversy see Nye, "N-rays."
8. Lagemann, "New Light on Old Rays," 283.
9. Russell Targ and Harold Puthoff, "Information Transmission under Conditions of Sensory Shielding," *Nature* 251 (18 October 1974): 602–607. For more on Targ and Puthoff, see Kaiser, *How the Hippies Saved Physics*, 69–70, 90–101.
10. [David Davies], "Investigating the Paranormal," *Nature* 251 (18 October 1974): 559–560.
11. David Marks and Richard Kammann, "Information Transmission in Remote Viewing Experiments," *Nature* 274 (17 August 1978): 680–681.
12. David Davies, interview by author, London, United Kingdom, 11 April 2012.
13. Sir Peter Scott and Robert Rines, "Naming the Loch Ness Monster," *Nature* 258 (11 December 1975): 466–468.
14. Davies, interview.
15. T. Imanishi-Kari et al., "Altered Repertoire of Endogenous Immunoglobulin Gene Expression in Transgenic Mice Containing a Rearranged Mu Heavy Chain Gene," *Cell* 45 (April 1986): 247–59.
16. On the investigation into the *Cell* paper, see Kevles, *Baltimore Case*.
17. John Maddox, "Can a Greek Tragedy Be Avoided?" *Nature* 333 (30 June 1988): 795–797.
18. "When to Believe the Unbelievable," *Nature* 333 (30 June 1988): 787.
19. For historical and anthropological sources on homeopathy, see Dean, "Homeopathy"; Degele, "On the Margins of Everything"; Kaufman, *Homeopathy in America*; Kenny, "Darker Shade of Green," 481–504; Weatherall, "Making Medicine Scientific."
20. "Editorial Reservation," *Nature* 333 (30 June 1989): 818.
21. A previous Benveniste paper was submitted in June 1986 and rejected. Peter Newmark (deputy editor of *Nature*) to Jacques Benveniste, 24 November 1986, personal papers of Brenda Maddox.
22. Walter W. Stewart to Peter Newmark (deputy editor of *Nature*), 15 July 1987, personal papers of Brenda Maddox.
23. Walter Gratzer, interview by author, 31 May 2012.
24. Charles Wenz, interview by author, 30 May 2012.
25. A technician named Jose Alvarez was also part of the team, but his name never appeared in *Nature*'s accounts of the investigation. See Picart, "Scientific Controversy as Farce," 12.
26. E.g., Antoine Danchin, "Explanation of Benveniste," *Nature* 334 (28 July 1988): 286; Ronald H. A. Plasterk, "Explanation of Benveniste," *Nature* 334 (28 July 1988): 285; David Taylor Reilly, "Explanation of Benveniste," *Nature* 334 (28 July 1988): 285.
27. John Maddox, James Randi, and Walter W. Stewart, "'High-Dilution' Experiments a Delusion," *Nature* 334 (28 July 1988): 287–290.
28. Ibid., 287.
29. Ibid.
30. Ibid., 290.
31. Jacques Benveniste, "Dr Jacques Benveniste Replies," *Nature* 334 (28 July 1988): 291.

32. Randi was also the author of a book on exposing various types of fraud. See Randi, *Flim-Flam!*.

33. Benveniste, "Dr. Jacques Benveniste Replies," 291.

34. M. J. Escribano, "Only the Smile Is Left," *Nature* 334 (4 August 1988): 376; P. M. Gaylarde, "Only the Smile Is Left," *Nature* 334 (4 August 1988): 375; J. Leslie Glick, "Only the Smile Is Left," *Nature* 334 (4 August 1988): 376; Kenneth S. Suslick, "Only the Smile Is Left," *Nature* 334 (4 August 1988): 375-376.

35. Henry Metzger and Stephen C. Dreskin, "Only the Smile Is Left," *Nature* 334 (4 August 1988): 375.

36. For letters from homeopaths and homeopathy supporters, see Peter Fisher, "Orthodoxy and Homeopathy," *Nature* 335 (22 September 1988): 292; Robin M. Gibson and Sheila L. M. Gibson, "Controversy Continues," *Nature* 335 (15 September 1988): 200. For letters proposing explanations, see Glick, "Only the Smile," 376; Ray M. Schilling, "More on Benveniste's Dilution Results," *Nature* 335 (13 October 1988): 584; Suslick, "Only the Smile," 375-376.

37. See, e.g., JeanClare Seagrave, "Evidence of Non-Reproducibility," *Nature* 334 (18 August 1988): 559; Sergio Bonini, Emilio Adriani, and Francesco Balsano, "Evidence of Non-Reproducibility," *Nature* 334 (18 August 1988): 559.

38. Gaylarde, "Only the Smile," 375.

39. Gregory A. Petsko, "Unreproducible Results," *Nature* 335 (8 September 1988): 109.

40. Peter Taylor, "Controversy Continues," *Nature* 335 (15 September 1988): 200.

41. Alexander M. Grimwade, "From Other Letters," *Nature* 335 (22 September 1988): 292.

42. G. J. Neville, "Controversy Continues," *Nature* 335 (15 September 1988): 200.

43. Jacques Benveniste, "Benveniste on the Benveniste Affair," *Nature* 335 (27 October 1988): 759.

44. John Maddox, "Waves Caused by Extreme Dilution," *Nature* 335 (27 October 1988): 760.

45. Ibid., 761.

46. Ibid.

47. Ibid., 762.

48. On Maddox's tactics and why they were ultimately successful at discrediting Benveniste, see Picart, "Scientific Controversy as Farce." For another interpretation of the Benveniste affair, see Fadlon and Lewin-Epstein, "Laughter Spreads." Fadlon and Lewin-Epstein argue that the Benveniste episode is best read not as scientific self-policing but as science responding to the threat of Benveniste making his claims in the popular press. However, as Picart argues in her rebuttal, Fadlon and Lewin-Epstein base this conclusion on Maddox's own claims that the Benveniste results might have reached the popular press, not on evidence that such publication was about to take place. See Picart, "Blurring Boundaries."

49. Deuterium, also called "heavy hydrogen," is a stable isotope of hydrogen with an extra neutron in its nucleus.

50. For a full account of the events leading up to Pons and Fleischmann's press conference, see Close, *Too Hot to Handle*; Collins and Pinch, *Golem*, chap. 3; Simon, *Undead Science*.

51. On the rhetoric employed in the press conference, see Gieryn, "(Cold) Fusion of Science."

52. On the early circulation of information about cold fusion, see Lewenstein, "From Fax to Facts"; Dearing, "Newspaper Coverage of Maverick Science."

53. "Disorderly Publication," *Nature* 338 (13 April 1989): 528.

54. Ibid., 527-528.

55. Ibid., 528.

56. "Cold Fusion in Print," *Nature* 338 (20 April 1989): 604.

57. Ibid..

58. Richard L. Garwin, "Consensus on Cold Fusion Still Elusive," *Nature* 338 (20 April 1989): 616-617.

59. S. E. Jones, E. P. Palmer, J. B. Czirr, D. L. Decker, G. L. Jensen, J. M. Thorne, S. F. Taylor,

and J. Rafelski, "Observation of Cold Nuclear Fusion in Condensed Matter," *Nature* 338 (27 April 1989): 737.

60. ^{6}Li is a natural isotope of lithium containing three protons and three neutrons. Lithium's most common isotope is ^{7}Li, which contains three protons and four neutrons.

61. Jones et al., "Observation," 740.

62. Ibid., 737.

63. Ibid., 740.

64. Vera Rich, "Prospect of Achieving Cold Fusion Tantalizes; Confirmation Reports Trickle In; Dispute over Primacy," *Nature* 338 (13 April 1989): 529.

65. R. D. Petrasso, X. Chen, K.W. Wenzel, R. R. Parker, C. K. Li, and C. Fiore, "Problems with the γ-Ray Spectrum in the Fleischmann *et al.* Experiments," *Nature* 339 (18 May 1989): 183-185.

66. Ibid., 183.

67. Ibid., 185.

68. Martin Fleischmann, Stanley Pons, and R. J. Hoffman, "Measurement of γ-Rays from Cold Fusion," *Nature* 339 (29 June 1989): 667.

69. R. D. Petrasso, X. Chen, K. W. Wenzel, R. R. Parker, C. K. Li, and C. Fiore, reply, *Nature* 339 (29 June 1989): 667-669.

70. Ibid., 667-668.

71. The team went on to say that "The estimated neutron flux in this experiment is at least a factor of 50 times smaller than that reported by Jones *et al.* and about one million times smaller than that reported by Fleischmann *et al.* The results suggest that a significant fraction of the observed neutron events are associated with cosmic rays." M. Gai, S. L. Rugari, R. H. France, B. J. Lund, Z. Zhao, A. J. Davenport, H. S. Isaacs, and K. G. Lynn, "Upper Limits on Neutron and γ-Ray Emission from Cold Fusion," *Nature* 340 (6 July 1989): 29.

72. D. E. Williams, D. J. S. Findlay, D. H. Craston, M. R. Sené, M. Bailey, S. Croft, B. W. Hooton, C. P. Jones, A. R. J. Kucernak, J. A. Mason, and R. I. Taylor, "Upper Bounds on 'Cold Fusion' in Electrolytic Cells," *Nature* 342 (23 November 1989): 375-384.

73. Ibid., 384.

74. M. H. Salamon, M. E. Wrenn, H. E. Bergeson, K. C. Crawford, W. H. Delaney, C. L. Henderson, Y. Q. Li, J. A. Rusho, G. M. Sandquist, and S. M. Seltzer, "Limits on the Emission of Neutrons, γ-Rays, Electrons and Protons from Pons/Fleischmann Electrolytic Cells," *Nature* 344 (29 March 1990): 401-405.

75. Ibid., 401.

76. Ibid., 404.

77. Ibid.

78. Ibid., 405.

79. David Lindley, "Cold Fusion Gathering Is Incentive to Collaboration; Some Evidence for Neutron Production; Claims of Energy Generation Mostly Ignored," *Nature* 325 (1 June 1989): 325.

80. David Swinbanks, "Cold Fusion: Efforts Abandoned in Japan," *Nature* 339 (18 May 1989): 167.

81. David Lindley, "Double Blow for Cold Nuclear Fusion; Harwell Investigation Stopped; Los Alamos Collaboration Fails," *Nature* 339 (22 June 1989): 567.

82. David Lindley, "Official Thumbs Down," *Nature* 342 (16 November 1989): 215.

83. David Lindley, "The Embarrassment of Cold Fusion," *Nature* 344 (29 March 1990): 375-376.

84. Lindley, "Cold Fusion Gathering," 325.

85. John Maddox, "What to Say about Cold Fusion," *Nature* 338 (27 April 1989): 701.

86. Ibid.

87. Gareth Morgan, "Peer Review?," *Nature* 339 (18 May 1989): 170.

88. John Maddox, "Can Journals Influence Science?," *Nature* 339 (29 June 1989): 657.

89. John Maddox, "End of Cold Fusion in Sight," *Nature* 340 (6 July 1989): 15.

90. "Farewell (Not Fond) to Cold Fusion," *Nature* 344 (29 March 1990): 365. If Maddox thought cold fusion deserved a "waspish footnote," we can only speculate on what he would have thought about the present chapter.

91. Lewenstein, "From Fax to Facts"; "Cold Fusion and Hot History."

92. For secondary sources on the cold fusion controversy, see nn. 50–52.

93. Gieryn, "(Cold) Fusion of Science," 186–187.

94. Ibid., 219.

95. Tim Beardsley, "Fusion Breakthrough?" *Scientific American*, May 1989, 28.

96. Tim Beardsley, "Playing with Fire," *Scientific American*, June 1989, 22.

97. Tim Beardsley, "Morning After," *Scientific American*, August 1989, 12, 16.

98. "Fusion Illusion?" *Time*, 8 May 1989.

99. Ibid.

100. Ibid.

101. "So You Want to Be a Star," *Economist*, 1 April 1989, 75.

102. For more on the disciplinary boundaries in the cold fusion controversy, see McAllister, "Competition."

103. "So You Want to Be a Star," *Economist*, 1 April 1989, 75.

104. "Reports of My Death . . . ," *Economist*, 13 May 1989, 90.

105. "Not What It Used to Be," *Economist*, 3 June 1989, 84.

106. "The Secret Life of Cold Fusion," *Economist*, 30 September 1989, 87.

107. Ibid.

108. See, e.g., Jerry E. Bishop, "Science: Heat Source in Fusion Find May Be Mystery Reaction," *Wall Street Journal*, 3 April 1989, B1, B3.

109. Jerry Bishop, "Development in Atom Fusion to Be Unveiled," *Wall Street Journal*, 23 March 23, B1, B8.

110. Jerry Bishop, "Second Fusion Discovery Comes to Light: Findings at Brigham Young Are Shrouded in Secrecy, Seem Less Controversial," *Wall Street Journal*, 29 March 1989, B4.

111. Richard L. Hudson, "Fusion Findings Withdrawn from Journal," *Wall Street Journal*, 20 April 1989, B4.

112. Ibid.

113. Richard L. Hudson, "British Journal Attacks Claims of 'Cold Fusion,'" *Wall Street Journal*, 27 April 1989, B1, B4.

114. See, e.g., Jerry E. Bishop, "'Cold Fusion' Still Generates Some Heat," *Wall Street Journal*, 8 September 1989, B1; "Cold Comfort on Cold Fusion Front," *Wall Street Journal*, 23 October 1989, B1, B5; "'Cold Fusion' Research Dispels Some Doubts," *Wall Street Journal*, 2 March 1990, B1; Jerry E. Bishop and Jacob M. Schlesinger, "Japan's 'Cold Fusion' Effort Produces Startling Claims of Bursts of Neutrons," *Wall Street Journal*, 1 December 1989, B4.

115. Jerry E. Bishop, "'Cold Fusion' Chemists Reiterate Claim; Other Scientists Report Similar Results," *Wall Street Journal*, 30 March 1990, B3.

116. Richard L. Hudson, "If You Read It First in *Nature*, It's Big and (Usually) True: Cold Fusion, Homeopathy and Hunchbacked Flies Enliven Scientific Journal," *Wall Street Journal*, 15 May 1989, A1, A6.

117. Maddox, quoted in Hudson, "If You Read It First in *Nature*," A6.

CONCLUSION

1. Although Maddox did not offer specific information about the subcommittee hearing where this observation was made, it was almost certainly part of Rep. John Dingell's hearings into scientific fraud. On Dingell's hearings, see Kevles, *Baltimore Case*.

2. John Maddox, "Can Journals Influence Science?" *Nature* 339 (29 June 1989): 657.

3. For histories of peer review and refereeing, see Baldwin, "Tyndall and Stokes"; Burnham, "Editorial Peer Review," "Journal Editors"; Rennie, "Editorial Peer Review."

4. See "*Nature*: Editorial Process," Nature Publishing Group, accessed 5 January 2014, http://www.nature.com/nature/authors/gta/1a_Editorial_process.pdf.

5. See "Availability of Data and Materials," Nature Publishing Group, accessed 4 June 2014, http://www.nature.com/authors/policies/availability.html.

6. See, e.g., "Gender Progress (?)," *Nature* 504 (11 December 2013): 188.

7. John Maddox, "Towards the Electric Journal?" *Nature* 344 (22 March 1990): 287. In 2015 dollars, this is roughly $1,800. In 2015, however, the cost of a 1 GB (1024 MB) external storage drive was approximately $5.

8. Alun Anderson, "Cautious Start for AAAS," *Nature* 344 (1 March 1990): 6.

9. On the closing of the *Online Journal of Current Clinical Trials*, see Tumber and Dickersin, "Publication of Clinical Trials."

10. Geoff Brumfiel, "Ousted Creationist Sues over Website," *Nature* 420 (12 December 2002): 597.

11. "Retribution Denied to Creationist Suing arXiv over Religious Bias," *Nature* 428 (1 April 2004): 458

12. "The arXiv Endorsement System," arXiv.org, accessed 4 June 2014, http://arxiv.org/help/endorsement.

13. John Maddox, "Electronic Journals Have a Future," *Nature* 356 (16 April 1992): 559; "Electronic Journals Are Already Here," *Nature* 365 (21 October 1993): 689.

14. John Maddox, "*Nature* on the Internet (at Last)!," *Nature* 377 (12 October 1995): 475.

15. Charles Wenz, interview by author, Toronto, Canada, 30 May 2012.

16. Owen, *Scientific Article*.

17. Wenz, interview.

18. On the effects of online commenting on scientific publishing, see Casper, "Mutable Mobiles."

19. On the 1997 redesign, see "In Pursuit of Usefulness and Distraction," *Nature* 385 (16 January 1997): 185.

20. John Maddox, "Towards the Electric Journal?," *Nature* 344 (22 March 1990): 287.

21. Meredith Wadman, "Libraries Offer Incentive for Web-Based Rivals to 'Costly' Journals," *Nature* 398 (25 March 1999): 272.

22. "Cost of Journals," University of Illinois Library, accessed 5 January 2014, http://researchguides.uic.edu/content.php?pid=43156&sid=1386845.

23. Philosopher and computer scientist Peter Suber is arguably the most visible advocate of open-access publishing. See Suber, *Open Access*; "Open Access: Other Ways," *Nature* 426 (6 November 2003): 15.

24. "*PLOS ONE* Journal Information," PLOS ONE, accessed 5 January 2014, http://www.plosone.org/static/information. See also "*PLOS ONE* Guidelines for Reviewers," PLOS ONE, accessed 5 January 2014, http://www.plosone.org/static/reviewerGuidelines, and Jim Giles, "Open-Access Journal Will Publish First, Judge Later," *Nature* 445 (4 January 2007): 9.

25. Steven Harnad, "The Self-Archiving Initiative," *Nature* 410 (26 April 2001): 1024–1025.

26. Ibid.; Michael Jubb, "Open Access: Let's Go for Gold," *Nature* 487 (19 July 2012): 302.

27. See, e.g., Raghavendra Gadakar, "Open-Access More Harm than Good in Developing World," *Nature* 453 (21 May 2008): 453; Christopher Smith, "Open Access: Hard on Lone Authors," *Nature* 487 (26 July 2012): 432. For Campbell's cost estimate, see Richard Van Noorden, "*Nature* Makes All Articles Free to View," Nature.com, 2 December 2014, accessed 3 December 2014, http://www.nature.com/news/nature-makes-all-articles-free-to-view-1.16460.

28. Van Noorden, "*Nature* Makes All Articles Free to View"; John Bohannon, "*Nature* Publisher

Hopes to End 'Dark Sharing' by Making Read-Only Papers Free," *ScienceInsider*, 2 December 2014, accessed 3 December 2014, http://news.sciencemag.org/scientific-community/2014/12/nature-publisher-hopes-end-dark-sharing-making-read-only-papers-free.

29. Alison McCook, "Researchers Boycott Cell Press," *Scientist*, 23 October 2003, accessed 15 October 2012, http://classic.the-scientist.com/?articles.view/articleNo/22548/title/Researchers-boycott-Cell-Press/.

30. E.g., Michael Eisen, "Elsevier-Funded NY Congresswoman Carolyn Maloney Wants to Deny Americans Access to Taxpayer Funded Research," *it is NOT junk*, 5 January 2012, accessed 18 October 2012, http://www.michaeleisen.org/blog/?p=807.

31. Timothy Gowers, "Elsevier: My Part in Its Downfall," *Gowers's Weblog*, 21 January 2012, accessed 5 January 2014, http://gowers.wordpress.com/2012/01/21/elsevier-my-part-in-its-downfall/.

32. Jennifer Howard, "U. of California Tries Just Saying No to Rising Journal Costs," *Chronicle of Higher Education*, 8 June 2010, accessed 15 October 2012, http://chronicle.com/article/U-of-California-Tries-Just/65823/.

33. Jennifer Howard, "Nature Publishing Group Defends Its Price Increase for U. of California," *Chronicle of Higher Education*, 9 June 2010, accessed 15 October 2012, http://chronicle.com/article/Nature-Publishing-Group/65848/.

34. Daniel Cressey, "Nature Publishing Group and the University of California Make Nice," *Nature News Blog*, 25 August 2010, accessed 15 October 2012, http://blogs.nature.com/news/2010/08/nature_and_california_make_nic_1.html.

35. Sarah Greaves, Joanna Scott, Maxine Clarke, Linda Miller, Timo Hannay, Annette Thomas, and Philip Campbell, "*Nature*'s Trial of Open Peer Review," *Nature.com*, December 2006, accessed 12 December 2013, http://www.nature.com/nature/peerreview/debate/nature05535.html.

36. "Peer Review and Fraud," *Nature* 444 (21/28 December 2006): 972.

37. For the peer-review debate, see "Peer Review: Debate," *Nature.com*, December 2006, accessed 12 December 2013, http://www.nature.com/nature/peerreview/debate/. For the archives of "Peer-to-Peer," see "Peer-to-Peer," *Nature.com*, December 2006–May 2010, accessed 12 December 2013, http://blogs.nature.com/peer-to-peer/.

38. See, e.g., a recent report on scientific peer review by the British Parliament, "Science and Technology Committee, Eighth Report: Peer Review in Scientific Publications," United Kingdom Parliament, 18 July 2011, accessed 13 June 2013, http://www.publications.parliament.uk/pa/cm201012/cmselect/cmsctech/856/85602.htm.

39. Of course, this is not a new problem—researchers have been complaining about an excess of literature for centuries. See, e.g., Blair, *Too Much to Know*.

Bibliography

PRIMARY SOURCE COLLECTIONS

Alfred Russel Wallace Papers. British Library. (ARWP:BL)
John Tyndall Papers. Royal Institution of Great Britain, London.
Macmillan Papers. British Library. (MP:BL)
Macmillan Papers. Special Collections. University of Reading Library. (MP:UR)
Nature 1–516 (1869–2014).
Norman Lockyer Papers. Special Collections. University of Exeter Library. (NLP)
Sir Richard Gregory Papers. Special Collections. University of Sussex Library. (SRGP)
Sir Richard Gregory Papers, Addition from the University of Sheffield. Special Collections. University of Sussex Library. (SRGP:SA)
Thomas Henry Huxley Collection. Records and Archives. Imperial College London Library. (THHC)

SECONDARY SOURCES

Abelson, Philip. "Scientific Communication." *Science* 209 (4 July 1980): 60–62.
Adams, Mark B. *The Wellborn Science: Eugenics in Germany, France, Brazil, and Russia.* Oxford: Oxford University Press, 1990.
Ammon, Ulrich, and Grant D. McConnell. *English as an Academic Language in Europe.* New York: Peter Lang, 2002.
Armytage, W. H. G. *Sir Richard Gregory: His Life and Work.* London: Macmillan, 1957.
Atkinson, Dwight. *Scientific Discourse in Sociohistorical Context: The Philosophical Transactions of the Royal Society of London, 1675–1975.* Mahwah, NJ: L. Erlbaum, 1999.
Badash, Lawrence. "British and American Views of the German Menace in World War I." *Notes and Records of the Royal Society of London* 34 (1979): 91–121.
———. *Ernest Rutherford and Theoretical Physics.* Cambridge, MA: MIT Press, 1987.
———. "The Influence of New Zealand on Rutherford's Scientific Development." In *Scientific Colonialism: A Cross-Cultural Comparison,* edited by Nathan Reingold and Marc Rothenberg, 379–389. Washington, DC: Smithsonian Institution Press, 1987.
———. *Kapitza, Rutherford, and the Kremlin.* New Haven, CT: Yale University Press, 1985.
———. "The Origins of Big Science: Rutherford at McGill." In *Rutherford and Physics at the Turn*

of the Century, edited by Mario Bunge and William R. Shea, 23-41. New York: Science History Publications, 1979.

———. "Radium, Radioactivity, and the Popularity of Scientific Discovery." *Proceedings of the American Philosophical Society* 122 (1978): 145-154.

———. *Radioactivity in America: Growth and Decay of a Science*. Baltimore: Johns Hopkins University Press, 1979.

———, ed. *Rutherford and Boltwood: Letters on Radioactivity*. New Haven, CT: Yale University Press, 1969.

Bailey, R. W. *Nineteenth-Century English*. Ann Arbor: University of Michigan Press, 1996.

Baldwin, Melinda. "Tyndall and Stokes: Correspondence, Referee Reports and the Physical Sciences in Victorian Britain." In *The Age of Scientific Naturalism: John Tyndall and His Contemporaries*, edited by Bernard Lightman and Michael Reidy, 171-186. London: Pickering and Chatto, 2014.

Barr, Alan P., ed. *Thomas Henry Huxley's Place in Science and Letters: Centenary Essays*. Athens: University of Georgia Press, 1997.

Bartholomew, James R. *The Formation of Science in Japan: Building a Research Tradition*. New Haven, CT: Yale University Press, 1989.

———. "Science in Twentieth-Century Japan." In *Science in the Twentieth Century*, edited by John Krige and Dominique Pestre, 879-896. Amsterdam: Harwood, 1997.

Barton, Ruth. "'Huxley, Lubbock, and Half a Dozen Others': Professionals and Gentlemen in the Formation of the X Club, 1851-1864." *Isis* 89 (1998): 410-444.

———. "'An Influential Set of Chaps': The X-Club and Royal Society Politics 1864-85." *British Journal for the History of Science* 23 (1990): 53-81.

———. "Just before *Nature*: The Purposes of Science and the Purposes of Popularization in Some English Popular Science Journals of the 1860s." *Annals of Science* 55 (1998): 1-33.

———. "'Men of Science': Language, Identity and Professionalization in the Mid-Victorian Scientific Community." *History of Science* 41 (2003): 73-119.

———. "Scientific Authority and Scientific Controversy in *Nature*: North Britain against the X Club." In *Culture and Science in the Nineteenth-Century Media*, edited by Louise Henson, G. N. Cantor, Gowan Dawson, Richard Noakes, Sally Shuttleworth, and Jonathan Topham, 223-235. Aldershot, UK: Ashgate, 2004.

———. "The X Club: Science, Religion, and Social Change in Victorian England." PhD diss., University of Pennsylvania, 1976.

Bateson, Beatrice. *William Bateson, F.R.S., Naturalist: His Essays and Addresses, Together with a Short Account of His Life*. Cambridge: Cambridge University Press, 1928.

Beer, Gillian. "The *Academy*: Europe in England." In *Science Serialized: Representations of the Sciences in Nineteenth-Century Periodicals*, edited by Geoffrey Cantor and Sally Shuttleworth, 181-198. Cambridge, MA: MIT Press, 2004.

Beyerchen, Alan D. *Scientists under Hitler: Politics and the Physics Community in the Third Reich*. New Haven, CT: Yale University Press, 1977.

Blair, Ann. *Too Much to Know: Managing Scholarly Information before the Modern Age*. New Haven, CT: Yale University Press, 2011.

Boag, J. W., P. E. Rubinin, and D. Schoenberg, eds. *Kapitza in Cambridge and Moscow: Life and Letters of a Russian Physicist*. Amsterdam: North Holland, 1990.

Boudia, Soraya. *Marie Curie et son laboratoire: Sciences et industrie de la radioactivité en France*. Paris: Éditions des archives contemporaines, 2001.

Bowler, Peter. *The Eclipse of Darwinism: Anti-Darwinian Evolution Theories in the Decades around 1900*. Baltimore: Johns Hopkins University Press, 1983.

———. *Science for All: The Popularization of Science in Early Twentieth-Century Britain*. Chicago: University of Chicago Press, 2009.

Brake, Laura, and Marysa Demoor, eds. *Dictionary of Nineteenth-Century Journalism in Great Britain and Ireland*. London: Academia Press, 2009.

Brian, Denis. *The Curies: A Biography of the Most Controversial Family in Science.* Hoboken, NJ: John Wiley and Sons, 2005.
Brock, William H. "Advancing Science: The British Association and the Professional Practice of Science." In *Parliament of Science,* edited by R. M. MacLeod and P. M. Collins, 89–117. London: Science Reviews, 1981.
———. "The *Chemical News,* 1859–1932." *Bulletin for the History of Chemistry* 12 (1992): 30–35.
———. "The Development of Commercial Science Journals in Victorian Britain." In *Development of Science Publishing in Europe,* edited by A. J. Meadows, 95–122. New York: Elsevier Science, 1980.
———, ed. *H.E. Armstrong and the Teaching of Science.* Cambridge: Cambridge University Press, 1973.
———. *William Crookes (1832–1919) and the Commercialization of Science.* Aldershot, UK: Ashgate, 2008.
Broman, Thomas. "The Habermasian Public Sphere and 'Science in the Enlightenment.'" *History of Science* 36 (1998): 123–149.
———. "Periodical Literature." In *Books and the Sciences in History,* edited by Marina Frasca-Spada and Nicholas Jardine, 225–238. Cambridge: Cambridge University Press, 2000.
Browne, E. Janet. *Charles Darwin: The Power of Place.* Princeton, NJ: Princeton University Press, 2003.
———. *Charles Darwin: Voyaging.* Princeton, NJ: Princeton University Press, 1996.
———. "'I Could Have Retched All Night': Charles Darwin and His Body." In *Science Incarnate: Historical Embodiments of Natural Knowledge,* edited by Christopher Lawrence and Steven Shapin, 240–287. Chicago: University of Chicago Press, 1998.
Buchanan, R. A. "The Rise of Scientific Engineering in Britain." *British Journal for the History of Science* 18 (1985): 218–233.
———. "Science and Engineering: A Case Study in British Experience in the Mid-Nineteenth Century." *Notes and Records of the Royal Society of London* 32 (1978): 215–223.
Bunge, Mario, and William R. Shea, eds. *Rutherford and Physics at the Turn of the Century.* New York: Science History Publications, 1979.
Burchfield, Joe D. *Lord Kelvin and the Age of the Earth.* New York: Science History Publications, 1975.
Burnham, John. "The Evolution of Editorial Peer Review." *Journal of the American Medical Association* 263 (1990): 1323–1329.
———. "How Journal Editors Came to Develop and Critique Peer Review Procedures." In *Research Ethics, Manuscript Review, and Journal Quality: Proceedings of a Symposium on the Peer Review–Editing Process,* edited by H. F. Maryland and R. E. Sojka, 55–62. Madison, WI: ACS Miscellaneous Publications, 1992.
Campbell, Philip. "Escape from the Impact Factor." *Ethics in Science and Environmental Politics* 8 (2008): 5–7.
Cantor, Geoffrey, ed. *Science in the Nineteenth-Century Periodical: Reading the Magazine of Nature.* Cambridge: Cambridge University Press, 2004.
Cantor, Geoffrey, and Sally Shuttleworth, eds. *Science Serialized: Representations of the Sciences in Nineteenth-Century Periodicals.* Cambridge, MA: MIT Press, 2004.
Carneiro, Robert, and Robert Perrin. "Herbert Spencer's Principles of Sociology: A Centennial Retrospective and Appraisal." *Annals of Science* 59 (2002): 221–261.
Casper, Christian Frederick. "Mutable Mobiles: Online Journals and the Evolving Genre Ecosystem of Science." PhD diss., North Carolina State University, 2009.
Cherny, Robert W. *Righteous Cause: The Life of William Jennings Bryan.* Boston: Little, Brown, 1985.
Chitty, Susan. *The Beast and the Monk: A Life of Charles Kingsley.* London: Hodder and Stoughton, 1974.

Churchill, Frederick B. "August Weismann and a Break from Tradition." *Journal of the History of Biology* 1 (1968): 91–112.

Churchill, Frederick B., and Helmut Risler, eds. *August Weismann: Ausgewahlte Briefe und Dokumente, Selected Letters and Documents*, 2 Vols. Freiburg: Universitätbibliothek, 1999.

Cleevely, R. J. "Bennett, Alfred William (1833–1902)." In *Oxford Dictionary of National Biography*, edited by H. C. G. Matthew and Brian Harrison. Oxford: Oxford University Press, 2004.

Close, Frank. *Too Hot to Handle: The Race for Cold Fusion*. Princeton, NJ: Princeton University Press, 1991.

Clow, Nani N. "The Laboratory of Victorian Culture: Experimental Physics, Industry, and Pedagogy in the Liverpool Laboratory of Oliver Lodge, 1881–1900." PhD diss., Harvard University, 1999.

Collins, Harry, and Trevor Pinch. *The Golem: What Everyone Should Know about Science*. Cambridge: Cambridge University Press, 1993.

Costabel, Pierre. "'L'à-peu-près n'est pas leur fait': Journal des sçavans 12 mai 1698." In *L'à-peu-près: Aspects anciens et modernes de l'approximation*, edited by Louis Frey, Gilles Lachaud, Charlotte Carcassonne, Jacques Le Goff, Jean-Luc Verley, Pierre Costabel, Bernard Bru, and Claude Corp Brezinski, 79–85. Paris: École des Hautes Études en Sciences Sociales, 1988.

Cox, Allan, ed. *Plate Tectonics and Geomagnetic Reversals*. New York: W. H. Freeman, 1973.

Cox, Stephen. "The Royal Society in Cold War Europe." *Notes and Records of the Royal Society of London* 64, no. S1 (2010): S131–S136.

Crawford, Elisabeth. "Internationalism in Science as a Casualty of the First World War: Relations between German and Allied Scientists as Reflected in Nominations for the Nobel Prizes in Physics and Chemistry." *Sociology of Science* 27 (1988): 163–201.

———. *Nationalism and Internationalism in Science, 1880–1939: Four Studies of the Nobel Population*. Cambridge: Cambridge University Press, 1992.

———. "The Universe of International Science, 1880–1939." In *Solomon's House Revisited: The Organization and Institutionalization of Science*, edited by Tore Frängsmyr, 251–269. Canton, MA: Science History Publications, 1990.

Crawford, Elisabeth, Terry Shinn, and Sverker Sörlin, eds. *Denationalizing Science: The Contexts of International Scientific Practice*. Dordrecht: Kluwer Academic, 1993.

Creager, Angela. "Tracing the Politics of Changing Postwar Research Practices: The Export of 'American' Radioisotopes to European Biologists." *Studies in History and Philosophy of Biological and Biomedical Sciences* 33C (2002): 367–388.

Creager, Angela, and Gregory J. Morgan. "After the Double Helix: Rosalind Franklin's Research on Tobacco Mosaic Virus." *Isis* 99 (2008): 239–272.

Crick, Francis. *What Mad Pursuit: A Personal View of Scientific Discovery*. New York: Basic Books, 1988.

Crosland, Maurice. "History of Science in a National Context." *British Journal for the History of Science* 10 (1977): 95–113.

Crystal, David. *English as a Global Language*. Cambridge: Cambridge University Press, 1997.

Csiszar, Alex. "Broken Pieces of Fact: The Scientific Periodical and the Politics of Search in Nineteenth-Century France and Britain." PhD diss., Harvard University, 2010.

Cunningham, Andrew, and Nicholas Jardine, eds. *Romanticism and the Sciences*. Cambridge: Cambridge University Press, 1990.

Darden, Lindley. "William Bateson and the Promise of Mendelism." *Journal of the History of Biology* 10 (1977): 87–106.

Daston, Lorraine. "The Ideal and Reality of the Republic of Letters in the Enlightenment." *Science in Context* 4 (1991): 367–386.

Daston, Lorraine, and Michael Otte, eds. "Styles of Science." Special issue, *Science in Context* 4 (1991).

Davis, J. L. "The Research School of Marie Curie in the Paris Faculty, 1907-14." *Annals of Science* 52 (1995): 321-355.

Dawson, Gowan. "The *Cornhill Magazine* and Shilling Monthlies in Mid-Victorian Britain." In *Science in the Nineteenth-Century Periodical*, edited by Geoffrey Cantor, Gowan Dawson, Grame Gooday, Richard Noakes, Sally Shuttleworth, and Jonathan Topham, 123-150. Cambridge: Cambridge University Press, 2004.

———. *Darwin, Literature and Victorian Respectability*. Cambridge: Cambridge University Press, 2007.

———. "The *Review of Reviews* and the New Journalism in Late-Victorian Britain." In *Science in the Nineteenth-Century Periodical*, edited by Geoffrey Cantor, Gowan Dawson, Grame Gooday, Richard Noakes, Sally Shuttleworth, and Jonathan R. Topham, 172-195. Cambridge: Cambridge University Press, 2004.

Dean, Michael Emmans. "Homeopathy and 'the Progress of Science.'" *History of Science* 39 (2001): 255-283.

Dearing, James W. "Newspaper Coverage of Maverick Science: Creating Controversy through Balancing." *Public Understanding of Science* 4 (1995): 341-361.

de Chadarevian, Soraya. *Designs for Life: Molecular Biology after World War II*. Cambridge: Cambridge University Press, 2002.

Degele, Nina. "On the Margins of Everything: Doing, Performing, and Staging Science in Homeopathy." *Science, Technology, and Human Values* 30 (2005): 111-136.

Desmond, Adrian. *Huxley*. 2 vols. London: Penguin, 1994.

———. "Redefining the X Axis: 'Professionals,' 'Amateurs' and the Making of Mid-Victorian Biology." *Journal of the History of Biology* 34 (2001): 3-50.

de Young, Ursula. *A Vision of Modern Science: John Tyndall and the Role of the Scientist in Victorian Culture*. London: Palgrave Macmillan, 2011.

Dibner, Bern. *Wilhelm Conrad Röntgen and the Discovery of X Rays*. New York: Watts, 1968.

Di Gregorio, Mario. *T. H. Huxley's Place in Natural Science*. New Haven, CT: Yale University Press, 1984.

Duhem, Pierre. *The Aim and Structure of Physical Theory*. Princeton, NJ: Princeton University Press, 1954.

Elliott, Paul. "Erasmus Darwin, Herbert Spencer, and the Origins of the Evolutionary Worldview in British Provincial Scientific Culture, 1770-1850." *Isis* 94 (2003): 1-29.

Endersby, Jim. *Imperial Nature: Joseph Hooker and the Practices of Victorian Science*. Chicago: University of Chicago Press, 2008.

England, Philip C., Peter Moinar, and Frank M. Richter. "Kelvin, Perry and the Age of the Earth." *American Scientist* 95 (2007): 342-349.

England, Richard, ed. *Design after Darwin, 1860-1900*. Bristol: Thoemmes, 2003.

Evans, Richard J. *The Third Reich in Power*. New York: Penguin, 2005.

Eve, A. S., ed. *Rutherford: Being the Life and Letters of the Rt. Hon. Lord Rutherford, O.M.* New York; London: The Macmillan Company, 1939.

Everett-Lane, Debra. "International Scientific Congresses, 1878-1913: Community and Conflict in the Pursuit of Knowledge." PhD diss., Columbia University, 2004.

Eyre, J. V., and E. H. Rodd. "Raphael Meldola." In *British Chemists*, edited by A. Findlay and W. H. Mills, 96-125. London: Chemical Society, 1947.

Fadlon, Judith, and Noah Lewin-Epstein. "Laughter Spreads: Another Perspective on Boundary Crossing in the Benveniste Affair." *Social Studies of Science* 27 (1997): 131-141.

Fan, Fa-ti. "Redrawing the Map: Science in Twentieth-Century China." *Isis* 98 (2007): 524-538.

Farrall, Lyndsay A. "Controversy and Conflict in Science: A Case Study: The English Biometric School and Mendel's Laws." *Social Studies of Science* 5 (1975): 269-301.

Feltes, N. N. *Modes of Production of Victorian Novels*. Chicago: University of Chicago Press, 1986.

Fölsing, Albrecht. *Albert Einstein: A Biography*. Translated by Ewald Osers. New York: Viking, 1997.

Forsdyke, Donald. *The Origin of Species, Revisited: A Victorian Who Anticipated Modern Developments in Darwin's Theory*. Kingston, ON: McGill-Queen's University Press, 2001.

Francis, Mark. *Herbert Spencer and the Invention of Modern Life*. Stocksfield, UK: Acumen, 2007.

Frankel, Henry. "The Development, Reception, and Acceptance of the Vine-Matthews-Morley Hypothesis." *Historical Studies in the Physical Sciences* 13 (1982): 1–39.

Fyfe, Aileen. "Conscientious Workmen or Booksellers' Hacks? The Professional Identities of Science Writers in the Mid-Nineteenth Century." *Isis* 96 (2005): 192–223.

———. *Science and Salvation: Evangelical Popular Science Publishing in Victorian Britain*. Chicago: University of Chicago Press, 2004.

———. *Steam-Powered Knowledge: William Chambers and the Business of Publishing, 1820–1860*. Chicago: University of Chicago Press, 2012.

Gay, Hannah. "Chemist, Entomologist, Darwinian, and Man of Affairs: Raphael Meldola and the Making of a Scientific Career." *Annals of Science* 67 (2010): 79–119.

Geison, Gerald. *Michael Foster and the Cambridge School of Physiology: The Scientific Enterprise in Late Victorian Society*. Princeton, NJ: Princeton University Press, 1992.

———. *The Private Science of Louis Pasteur*. Princeton, NJ: Princeton University Press, 1995.

Gieryn, Thomas F. "The (Cold) Fusion of Science, Mass Media, and Politics." In *Cultural Boundaries of Science*, 183–232. Chicago: University of Chicago Press, 1999.

Gieryn, Thomas F., George M. Bevins, and Stephen G. Zehr. "Professionalization of American Scientists: Public Science in the Creation/Evolution Trials." *American Sociological Review* 50 (1985): 392–409.

Gillespie, N. C. "The Duke of Argyll, Evolutionary Anthropology, and the Art of Scientific Controversy." *Isis* 68 (1977): 40–54.

Gingras, Yves. "Revisiting the 'Quiet Debut' of the Double Helix: A Bibliometric and Methodological Note on the 'Impact' of Scientific Publications." *Journal of the History of Biology* 43 (2010): 159–181.

Glasser, Otto. *Wilhelm Conrad Röntgen and the Early History of the Roentgen Rays*. Springfield, IL: Thomas, 1934.

Goldgar, Anne. *Impolite Learning: Conduct and Community in the Republic of Letters, 1680–1750*. New Haven, CT: Yale University Press, 1995.

Gooday, Graeme. "The Morals of Energy Metering." In *The Values of Precision*, edited by M. Norton Wise, 239–282. Princeton, NJ: Princeton University Press, 1995.

———. *The Morals of Measurement: Accuracy, Irony and Trust in Late Victorian Electrical Practice*. Cambridge: Cambridge University Press, 2004.

Goodman, Dena. *The Republic of Letters: A Cultural History of the French Enlightenment*. Ithaca, NY: Cornell University Press, 1996.

Gordin, Michael. *The Pseudoscience Wars: Immanuel Velikovsky and the Birth of the Modern Fringe*. Chicago: University of Chicago Press, 2012.

———. *A Well-Ordered Thing: Dmitrii Mendeleev and the Shadow of the Periodic Table*. New York: Basic Books, 2004.

Graves, Charles L. *Life and Letters of Alexander Macmillan*. London: Macmillan, 1910.

Gross, Alan G., Joseph E. Harmon, and Michael Reidy. *Communicating Science: The Scientific Article from the 17th Century to the Present*. Oxford: Oxford University Press, 2002.

Gross, John. *The Rise and Fall of the Man of Letters: Aspects of English Literary Life since 1800*. London: Weidenfeld and Nicolson, 1969.

Habermas, Jürgen. *The Structural Transformation of the Public Sphere*. Translated by Thomas Burger. Cambridge, MA: MIT Press, 1991.

Hamblin, Jacob D. "Science in Isolation: American Marine Geophysics Research, 1950–1968." *Physics in Perspective* 2 (2000): 293–312.

Harman, Oren Solomon. "C.D. Darlington and the British and American Reaction to Lysenko and the Soviet Conception of Science." *Journal of the History of Biology* 36 (2003): 309-352.

Harwood, Jonathan. "Engineering Education between Science and Practice: Rethinking the Historiography." *History and Technology* 22 (2006): 53-79.

———. "German Science and Technology under National Socialism." *Perspectives on Science* 5 (1997): 128-151.

———. "National Styles in Science: Genetics in Germany and the United States between the World Wars." *Isis* 78 (1987): 390-414.

———. *Styles of Scientific Thought: The German Genetics Community, 1900-1933*. Chicago: University of Chicago Press, 1993.

Heilbron, J. L. *H.G.J. Moseley: The Life and Letters of an English Physicist, 1887-1915*. Berkeley: University of California Press, 1974.

———. "Physics at McGill in Rutherford's Time." In *Rutherford and Physics at the Turn of the Century*, edited by Mario Bunge and William R. Shea, 42-73. New York: Science History Publications, 1979.

Henson, Louise, Geoffrey Cantor, Gowan Dawson, Richard Noakes, Sally Shuttleworth, Jonathan R. Topham, Aileen Fyfe, et al., eds. *Culture and Science in the Nineteenth-Century Media*. Aldershot, UK: Ashgate, 2004.

Hossfeld, Uwe, and Lennart Olsson. "*Nature* and Hitler." Accessed January 1, 2014. http://www.nature.com/nature/history/full/nature06242.html.

Houghton, Walter, ed. *Wellesley Index to Victorian Periodicals*. New York: Routledge, 1999.

Hughes, Jeff. "The Radioactivists: Community, Controversy, and the Rise of Nuclear Physics." PhD diss., University of Cambridge, 1993.

Hughes, Linda K., and Michael Lund. *The Victorian Serial*. Charlottesville: University Press of Virginia, 1991.

Hunt, Bruce. "Experimenting on the Ether: Oliver J. Lodge and the Great Whirling Machine." *Historical Studies in the Physical and Biological Sciences* 16 (1986): 111-134.

———. *The Maxwellians*. Ithaca: Cornell University Press, 2005.

Huxley, Leonard, ed. *Life and Letters of Thomas Henry Huxley*. London: Macmillan, 1900.

Hynes, Samuel. *A War Imagined: The First World War and British Culture*. London: Bodley Head, 1990.

Isaacson, Walter. *Einstein: His Life and Universe*. New York: Simon and Schuster, 2007.

Ivanov, Konstantin. "Science after Stalin: Forging a New Image of Soviet Science." *Science in Context* 15 (2002): 317-338.

James, Elizabeth, ed. *Macmillan: A Publishing Tradition from 1843*. New York: Palgrave, 2002.

James, Frank A. J. L. "Reporting Royal Institution Lectures, 1826-1967." In *Science Serialized: Representations of the Sciences in Nineteenth-Century Periodicals*, edited by Geoffrey Cantor and Sally Shuttleworth, 67-79. Cambridge, MT: MIT Press, 2004.

Jenson, J. V. "The X Club: Fraternity of Victorian Scientists." *British Journal for the History of Science* 5 (1970): 63-72.

Johns, Adrian. "Miscellaneous Methods: Authors, Societies and Journals in Early Modern England." *British Journal for the History of Science* 33 (2000): 159-186.

———. *The Nature of the Book: Print and Knowledge in the Making*. Chicago: University of Chicago Press, 1998.

Jolly, William P. *Sir Oliver Lodge*. London: Constable, 1974.

Jones, Greta, and Robert Peel, eds. *Herbert Spencer: The Intellectual Legacy*. London: Galton Institute, 2004.

Jordan, John O., and Robert L. Patten. *Literature in the Marketplace: Nineteenth-Century British Publishing and Reading Practices*. Cambridge: Cambridge University Press, 1995.

Jungnickel, Christa, and Russell McCormmach. *Intellectual Mastery of Nature: Theoretical Physics from Ohm to Einstein*. 2 vols. Chicago: University of Chicago Press, 1986.

Kaiser, David. *American Physics and the Cold War Bubble.* Chicago: University of Chicago Press, forthcoming.

———. "Cold War Requisitions, Scientific Manpower, and the Production of American Physicists after World War II." *Historical Studies in the Physical and Biological Sciences* 33 (2002): 131–159.

———. *How the Hippies Saved Physics: Science, Counterculture, and the Quantum Revival.* Chicago: University of Chicago Press, 2012.

Kauffman, G. B., ed. *Frederick Soddy, 1877–1956.* Boston: D. Reidel, 1985.

Kaufman, Martin. *Homeopathy in America: The Rise and Fall of a Medical Heresy.* Baltimore: Johns Hopkins University Press, 1971.

Kenny, Michael G. "A Darker Shade of Green: Medical Botany, Homeopathy, and Cultural Politics in Interwar Germany." *Social History of Medicine* 15 (2002): 481–504.

Kent, Susan Kingsley. *Aftershocks: Politics and Trauma in Britain, 1918–1931.* New York: Palgrave Macmillan, 2009.

Kevles, Daniel. *The Baltimore Case: A Trial of Science, Politics, and Character.* New York: W. W. Norton, 1998.

———. "Genetics in the United States and Great Britain, 1890–1930: A Review with Speculations." *Isis* 71 (1980): 441–455.

———. *In the Name of Eugenics: Genetics and the Uses of Human Heredity.* Cambridge, MA: Harvard University Press, 1985.

———. "'Into Hostile Political Camps': The Reorganization of International Science in World War I." *Isis* 62 (1971): 47–60.

Kim, Stephen. *John Tyndall's Transcendental Materialism and the Conflict between Religion and Science in Victorian England.* Lewiston, NY: Mellen University Press, 1996.

Kjærgaard, Peter C. "'Within the Bounds of Science': Redirecting Controversies to *Nature*." In *Culture and Science in the Nineteenth-Century Media,* edited by Louise Henson, G. N. Cantor, Gowan Dawson, Richard Noakes, Sally Shuttleworth, and Jonathan Topham, 211–221. Aldershot, UK: Ashgate, 2004.

Knight, David. "Science and Culture in Mid-Victorian Britain: The Reviews, and William Crookes' *Quarterly Journal of Science*." *Nuncius* 11 (1996): 43–54.

Koenig, Louis William. *Bryan: A Political Biography of William Jennings Bryan.* New York: Putnam, 1971.

Kohlstedt, Sally Gregory. "*Science*: The Struggle for Survival, 1880 to 1894." *Science* 209 (4 July 1980): 33–42.

Kozhevnikov, Aleksei. "Piotr Kapitza and Stalin's Government: A Study in Moral Choice." *Historical Studies in the Physical and Biological Sciences* 22 (1991): 131–164.

Kragh, Helge. "Anatomy of a Priority Conflict: The Case of Element 72." *Centaurus: International Magazine of the History of Mathematics, Science, and Technology* 23 (1979–1980): 275–301.

Krementsov, Nikolai. *International Science between the World Wars: The Case of Genetics.* London: Routledge, 2005.

———. "In the Shadow of the Bomb: US-Soviet Biomedical Relations in the Early Cold War, 1944–1948." *Journal of Cold War Studies* 9 (2007): 41–67.

———. "A 'Second Front' in Soviet Genetics: The International Dimension of the Lysenko Controversy, 1944–1947." *Journal of the History of Biology* 29 (1996): 229–250.

———. *Stalinist Science.* Princeton, NJ: Princeton University Press, 1996.

Krige, John. *American Hegemony and the Postwar Reconstruction of Science in Europe.* Cambridge, MA: MIT Press, 2006.

Laeven, A. H. *The "Acta Eruditorum" under the Editorship of Otto Mencke (1644–1707): The History of an International Learned Journal between 1682 and 1707.* Translated by Lynne Richards. Amsterdam: APA-Holland University Press, 1990.

Lagemann, Robert T. "New Light on Old Rays: N Rays." *American Journal of Physics* 45 (March 1977): 281–284.

Larson, Edward. *Summer for the Gods: The Scopes Trial and America's Continuing Debate over Science and Religion*. New York: Basic Books, 1997.
Lavine, Matthew. "A Cultural History of Radiation and Radioactivity in the United States, 1895-1945." PhD diss., University of Wisconsin-Madison, 2008.
Lear, John. "Canada's Unappreciated Role as Scientific Innovator." *Saturday Review* (2 September 1967): 45-50.
LeGrand, H. E. *Drifting Continents and Shifting Theories: The Modern Revolution in Geology and Scientific Change*. Cambridge: Cambridge University Press, 1988.
Lesch, John E. "The Role of Isolation in Evolution: George J. Romanes and John T. Gulik." *Isis* 66 (1975): 483-503.
Lester, Joseph. *E. Ray Lankester and the Making of British Biology*. Oxford: Alden, 1995.
Lewenstein, Bruce V. "Cold Fusion and Hot History." *Osiris* 7 (1992): 135-163.
———. "From Fax to Facts: Communication in the Cold Fusion Saga." *Social Studies of Science* 25 (1995): 403-436.
Lienesch, Michael. *In the Beginning: Fundamentalism, the Scopes Trial, and the Making of the Antievolution Movement*. Chapel Hill: University of North Carolina Press, 2007.
Lightman, Bernard. "*Knowledge* Confronts *Nature*: Richard Proctor and Popular Science Periodicals." In *Culture and Science in the Nineteenth-Century Media*, edited by Louise Henson, G. N. Cantor, Gowan Dawson, Richard Noakes, Sally Shuttleworth, and Jonathan Topham, 199-210. Aldershot, UK: Ashgate, 2004.
———. "Marketing Knowledge for the General Reader: Victorian Popularizers of Science." *Endeavour* 24, no. 3 (2000): 100-106.
———. *Victorian Popularizers of Science: Designing Nature for New Audiences*. Chicago: University of Chicago Press, 2007.
———. "The Visual Theology of Victorian Popularizers of Science: From Reverent Eye to Chemical Retina." *Isis* 91 (2000): 650-680.
Lightman, Bernard, George Levine, Barbara T. Gates, and Frank M. Turner, eds. *Victorian Science in Context*. Chicago: University of Chicago Press, 1997.
Lockyer, Thomazine Mary Browne, Winifred Lucas Lockyer, and Herbert Dingle, eds. *Life and Work of Sir Norman Lockyer*. London: Macmillan, 1928.
Lucier, Paul. "The Professional and the Scientist in Nineteenth-Century America." *Isis* 100 (2009): 699-732.
Lux, David, and Harold Cook. "Closed Circles or Open Networks? Communicating at a Distance during the Scientific Revolution." *History of Science* 36 (1998): 179-211.
Mackenzie, Norman, and Jeanne Mackenzie. *The Time Traveller: The Life of H.G. Wells*. 2nd ed. London: Hogarth, 1987.
MacLeod, R. M. "Is It Safe to Look Back?" *Nature* 224 (1 November 1969): 417-461.
———. "Science for Imperial Efficiency and Social Change: Reflections on the British Science Guild, 1905-1936." *Public Understanding of Science* 3 (1994): 155-193.
———. "The X-Club: A Social Network of Science in Late-Victorian England." *Notes and Records of the Royal Society of London* 24 (1970): 305-322.
Maddox, Brenda. *Rosalind Franklin: The Dark Lady of DNA*. New York: Harper Perennial, 2003.
Maddox, John. "Introduction." In *Nature: 1869-1879, Index to Volumes 1-20*, 1-19. London: Palgrave Macmillan Archive Press, 2002.
Magnello, M. E. "Karl Pearson's Gresham Lectures: W.F.R. Weldon, Speciation and the Origins of Pearsonian Statistics." *British Journal for the History of Science* 29 (1996): 43-63.
———. "Karl Pearson's Mathematization of Inheritance: From Ancestral Heredity to Mendelian Genetics (1895-1909)." *Annals of Science* 55 (1998): 35-94.
Manlove, Colin. "Charles Kingsley, H.G. Wells, and the Machine in Victorian Fiction." *Nineteenth-Century Literature* 48 (1993): 212-239.
Marchant, James, ed. *Raphael Meldola: Reminiscences of His Worth and Work by Those Who Knew*

Him, Together with a Chronological List of His Publications, 1869-1915. London: Williams and Norgate, 1916.

Marsden, Ben, and Crosbie Smith. *Engineering Empires: A Cultural History of Technology in Nineteenth-Century Britain.* London: Palgrave Macmillan, 2005.

Mason, Joan. "Hertha Ayrton (1854-1923) and the Admission of Women to the Royal Society of London." *Notes and Records of the Royal Society of London* 45 (1991): 201-220.

Matricon, Jean, and Georges Waysand. *Cold Wars: A History of Superconductivity.* New Brunswick, NJ: Rutgers University Press, 2003.

Matthew, H. C. G. "Campbell, George Douglas, Eighth Duke of Argyll in the Peerage of Scotland, and First Duke of Argyll in the Peerage of the United Kingdom (1823-1900)." In *Oxford Dictionary of National Biography*, edited by H. C. G. Matthew, Brian Harrison, and Lawrence Goldman. Oxford: Oxford University Press, 2004.

Mayer, Anna-Katherine. "Reluctant Technocrats: Science Promotion in the Neglect-of-Science Debate of 1916-1918." *History of Science* 43 (2005): 139-159.

Mayhew, Robert. "Mapping Science's Imagined Community: Geography as a Republic of Letters, 1600-1800." *British Journal for the History of Science* 38 (2005): 73-92.

McAllister, James W. "Competition among Scientific Disciplines in Cold Nuclear Fusion Research." *Science in Context* 5 (1992): 17-49.

McElheny, Victor K. *Watson and DNA: Making a Scientific Revolution.* Cambridge, MA: Perseus, 2003.

McLean, Roderick. "Kaiser Wilhelm II and the British Royal Family: Anglo-German Dynastic Relations in Political Context, 1890-1914." *History* 86 (2001): 478-502.

Meadows, A. Jack. *Communication in Science.* London: Butterworths, 1974.

———. *Science and Controversy: A Biography of Sir Norman Lockyer.* Cambridge, MA: MIT Press, 1972.

———. *Victorian Scientist: The Growth of a Profession.* London: British Library, 2005.

Mendel, Gregor. *Experiments in Plant Hybridisation*, edited by J. H. Bennett. London: Oliver and Boyd, 1965.

Merricks, Linda. *The World Made New: Frederick Soddy, Science, Politics, and Environment.* Oxford: Oxford University Press, 1996.

Milner, Richard. "Huxley's Bulldog: The Battles of E. Ray Lankester (1846-1929)." *The Anatomical Record* 257 (1999): 90-95.

Moran, Jeffrey P. *The Scopes Trial: A Brief History with Documents.* New York: Palgrave, 2002.

Morange, Michel. *A History of Molecular Biology.* Cambridge, MA: Harvard University Press, 1998.

Morgan, B. T. *Histoire du* Journal des Sçavans *depuis 1665 jusqu'en 1701.* Paris: Presses Universitaires de France, 1928.

Morgan, Charles. *The House of Macmillan (1843-1943).* New York: Macmillan, 1944.

Morley, Lawrence. "The Zebra Pattern." In *Plate Tectonics: An Insider's History of the Modern Theory of the Earth*, edited by Naomi Oreskes, 67-85. Cambridge, MA: Westview, 2001.

Morrell, Jack. "Professionalisation." In *Companion to the History of Modern Science*, edited by Robert C. Olby, 980-989. London: Routledge, 1990.

Morrell, Jack, and Arnold Thackray. *Gentlemen of Science: Early Years of the British Association for the Advancement of Science.* Oxford: Oxford University Press, 1981.

Morrison, Mark. *Modern Alchemy: Occultism and the Emergence of Atomic Theory.* Oxford: Oxford University Press, 2007.

Mussell, James. *Science, Time and Space in the Late Nineteenth-Century Periodical Press.* Aldershot, UK: Ashgate, 2007.

Needham, Joseph. *The Grand Titration: Science and Society in East and West.* London: Allen and Unwin, 1969.

Noakes, Richard. "The 'Bridge Which Is between Physical and Psychical Research': William Fletcher Barrett, Sensitive Flames, and Spiritualism." *History of Science* 42 (2004): 419-464.

———. "Ethers, Religion and Politics in Late-Victorian Physics: Beyond the Wynne Thesis." *History of Science* 43 (2005): 415-455.

Nova. "The Secret of Photo 51." PBS Home Video, 2003.

Nowotny, Helga, and Ulrike Felt. *After the Breakthrough: The Emergence of High-Temperature Superconductivity as a Research Field*. New York: Cambridge University Press, 1997.

Nye, Mary Jo. *Before Big Science: The Pursuit of Modern Chemistry and Physics, 1800-1940*. Cambridge, MA: Harvard University Press, 1996.

———. "N-Rays: An Episode in the History and Psychology of Science." *Historical Studies in the Physical Sciences* 11 (1980): 125-156.

Ogilvie, Marilyn Bailey. *Marie Curie: A Biography*. Westport, CT: Greenwood, 2004.

Olby, Robert. "The Dimensions of Scientific Controversy: The Biometric-Mendelian Debate." *British Journal for the History of Science* 22 (1989): 299-320.

———. *The Path to the Double Helix*. Seattle: University of Washington Press, 1974.

———. "Quiet Debut for the Double Helix." *Nature* 421 (23 January 2003): 402-405.

———. "William Bateson's Introduction of Mendelism to England: A Reassessment." *British Journal for the History of Science* 20 (1987): 399-420.

Oppenheim, Janet. *The Other World: Spiritualism and Psychical Research in England, 1850-1914*. Cambridge: Cambridge University Press, 1985.

Oreskes, Naomi. "The Rejection of Continental Drift." *Historical Studies in the Physical and Biological Sciences* 18 (1988): 311-348.

Oreskes, Naomi, and Erik Conway. *Merchants of Doubt: How a Handful of Scientists Obscured the Truth on Issues from Tobacco Smoke to Global Warming*. New York: Bloomsbury, 2010.

Owen, Alex. *The Place of Enchantment: British Occultism and the Culture of the Modern*. Chicago: University of Chicago Press, 2004.

Owen, J. S. Mackenzie. *The Scientific Article in the Age of Digitization*. Dordrecht: Springer, 2007.

Paradis, James G. *T. H. Huxley: Man's Place in Nature*. Lincoln: University of Nebraska Press, 1978.

Peierls, Rudolf. "Rutherford and Bohr." *Notes and Records of the Royal Society of London* 42 (1988): 229-241.

Perutz, Max. "DNA Helix." *Science* 164 (27 June 1969): 1537-1539.

Pestre, Dominique. "The Moral and Political Economy of French Scientists in the First Half of the XXth Century." *History and Technology* 13 (1997): 241-248.

Picart, Caroline Joan. "Blurring Boundaries: A Reply to Fadlon and Lewin-Epstein." *Social Studies of Science* 27 (1997): 142-146.

———. "Scientific Controversy as Farce: The Benveniste-Maddox Counter Trials." *Social Studies of Science* 24 (1994): 7-37.

Pole, William. *The Life of Sir William Siemens*. London: John Murray, 1888.

Porter, Roy. "Gentlemen and Geology: The Emergence of a Scientific Career, 1660-1920." *Historical Journal* 21 (1978): 809-836.

Porter, Theodore M. "The Fate of Scientific Naturalism." In *Victorian Scientific Naturalism: Community, Identity, Continuity*, edited by Gowan Dawson and Bernard Lightman, 265-287. Chicago: University of Chicago Press, 2014.

———. *Karl Pearson: The Scientific Life in a Statistical Age*. Princeton, NJ: Princeton University Press, 2004.

Pyenson, Lewis. "The Incomplete Transmission of a European Image: Physics at Greater Buenos Aires and Montreal, 1890-1920." *Proceedings of the American Philosophical Society* 122 (1978): 92-114.

Quinn, Susan. *Marie Curie: A Life*. New York: Simon and Schuster, 1995.

Raby, Peter. *Alfred Russel Wallace: A Life*. Princeton, NJ: Princeton University Press, 2001.

Randi, James. *Flim-Flam! Psychics, ESP, Unicorns and Other Delusions*. New York: Prometheus, 1982.
Reingold, Nathan. "National Style in the Sciences: The United States Case." In *Human Implications of Scientific Advance*, edited by Eric G. Forbes, 163-173. Edinburgh: Edinburgh University Press, 1978.
Renneberg, Monika, and Mark Walker. *Science, Technology, and National Socialism*. Cambridge: Cambridge University Press, 1994.
Rennie, Drummond. "Editorial Peer Review: Its Development and Rationale." In *Peer Review in Health Sciences*, edited by Fiona Godlee and Tom Jefferson, 1-13. London: BMJ, 1999.
Richards, Robert J. *The Romantic Conception of Life: Science and Philosophy in the Age of Goethe*. Chicago: University of Chicago Press, 2002.
———. *The Tragic Sense of Life: Ernst Haeckel and the Struggle over Evolutionary Thought*. Chicago: University of Chicago Press, 2008.
Richmond, Marsha. "Women in the Early History of Genetics: William Bateson and the Newnham College Mendelians, 1900-1910." *Isis* 92 (2001): 55-90.
Ridley, Mark. "Coadaptation and the Inadequacy of Natural Selection." *British Journal for the History of Science* 15 (1982): 45-68.
Roll-Hansen, Diderik. *The Academy, 1869-1879: Victorian Intellectuals in Revolt*. Copenhagen: Rosenkilde and Bagger, 1957.
Roos, David A. "The 'Aims and Intentions' of *Nature*." In *Victorian Science and Victorian Values: Literary Perspectives*, edited by James Paradis and Thomas Postlewait, 159-180. New Brunswick, NJ: Rutgers University Press, 1981.
Ross, Sydney. "Scientist: The Story of a Word." *Annals of Science* 18 (1962): 65-85.
Rossianov, Kirill O. "Editing Nature: Joseph Stalin and the 'New' Soviet Biology." *Isis* 84 (1993): 728-745.
Rowlands, Peter. *Oliver Lodge and the Liverpool Physical Society*. Cambridge: Cambridge University Press, 1990.
Rowlands, Peter, and J. Patrick Wilson, eds. *Oliver Lodge and the Invention of Radio*. Liverpool: PD Publications, 1994.
Ryan, Frank X. *Darwinism and Theology in America*. Vol. 4, *Science, Humanism and the Scopes Trial*. Bristol: Thoemmes, 2002.
Rylance, Rick. *Victorian Psychology and British Culture 1850-1880*. Oxford: Oxford University Press, 2000.
Sapp, Jan. "The Struggle for Authority in the Field of Heredity, 1900-1932: New Perspectives on the Rise of Genetics." *Journal of the History of Biology* 16 (1983): 311-342.
Sayre, Anne. *Rosalind Franklin and DNA*. New York: W. W. Norton, 1975.
Schroeder-Gudehus, Brigitte. "Challenge to Transnational Loyalties: International Scientific Organizations after the First World War." *Science Studies* 3 (1973): 93-118.
———. "Nationalism and Internationalism." In *Companion to the History of Modern Science*, edited by Robert Olby, 909-919. New York: Routledge, 1990.
Schwartz, Joel S. "George John Romanes's Defense of Darwinism: The Correspondence of Charles Darwin with His Chief Disciple." *Journal of the History of Biology* 28 (1995): 281-316.
———. "Out from Darwin's Shadow: George John Romanes's Efforts to Popularize Science in *Nineteenth Century* and Other Victorian Periodicals." *Victorian Periodicals Review* 35 (2002): 133-159.
Scott, J. D. *Siemens Brothers, 1858-1958*. London: Weidenfeld and Nicolson, 1958.
Secord, James A. "The Geological Survey of Great Britain as a Research School, 1839-1855." *History of Science* 24 (1986): 223-275.
———. "How Scientific Conversation Became Shop Talk." In *Science in the Marketplace: Nineteenth-Century Sites and Experiences*, edited by Bernard Lightman and Aileen Fyfe, 23-59. Chicago: University of Chicago Press, 2007.

———. "Science, Technology and Mathematics." In *The Cambridge History of the Book in Britain*. Vol. 6, *1830–1914*, edited by David McKitterick, 444–474. Cambridge: Cambridge University Press, 2009.

———. *Victorian Sensation: The Extraordinary Publication, Reception, and Secret Authorship of Vestiges of the Natural History of Creation*. Chicago: University of Chicago Press, 2000.

———. *Visions of Science: Books and Readers at the Dawn of the Victorian Age*. Chicago: University of Chicago Press, 2014.

Segrè, Emilio. *Enrico Fermi, Physicist*. Chicago: University of Chicago Press, 1970.

Shapin, Steven. *A Social History of Truth: Civility and Science in Seventeenth-Century England*. Chicago: University of Chicago Press, 1994.

Sibum, H. Otto. "Experimentalists in the Republic of Letters." *Science in Context* 16 (2003): 89–120.

Sime, Ruth Lewin. *Lise Meitner: A Life in Physics*. Berkeley: University of California Press, 1996.

Simon, Bart. *Undead Science: Science Studies and the Afterlife of Cold Fusion*. New Brunswick, NJ: Rutgers University Press, 2002.

Slotten, Ross A. *The Heretic in Darwin's Court: The Life of Alfred Russel Wallace*. New York: Columbia University Press, 2004.

Smith, Crosbie, and M. Norton Wise. *Energy and Empire: A Biographical Study of Lord Kelvin*. Cambridge: Cambridge University Press, 1989.

Smith, C. U. M. "Evolution and the Problem of Mind. Part I. Herbert Spencer." *Journal of the History of Biology* 15 (1982): 241–262.

Smith, David C. *H. G. Wells: Desperately Mortal*. New Haven, CT: Yale University Press, 1986.

Snyder, Laura. *The Philosophical Breakfast Club: Four Remarkable Friends Who Transformed Science and Changed the World*. New York: Broadway Books, 2011.

Soifer, Valerii, Leo Gruliow, and Rebecca Gruliow. *Lysenko and the Tragedy of Soviet Science*. New Brunswick, NJ: Rutgers University Press, 1994.

Sokal, Michael M. "Science and James McKeen Cattell, 1894 to 1945." *Science* 209 (4 July 1980): 43–52.

Sponsel, Alistair. "Constructing a 'Revolution in Science': The Campaign to Promote a Favourable Reception for the 1919 Solar Eclipse Experiments." *British Journal for the History of Science* 35 (2002): 439–467.

———. "Coral Reef Formation and the Sciences of the Earth, Life, and Sea, 1770–1952." PhD diss., Princeton University, 2009.

Stanley, Matthew. "'An Expedition to Heal the Wounds of War': The 1919 Eclipse and Eddington as Quaker Adventurer." *Isis* 94 (2003): 57–89.

———. "Where Naturalism and Theism Met: The Uniformity of Nature." In *Victorian Scientific Naturalism: Community, Identity, Continuity*, edited by Bernard Lightman and Gowan Dawson, 242–264. Chicago: University of Chicago Press, 2014.

Stewart, John A. *Drifting Continents and Colliding Paradigms: Perspectives on the Geoscience Revolution*. Bloomington: Indiana University Press, 1990.

Stone, Dan. *Responses to Nazism in Britain, 1933–1939: Before the War and Holocaust*. London: Palgrave Macmillan, 2002.

Straley, Jessica. "Of Beasts and Boys: Kingsley, Spencer, and the Theory of Recapitulation." *Victorian Studies* 49 (2007): 583–609.

Strasser, Bruno J. "Who Cares About the Double Helix." *Nature* 422 (4 April 2003): 803–804.

Stueland Yruma, Jeris. "How Experiments Are Remembered: The Discovery of Nuclear Fission, 1938–1968." PhD diss., Princeton University, 2008.

Suber, Peter. *Open Access*. Cambridge, MA: MIT Press, 2012.

Szöllösi-Janze, Margit. *Science in the Third Reich*. New York: Berg, 2001.

Taylor, M. W. *Man Versus the State: Herbert Spencer and Late Victorian Individualism*. Oxford: Oxford University Press, 1992.

Teich, Mikulás. "Haldane and Lysenko Revisited." *Journal of the History of Biology* 40 (2007): 557–563.

Tennyson, G. B. "The Sacramental Imagination." In *Nature and the Victorian Imagination*, edited by U. C. Knoepflmacher and G. B. Tennyson, 370–390. Berkeley: University of California Press, 1977.

Tharp, Marie. "Connect the Dots: Mapping the Seafloor and Discovering the Mid-Ocean Ridge." In *Lamont-Doherty Earth Observatory of Columbia: Twelve Perspectives on the First Fifty Years 1949–1999*, edited by Laurence Lippsett, 31–37. New York: Columbia University Press, 1999.

Thomason, Bernard. "Dyer, Sir William Turner Thiselton (1843–1928)." In *Oxford Dictionary of National Biography*, edited by H. C. G. Matthew and Brian Harrison. Oxford: Oxford University Press, 2004.

Tumber, M. B. and K. Dickersin. "Publication of Clinical Trials: Accountability and Accessibility." *Journal of Internal Medicine* 256 (2004): 278–279.

Turner, Frank M. *Between Science and Religion: The Reaction to Scientific Naturalism in Late Victorian England*. New Haven, CT: Yale University Press, 1974.

———. *Contesting Cultural Authority: Essays in Victorian Intellectual Life*. Cambridge: Cambridge University Press, 1993.

———. "Public Science in Britain, 1880–1919." *Isis* 71 (1980): 589–608.

———. "The Victorian Conflict between Science and Religion: A Professional Dimension." *Isis* 69 (1978): 356–376.

van Ardsel, Rosemary T. "Macmillan Family (per. c.1840–1986)." In *Oxford Dictionary of National Biography*, edited by H. C. G. Matthew and Brian Harrison. Oxford: Oxford University Press, 2004.

Vandome, Robin. "The Advancement of *Science*: James McKeen Cattell and the Networks of Prestige and Authority, 1894–1915." *American Periodicals* 23 (2013): 172–187.

Vine, Frederick J. "Reversals of Fortune." In *Plate Tectonics: An Insider's History of the Modern Theory of the Earth*, edited by Naomi Oreskes, 46–66. Cambridge, MA: Westview, 2001.

von Gizycki, Rainald. "Centre and Periphery in the International Scientific Community: Germany, France and Great Britain in the 19th Century." *Minerva* 11 (1973): 474–494.

Walker, Mark. *German National Socialism and the Quest for Nuclear Power, 1939–1949*. Cambridge: Cambridge University Press, 1989.

———. *Nazi Science: Myth, Truth, and the German Atomic Bomb*. New York: Perseus, 1995.

Warwick, Andrew. *Masters of Theory: Cambridge and the Rise of Mathematical Physics*. Chicago: University of Chicago Press, 2003.

Watson, James. *The Double Helix: A Personal Account of the Discovery of the Structure of DNA*. New York: W. W. Norton, 1980.

Watts, Iain. "'We Want No Authors': William Nicholson and the Contested Role of the Scientific Journal in Britain, 1797–1813." *British Journal for the History of Science* 47 (2014): 397–419.

Weatherall, Mark W. "Making Medicine Scientific: Empiricism, Rationality, and Quackery in Mid-Victorian Britain." *Social History of Medicine* 9 (1996): 175–194.

Webb, K.R. "Raphael Meldola, 1849–1915." *Chemistry in Britain* 13 (1977): 345–348.

Weiner, Douglas R. "The Roots of 'Michurinism': Transformist Biology and Acclimatization as Currents in the Russian Life Sciences." *Annals of Science* 42 (1985): 243–260.

Wells, H. G. *Experiment in Autobiography: Discoveries and Conclusions of a Very Ordinary Brain*. 2 vols. London: Macmillan, 1934.

Werskey, Gary. "*Nature* and Politics between the Wars." *Nature* 224 (1 November 1969): 462–472.

———. *The Visible College: A Collective Biography of British Scientists and Socialists of the 1930s*. New York: Viking, 1978.

White, Paul. *Thomas Huxley: Making the "Man of Science."* Cambridge: Cambridge University Press, 2002.

Wilkins, Maurice. *The Third Man of the Double Helix: The Autobiography of Maurice Wilkins*. Oxford: Oxford University Press, 2003.

Wilson, David. *Kelvin and Stokes: A Comparative Study in Victorian Physics*. Bristol, UK: Adam Hilger, 1987.

———. "P.G. Tait and Edinburgh Natural Philosophy, 1860-1901." *Annals of Science* 48 (1991): 267-287.

———. "A Physicist's Alternative to Materialism: The Religious Thought of George Gabriel Stokes." In *Energy and Entropy: Science and Culture in Victorian Britain*, edited by Patrick Brantlinger, 177-204. Bloomington: Indiana University Press, 1989.

———. *Rutherford: Simple Genius*. Cambridge, MA: MIT Press, 1983.

———. "The Thought of the Late Victorian Physicists: Oliver Lodge's Ethereal Body." *Victorian Studies* 15 (1971): 29-48.

Winter, Alison. *Mesmerized: Powers of Mind in Victorian Britain*. Chicago: University of Chicago Press, 1998.

Winter, Jay. *The Great War and the British People*. London: Macmillan, 1985.

Wolfe, Audra. "What Does It Mean to Go Public? The American Response to Lysenkoism, Reconsidered." *Historical Studies in the Natural Sciences* 40 (2010): 48-78.

Wolfle, Dael. "*Science*: A Memoir of the 1960's and 1970's." *Science* 209 (4 July 1980): 57-60.

Zeitlin, Jonathan, and Gary Herrigel, eds. *Americanization and Its Limits: Reworking US Technology and Management in Post-War Europe and Japan*. Oxford: Oxford University Press, 2000.

Ziman, John. *Public Knowledge: The Social Dimension of Science*. Cambridge: Cambridge University Press, 1968.

Index

Abel, Frederick, 31
Abelson, Philip, 187–88
Abney, William, 115
Académie des sciences, 102, 105, 107
Academy (periodical), 29–30, 38, 91
Acta Eruditorum, 11
Adonis project, 231
age of the earth debate, 60–61
Airy, George, 31
Alichanian, A. I., 190
Alichanow, A. I., 190, 191
Allman, George J., 24
American Association for the Advancement of Science, 143, 231
Ångström, Anders, 36
Annalen der Physik und Chemie, 120, 121
"Appeal to the Civilized World," 125
Appleton, Charles, 29–30
Argyll, eighth Duke of, 2, 16, 56, 74–75, 79–83, 86, 88, 89, 90
Armstrong, Henry, 7, 92, 94, 126
Arrhenius, Svante, 126
arXiv.org (formerly xxx.lanl.gov), 231–32
Athenaeum (periodical), 28
Athenaeum (social club), 48, 64, 69, 96, 154, 161
Ayrton, Acton Smee, 40
Ayrton, Hertha, 78, 92, 94, 98
Ayrton, W. E., 52, 92, 95

Baker, John R., 193
Baltimore, David, 204, 205

Baltimore Affair, the, 204–5, 206
Bassett, A. B., 126
Bateson, William, 117–18
Beardsley, Tim, 221
Beauvais, Francis, 207
Becquerel, Henri, 102, 105, 106–7
Becquerel rays. *See* radioactivity
Belousov, V. V., 195
Bémont, Gustave, 102
Bendyshe, Thomas, 25
Bennett, Alfred, 31, 33, 41, 252n78
Benveniste, Jacques, 18, 201, 205–11, 213, 214, 219, 225, 226
Berichte der deutschen botanischen Gesellschaft, 116
Berlin Academy of Sciences, 125
Berliner, Arnold, 138
Bernal, J. D., 140, 193
Berthelot, Maurice, 114
Biometrika, 117, 118
Bishop, Jerry, 222, 223, 224
Blackman, V. H., 136
Blackwell (publisher), 231
Blondlot, René, 202
Bohr, Niels, 129–130, 133
Boltwood, Bertram B., 14, 16, 100, 101, 112–14, 121, 122, 127, 233
Bonney, T. G., 81, 82
Bottomley, J. T., 104
Bragg, Lawrence, 152, 154, 155, 167
Brauner, Bohuslav, 134–35

302 INDEX

Brigham Young University, 200, 211, 212, 214, 218
Brightman, Rainald, 147-48, 149-50, 151, 165, 175, 192-93
Brimble, L. J. F., 13, 14, 192, 193, 196, 198, 229; career before *Nature*, 146, 164; death, 172, 174; and DNA papers, 154; editorial style, 17, 145-46, 148-52, 155, 160-61, 165, 167-69, 170, 175, 176, 185, 202; as Gregory's assistant, 146-47; and plate tectonics, 156, 158, 159, 160; promoted to editor in chief, 147; and World War II, 147-48
Bristol College, 52
Bristol Trade School, 90
British Association for the Advancement of Science (BA), 5, 15, 25, 64, 66, 69, 70, 72, 74, 87, 88, 94, 126, 149
British Quarterly Review, 28, 43-44, 55, 63, 66
British Science Guild, 94-95, 125, 135, 149
Brookhaven National Laboratory, 216
Brown, Patrick, 236
Brunel, I. K., 86
Bryan, William Jennings, 136
Bullard, Theodore ("Teddy"), 161
Bulletin of the Geological Society of America, 157, 163
Bunsen, Robert, 124, 125
Byam Shaw, Nicholas, 179

California, University of (system), 175-76, 236, 237, 238
Cambridge, University of, 51, 52, 53, 61, 62, 102, 107, 117, 130, 142, 146, 152, 154, 156, 157, 158, 159, 160, 166, 179, 238
Cambridge University Gazette, 29
Campbell, George Douglas. *See* Argyll, eighth Duke of
Campbell, George Granville, 80
Campbell, Norman, 4-5, 129
Campbell, Philip, 13, 14, 177, 230, 233, 235, 237
Carlyle, Thomas, 77
Carpenter, William, 84
Cattell, J. McKeen, 137, 163
Cavendish Laboratory, 102, 107, 108, 110, 129, 130, 152, 154
Cell (journal), 204, 229, 240
Chadwick, James, 130
Chambers' Edinburgh Journal, 26

Chase, Martha, 155
Chemical News, 27-28, 64-65, 70, 84, 104-6, 109, 110
Chemical Society, 78-79, 91
Chladek, Stanislav, 191
Clausius, Rudolf, 43
Clay and Sons (printer), 31-32
Clerke, Agnes M., 91-92
cold fusion, 14, 18, 171, 200-202, 211-15; scientific criticism of, 215-20; popular coverage of, 220-24
Cold War, 18, 171-72, 192-96, 198-99
Collection of Czechoslovak Chemical Communications (journal), 191
Collier, James, 45
Columbia University, 111, 156, 159
Comptes Rendus, 104, 106, 107, 108
Copenhagen, University of, 129. *See also* Institute for Theoretical Physics
coral reef formation, 2, 74-75, 80-81
Cornell University, 232
Cornhill Magazine, 26
Correns, Carl, 116
Coster, Dirk, 130
Cox, Allan, 157, 163
Crick, Francis, 2, 14, 17, 152-55, 168, 169, 222
Crookes, William, 27, 28, 64-65, 70, 84, 85, 86, 88, 105, 110, 167
Cunningham, J. T., 118
Curie, Irène, 130, 133
Curie, Marie, 100, 102, 103, 106, 107, 108, 109, 114, 119, 121, 130, 152
Curie, Pierre, 100, 102, 103, 106, 107, 108, 109, 114, 121, 130, 152

Dalrymple, Brent, 157
Dana, James Dwight, 80
Danne, Jacques, 120, 121
Danysz, Jean, 121
Darwin, Charles, 2, 11, 28, 51, 53, 55, 56, 63, 74, 76, 77, 79, 82, 93; papers in *Nature*, 54, 67; posthumous image in *Nature*, 67-69
Darwin, Francis, 60, 68
Davenas, Elisabeth, 207, 209
Davenport, Charles, 118
Davies, David, 13, 14, 150, 151, 164-65, 197, 234; approach to editorials, 181, 184-85, 194-95; career before *Nature*, 176, 177,

INDEX 303

178, 179-80; changes to *Nature*, 17-18, 170, 180-84, 187, 189, 198; editorial style, 170, 181, 184-85, 194-95, 202-4, 229; hired at *Nature*, 180, 186-87; left *Nature*, 196; and nuclear disarmament, 194-95; portrait, 181; and unusual papers, 202-4, 226
de Gray, Thomas, 128
Delbrück, Max, 155
de Vries, Hugo, 116, 117
Dietz, Robert, 157, 159, 160, 162, 163, 164
Dingell, John, 204-5
Dingle, Herbert, 7
Dixon, Edward, 84-85, 86
Dixon, Herbert, 129
DNA, discovery of the structure of, 152-53, 168, 168; papers in *Nature*, 153-55, 168, 169
Doell, Richard, 157
Doklady Armenian S.S.R., 190
Draper, Henry, 64, 101
Dreskin, Stephen C., 208-9

École Municipale de Physique, 102
École Polytechnique, 102
Economist, 202, 221, 222
Eddington, Arthur, 129
Edinburgh, University of, 110
Edison, Thomas, 71, 257n86
Edward VII, King, 115, 134
Education and Training, 177
Einstein, Albert, 129, 138, 275n92
Eisen, Michael, 236
Electrician, 104-6, 109
Elsevier (publisher), 171, 231, 237-38
Engineer, 87
engineering, 86-88
English Mechanic, 64
Eve, Arthur S., 100, 115
evolutionary theory, 25, 38, 41, 56-60, 67-69, 74, 76-78, 82-83, 92-93, 117, 136-37
Ewing, Maurice, 159, 161, 167, 174

Faraday, Michael, 95
Feather, Norman, 133
Fermi, Enrico, 130; rejected *Nature* paper, 130
Financial Times, 212
Finsbury Technical College, 52, 61
Finsen, Niels, 121
Fiske, John, 25

Fleischmann, Martin, 18, 19, 200-202, 211-20; image in the popular press, 220-25
Fortnightly Review, 28, 44, 58, 60, 63, 65, 66-67, 91
Foster, Michael, 30, 31, 34, 50, 51
Franklin, Rosalind, 152-53, 168, 169
Frisch, Otto, 132-33

Gale, A. J. V., 13, 14, 192, 193, 198, 229; career before *Nature*, 146, 164; and DNA papers, 154; editorial style, 17, 145-46, 148-52, 155, 159, 160-61, 165, 167-69, 170, 172, 174, 175, 176, 185, 202; as Gregory's assistant, 146-47; promoted to editor in chief, 147; retirement, 156, 176; and World War II, 147-48
Galton, Francis, 117
Gardener's Chronicle, 67
Garfield, Eugene, 188
Garwin, Richard, 214, 221
Gates, Reginald Ruggles, 99
Gaylarde, P. M., 209
Geikie, Archibald, 32
Geller, Uri, 203, 208
Genetics, 101, 116-19, 131, 135-36
Gentry, Robert, 232
Geological Society, 79, 81
Geophysical Journal of the Royal Astronomical Society, 180
George V, King, 134
Gerard, John, 76-77
Ginsparg, Paul, 231-32
Glasgow, University of, 52, 53, 61, 74, 146
Godwin-Austen, H. H., 128
Gosling, Raymond, 152, 153
Göttingen, University of, 141, 142
Goudsmit, Samuel, 164
Gould, Barbara Ayrton, 92
Gowers, Timothy, 238
Graham, Thomas, 33
Gratzer, Walter, 161, 167, 174, 176, 206
Gray, Andrew, 88
Greenhill, A. G., 88
Gregory, John, 90
Gregory, Kate Florence, 90
Gregory, Richard, 13, 14, 131, 132, 134, 135, 149, 151, 163, 165, 171, 193, 196, 234; becomes editor of *Nature*, 16, 17, 93-94, 98-99; and British Association for the Advancement of Science, 94; and British

Gregory, Richard (*continued*)
  Science Guild, 94-95; *Discovery; or, the Spirit and Service of Science*, 95-96, 134; early life, 90; editorial style, 98-99, 151, 160, 161, 167; elected FRS, 99; hired at *Nature*, 91; knighted, 96; as Lockyer's assistant, 75, 90, 91-93, 96, 114; and *Nature*'s coverage of Germany, 123, 124, 125, 128, 141, 143-44; opinions on women, 91-92; portrait, 97; relationship with Macmillan and Company, 91, 96, 145-46, 177; retirement, 145-46, 147; science writing, 91; and *scientist* debate, 4-5, 7-8
Guardian (newspaper), 99, 149, 150, 171-72, 175, 178
Guthrie, Frederick, 45

Haeckel, Ernst, 50
Hahn, Otto, 16, 100, 101, 114, 115-16, 119, 121, 127, 132-33, 143-44
Haldane, J. B. S., 140, 141
Hamburg, University of, 133
Hampson, George, 128
Happer, William, Jr., 221
Harnad, Steven, 237
Harwell National Laboratory, 216, 217
Hawkins, Marvin, 212
Hayward, Robert B., 45
Heezen, Bruce, 159
Heidelberg, University of, 124, 125, 139, 141, 142
Heirtzler, James, 162
Helmholtz, Hermann von, 120
Hershey, Alfred, 155
Hertz, Heinrich, 120
Hess, Harry, 156-57, 162, 163
Hevesy, George, 130
Hill, A. V., 139, 140, 141
Hill, Maurice, 158, 159
Hirst, Thomas, 24, 50
Hitler, Adolf, 133, 138, 141
Hogben, Lancelot, 140
homeopathy, 18, 206, 208, 209, 224
Hooker, Joseph D., 21, 25, 29, 30, 31, 39-41, 43, 49, 51, 53, 54, 63, 74, 89, 252n78
Hudson, Richard L., 223, 224, 225
Hughes, Jenny, 179, 180, 196
Hughes, Norman, 156
Hughes, Thomas, 24
Huth, Edward, 231

Huxley, Julian, 138, 149
Huxley, Thomas H., 38, 69, 74, 95; articles for literary periodicals, 28, 65, 66, 80-81; articles for *Nature*, 31, 33, 35, 39, 46, 53, 54; early support of *Nature*, 26-27, 29, 30; and *Reader*, 25; as scientific mentor, 50-51, 53, 62-63; views on women, 27, 78, 91; and X Club, 40-41, 49, 50

Imanishi-Kari, Thereza, 204
Imperial Institute of Physics and Technology, 140
Institute for Theoretical Physics, 129-30, 133. *See also* Copenhagen, University of

Jahrbuch der Radioaktivität und Elektronik, 101
Jenkin, Fleeming, 86
Joliot, Frederic, 130
Jones, Steven, 18-19, 200, 201, 211-15, 217, 218, 222
Joseph Hughes (publisher), 91
Joule, James, 33
Journal des Sçavans, 11
Journal of Electroanalytical Chemistry, 18, 212, 213, 214
Journal of General Physiology, 155
Journal of Geophysical Research, 158, 159, 187
Journal of Human Genetics, 235
Journal of Internal Medicine, 231
Journal of Physics, 190, 191
Journal of Physiology, 141
Journal of the Chemical Society, 37
Journal of the Society of Arts, 29
journals: history of, 11-12; and knowledge claims, 11, 19, 228; online, 230-31, 233-36; and open-access movement, 236-37; as primary form of scientific publishing, 11, 19, 229; and priority, 65, 108-9, 151-53, 160, 212, 239; and social media, 233-34

Kaiser-Wilhelm-Institut, 132
Kant, Immanuel, 45
Kapitza, Peter, 135
Kew Gardens, 21, 39-41, 51
King's College London, 152, 153, 154, 171
Kingsley, Charles, 13, 37-38
Kirchoff, Gustav, 120
Klyueva, Nina, 189
Knowledge, 9, 64, 71-72, 91
Korenchevsky, Vladimir, 135

Laborde, Albert, 103
Lamont Geological Observatory, 156, 159, 162, 174
Lankester, Edwin, 50
Lankester, E. Ray, 53, 63, 65, 66, 67, 126; biography, 50-51, 52; contributions to *Nature*, 6, 7, 54, 57, 58-60, 67-69, 111, 118, 137
Lankester, Phebe Pope, 50
Lattes, C. H. G., 190
Leipzig, University of, 124
Leisure Hour, 91
Lenard, Philipp, 140
Lenin, Vladimir, 134
Lenin Academy of Agricultural Sciences, 136
Lidgett, J. Scott, 137
Lindley, David, 202, 217-18
Lindley, Mary, 186
Linnean Society, 56
Loch Ness Monster, 203-4
Lock, R. H., 118
Lockyer, Norman, 13, 14, 47, 51, 62, 71, 74, 84, 95, 145, 175, 196, 229, 234; astronomical work, 22-23; correspondence about *Nature*, 30-31, 59-60, 64, 111, 114; early goals for *Nature*, 6, 9, 27-28, 35-36, 38-39; early life, 22; editorial style, 39-46, 48-49, 59-60, 69-70, 92, 114, 124, 125, 149, 151, 160, 167; foundation of *Nature*, 15, 21-22, 26-28, 30-31, 65; and Gregory, 75, 90, 91, 92; and *Reader*, 24-25; relationship with Macmillan and Company, 25-26, 93-94, 96, 177; retirement, 16, 75, 93-94, 96, 98; supporter of women's rights, 78, 92
Lockyer, Thomazine Mary Broadhurst, 78
Lockyer, Winifred James, 70
Lodge, Oliver, 6, 50, 52, 53, 54, 72-73, 84-85, 86, 88, 129, 234
London Medical College, 118
London Review, 24
Lorentz, Hendrik, 129
Los Alamos National Laboratory, 218, 231
Lubbock, John, 25, 40, 50, 95
Luria, Salvador, 155
Lysenko, Trofim, 135-36, 192, 193, 194

MacDonald, James Ramsay, 99
Macdonald, William, 115
Macdonald Physics Laboratory, 108, 115-16
Macmillan, Alexander, 21, 25, 26, 31, 46
Macmillan, Daniel, 145, 147

Macmillan, Frederick, 70, 93-94, 96
Macmillan, George, 70
Macmillan, Harold, 180, 204
Macmillan, Maurice (1863-1936), 96
Macmillan, Maurice (1921-1984), 172, 276n4
Macmillan and Company, 109, 123, 180, 192, 204, 237; as book publisher, 25, 32, 70, 91, 95, 149, 177, 194; existing archives, 9, 13; and foundation of *Nature*, 15, 21, 26, 27, 29, 30, 31-32; and management of *Nature*, 46, 70, 93-94, 96, 97, 146, 147, 170, 172, 177, 179, 196-97, 230, 235; nineteenth-century success, 26
Maddox, Arthur, 171
Maddox, Brenda, 175, 177, 179, 196
Maddox, John, 12, 13, 14, 64, 150, 186, 222, 223, 234; and Baltimore affair, 205; and Benveniste incident, 18-19, 201-2, 205-11, 213-14; changes to *Nature*, 17-18, 170, 174-76, 185, 186, 189, 235; and cold fusion, 18-19, 201, 213-15, 218-20, 226; early life, 171; editorial style, 164, 168, 169, 174-79, 182, 184, 194, 195, 198, 202, 204, 205, 225, 228; and electronic journals, 232-33; hired at *Nature*, 172; left *Nature* in 1973, 179-80; and Nuffield Science Teaching Project, 172, 174, 196; portrait, 173; relationship with Macmillan and Company, 177, 196-97; retirement, 230; returned to *Nature*, 196-97; science journalism, 171-72; three-journals plan for *Nature*, 177-79, 187
Manchester, University of, 100, 109, 110, 111, 127, 129, 130, 146, 171
man of science (term), 2, 75-76, 78, 83, 88-90, 95, 96-97, 151
Marxism, 140
Massachusetts Institute of Technology (MIT), 177, 180, 186, 195, 204, 209, 215, 216
Masson et Compagnie (publisher), 120
Matthews, Drummond, 156, 157-58, 159, 160, 161, 162, 164
Maxwell, James Clerk, 36, 39, 52
McGill University, 102-3, 107, 108, 110, 112, 114, 115
Medvedev, Zhores, 192, 194
Meitner, Lise, 132-33, 268n48
Meldola, Raphael, 50, 51-52, 53, 54, 58, 59, 60, 63, 65, 66, 68, 95; book reviews for *Nature*, 76-78

Mellor, J. W., 134
Mendel, Gregor, 116, 117, 118, 119
Mendeleev, Dmitri, 103
Metzger, Henry, 208-9
Meyer, Kuno, 127
Meyer, Stefan, 100, 119
Michels, John, 71
Michurin, Ivan, 135, 136
molecular biology, 153-54, 155, 161
Morley, John, 31
Morley, Lawrence, 156, 158-60, 167, 168-69
Moseley, Henry, 127, 130
Moulton, J. F., 43-45
Mountford, James, 149, 150
Mudge, George P., 118
Murray, John, 74-75, 80-81

National Socialism (Nazism), 2, 16-17, 123, 124, 137-44, 147
nationalism, 121-22
Nature
    acceptance rate, 184, 198, 229
    advertisements for, 28-29
    banned in Germany, 16-17, 123-24
    book reviews, 33, 34, 35, 36-37, 38, 53, 76-78, 86, 118, 146, 148, 175, 178
    British orientation, 9-10, 114-17, 123-28, 133-34, 170, 175, 187, 197-98
    and British scientific community, 49, 50, 52-55, 61-63, 72-73, 75-76, 78-79, 81-90, 114-19, 124-28, 140, 145, 154, 165, 168, 172
    changes in editorship, 93-94, 96-98, 145-47, 172-74, 179-80, 196-97, 230
    and cold fusion, 14, 18, 171, 200-202, 212-15
    competitors, 26-30, 53-55, 63-67, 104-6, 109-10, 162-65, 229, 240
    contributors, 2, 5, 6, 8-9, 14-15, 16-17, 18, 30-31, 38-39, 48-49, 50-55, 63-65, 81-83, 85-90, 114-19, 123-24, 129-33, 134-42, 161, 165-66, 170, 177, 184-87, 189, 197, 198-99, 225-26, 229, 240-41
    correspondence column, 13, 176, 191, 208, 234
    covers, 180, 182, 183, 227, 234
    criticism of British government in, 10, 16, 93, 98, 115, 123-25, 194-95
    criticism of foreign governments in, 2, 16-17, 125-27, 133-42, 192-96, 204-5
    current scholarship on, 12
    and DNA papers, 153-55, 168, 169
    debates in, 39-46, 56-60, 62-63, 81-83, 84-88, 92, 113-14, 117-19, 140-41, 149, 176, 205-11, 215-16
    editorials, 7, 10, 27, 28, 33, 36, 37, 54, 62, 78, 79, 81, 86-87, 98, 115, 124, 125-26, 127, 128, 136, 137, 138, 139, 143, 147-48, 149-50, 151-52, 175, 181, 184-85, 186, 192, 193-95, 196, 200-201, 202, 205-6, 213-14, 218-20, 222, 228, 233, 234, 239
    editors, 13-14. *See also* Brimble, L. J. F.; Campbell, Philip; Davies, David; Gale, A. J. V.; Gregory, Richard; Lockyer, Norman; Maddox, John
    and electronic publishing, 230-35, 240-41
    and evolutionary theory, 25, 38, 41, 56-58, 67-69, 74, 76-78, 82-83, 92-93, 117, 136-37
    foundation, 21-22, 25-26
    and genetics, 101, 116-19, 131, 135-36
    and Germany, 10, 16-17, 123-28, 137-44
    images in, 32, 115-16, 180
    impact factor, 188
    internationalism in, 16, 100-101, 114-16, 119-22, 123-24, 131-32, 165-66, 168, 170-71, 175-76, 185-88, 189-92, 197-99
    and Japan, 131, 189, 197
    Letters to the Editor column, 2, 4-7, 9, 13, 16, 17, 28, 36, 37, 41-46, 55, 61, 71, 81, 84-88, 97, 98, 101, 109, 111-12, 117, 118-19, 123-24, 128-33, 134, 148, 151-52, 158, 165, 168, 175, 185, 188, 189, 197, 215-16, 232, 234
    and Macmillan and Company, 15, 21, 26, 27, 29, 30, 31-32, 46, 70, 93-94, 96, 97, 146, 147, 170, 172, 177, 179, 196-97, 230, 235
    masthead, 33-34, 99, 148, 150-51
    and molecular biology, 153-54, 155, 161
    News and Views column, 98, 131-32, 135, 136, 138, 139, 142, 143, 148, 149, 150, 172, 175, 176, 177, 180, 187, 205, 214, 219, 228, 234
    news articles, 18, 33-37, 124, 138, 168, 170, 175-76, 185, 187, 195-96, 204, 217-18, 232
    Notes column, 27, 37, 98, 103, 105, 106, 128

and open-access trial program, 237
papers rejected by, 9, 13, 130, 158-61,
    162-64
peer review procedures, 18, 146, 160-61,
    167, 169, 174, 187, 229, 239
and plate tectonics, 17, 155-56, 157-61
printing of, 31-32, 148, 180, 186
publication speed, 63-64, 104, 105, 109-
    10, 151, 153, 172, 174, 215, 240
and radioactivity, 16, 86, 105-7, 112-14,
    119, 129, 130, 165, 166, 232, 240
readers, 2, 5, 8, 9, 13-14, 27, 28-29, 30, 35,
    37-38, 39, 64-65, 70, 71, 114, 122, 143,
    167, 178, 180, 197, 225-26, 229, 233-34
and scientific community, 1-3, 5-6, 8-10,
    11, 47, 63-67, 72-73, 75-76, 94-98, 124-
    28, 131-34, 143-44, 151-52, 154-55, 161-
    62, 165-69, 175-76, 184-85, 197-99,
    211, 224-26, 229-30, 239-41
and scientific priority, 65, 108-9, 151-53,
    160, 212, 239
sister publications, 178, 235
Societies and Academies column, 28, 37,
    105, 106
subscribers, 26, 28, 64, 96, 114, 148, 177,
    186, 188, 197, 224, 233, 234
and the United States, 10, 17, 64, 71, 112-
    14, 124, 135, 136-37, 159, 166, 170, 176,
    185-87, 197, 203
and the USSR, 17, 18, 124, 134-36, 171,
    189-96, 198-99, 229
Washington office, 170, 185-86, 197, 203
website, 230, 232-33, 234, 235, 240
and women in science, 4, 27, 78-79, 91-
    92, 230
and X Club, 15, 25, 26, 29, 31, 39-46, 47,
    49-50, 51, 53, 54, 60, 63, 77, 114
and X rays (Röntgen rays), 86, 101-2, 103-
    5, 106, 112
Nature New Biology, 177-79, 180
Nature Physical Science, 177-79, 180
Nature Publishing Group (NPG), 235, 237, 238
Naturwissenschaften, Die (journal), 138
neutron, discovery of, 130
Neville, G. J., 209-10
New Scientist, 171
Newton, Isaac, 11, 44-45
Newton, Mary, 108
New York Times (newspaper), 144, 186
Nineteenth Century, 28, 55, 66, 74-75, 80

Nobel Institute, 122, 133, 188
Nobel Prize, 16, 106, 107, 109, 114, 122, 129,
    140, 153, 155, 204, 222
N rays, 202, 206
nuclear fission, discovery of, 132-33

Occhialini, G. P. S., 190
Ohm, Georg, 120
Oldenburg, Henry, 166-67
Online Computer Library Center, 231
Online Journal of Current Clinical Trials, 231
open-access publishing, 236-37
Ostwald, Wilhelm, 124, 125
O'Toole, Margot, 204, 205
Owen, Richard, 40-41, 43, 63
Oxford University, 29, 50-51, 52, 63, 70, 103,
    127, 142, 159, 171, 197
Oxford University Extension Delegacy, 90, 91

Paris Academy of Sciences, 79
Parson, James, 90
Pascoe, Francis, 77-78
Pasteur, Louis, 124
Pauling, Linus, 152
Pearl, Raymond, 118
Pearson, Karl, 85, 117-18
peer review, 18, 146, 160-61, 166-67, 169,
    187-88, 201, 214, 223, 225, 229, 232,
    234, 236, 239
Pengelly, William, 31
Pergamon (publisher), 231
periodicals: popular science, 11-12, 21, 26-30;
    specialist scientific (*see* journals); Victo-
    rian, 15, 21, 26-30, 31, 37, 46, 54, 55, 63,
    66-67, 91, 230
Perry, John, 50, 52, 53, 54, 60-61, 63, 66, 67,
    111
Petrasso, Richard, 215-16, 221
Petrograd University, 134
Petsko, Gregory, 209
Philosophical Magazine, 104, 105, 106, 108,
    109, 112
*Philosophical Transactions of the Royal Society
    of London*, 11, 34, 37, 66, 72-73, 104,
    106, 166
Physical Review, 164
Physical Review Letters, 164-65, 168
Pitman, Walter, 162
Planck, Max, 120, 275n92
plate tectonics, 17, 156-58, 162-64, 192, 199

PLOS ONE, 236, 238-39
Poggendorff, Christian, 120
Pons, Stanley, 18, 19, 200-202, 211-20; image in popular press, 220-25
Poulton, Edward B., 118
Powell, C. F., 190-91
Princeton University, 131, 156, 166, 221
Pritchard, Charles, 31
Proceedings (journal of Society for Psychical Research), 86
Proceedings of the National Academy of Science, 225
Proceedings of the Royal Society of Edinburgh, 80
Proceedings of the Royal Society of London, 104, 106, 109, 112
Proctor, Richard, 2, 9, 71-72
professionalization, 10, 89-90
Public Library of Science (PLOS), 236
PubMed Central, 236, 237
Puthoff, Harold, 203, 226, 229

Queen's College (Belfast), 52

radioactivity, 16, 86, 100, 102-3, 105-7, 112-14, 119, 120-21, 129, 130, 165, 166, 232, 240
Radium, Le, 101, 107, 120-21
Ramsay, William, 95, 103, 113, 124, 125-26, 127
Randall, John, 154, 155
Randi, James, 207, 208, 210
Rankine, William J. M., 86
Reade, T. Mellard, 81-82
Reader, 24-25, 26, 28, 29, 30
Reagan, Ronald, 175
Reid, G. Archdall, 118, 119
relativity, 129
Republic of Letters, 8
Rich, Vera, 195-96, 215
Rines, Robert, 203
Romanes, George John, 14, 50, 54, 64, 65, 111, 118; biography, 51, 52, 53; and Darwin, 55-56, 63, 67-69; evolutionary theories of, 56-60; and literary periodicals, 63, 66
Röntgen, Wilhelm Conrad, 101-2, 103-4
Röntgen rays. *See* X-rays
Root, Charles, 45
Roscoe, Henry E., 124
Roskin, Grigorii, 189

Royal Agricultural College, 51
Royal Artillery Officers academy, 88
Royal Astronomical Society, 22
Royal College of Chemistry, 51
Royal College of Science, 51, 90
Royal Institution, 5, 52, 139
Royal Society of Edinburgh, 74
Royal Society of London, 5, 16, 24, 34, 49, 51, 52, 61-63, 73, 78, 79, 83, 94, 99, 115, 117, 126, 166
Rubens, Heinrich, 121, 202
Rügemer, H., 142-43
Rust, Bernhard, 123, 139, 142, 143
Rutherford, Ernest, 14, 100, 121, 122, 126-27, 152, 163, 233; as mentor, 119, 129-30, 133, 135; picture from *Nature*, 115-16; publishing in *Nature*, 16, 101, 109-14, 118; radioactivity research, 102-3, 107-9

Salamon, Michael, 217, 222, 224
Savitch, Paul, 133
Scallan, Mary, 186
Schuster, Arthur, 103, 126
Science (journal), 71, 110, 137, 162, 163-64, 168, 187-88, 197, 225, 229, 231, 240
science journalism, 26-27, 38-39, 220-25
Scientific American, 221, 235
scientific internationalism, 16, 100-101, 114-16, 119-22, 123-24, 131-32, 165-66, 168, 170-71, 175-76, 185-88, 189-92, 197-99
Scientific Opinion, 31
scientist (word), 4-8
Scopes, J. T., 17, 124, 136-37
Scott, Peter, 203
Sheehan, Mary, 150, 164, 172, 174, 178, 186, 196
Siemens, C. W., 48
Society for Psychical Research (SPR), 84-86
Soddy, Frederick, 16, 103, 108, 109, 113, 114, 121
Sokoloff, Boris, 134
Spectator, 24
Spencer, Herbert, 25, 28, 43-46, 49, 50, 53, 54, 65, 66, 68-69
spiritualism, 83-86
Spottiswoode, William, 40, 49
Springer (publisher), 231
Stalin, Joseph, 135, 189, 191, 192, 194
St. Andrews, University of, 74, 137

Stanford University, 203, 236
Stanton, Arthur, 104
Stark, Johannes, 140-41
Stewart, Balfour, 62-63
Stewart, Walter, 205, 206, 207, 208, 210
Stirling, James Hutchison, 76
Stokes, George Gabriel, 61-63, 83
Strassmann, Fritz, 132-33
Swann, Michael, 197
Swinbanks, David, 217
Swinton, A. A. C., 104
Syme, David, 76, 77

Tait, Peter Guthrie, 41-43, 44-46, 52-53, 54, 60-61, 65, 87, 110
Tansley, A. G., 193
Targ, Russell, 203, 226, 229
Tharp, Marie, 156
Thiselton-Dyer, W. T., 50, 51, 52, 53, 54, 59, 62-63, 65, 66, 67, 82, 117, 118
Thompson, D'Arcy Wentworth, 6, 137
Thomson, J. J., 16, 102, 103, 105, 108, 109, 110, 129
Thomson, William (Lord Kelvin), 2, 41, 44, 52, 63, 87, 88, 95, 103, 104; and age of the earth controversy, 60-61; publishing in *Nature*, 65, 70-71
Thomson, Wyville, 46
Time (magazine), 202, 221-22
Times (newspaper), 29, 99, 124
Timiryazev Academy, 194
Transactions of the Linnean Society, 37
Tschermak, Erich, 116
Tyndall, John, 25, 30, 40, 41-43, 45, 49, 50, 52, 69-70

University College, Liverpool, 52
University College London, 51, 52, 103
University College of North Wales, 88
University College of Reading, 146
University Correspondence College, 90
University of Western Australia, 111
Utah, University of, 18, 200, 211, 212, 213, 217, 218, 222, 224

Varmus, Harold, 236
Vavilov, Nikolai, 135, 136, 269n67
Victoria, Queen, 115, 134

Victoria University College, 111
Villard, Paul, 102
Vine, Frederick, 156, 157-60, 161-62, 164, 166, 167, 168

Wade, Nicholas, 186
Wallace, Alfred Russel, 28, 30, 32, 53, 56, 63, 65, 66, 67, 74; book reviews for *Nature*, 76-77; contributions to *Nature*, 53, 54, 60, 84-86; and evolutionary theory, 58-59, 67-68, 69
Wall Street Journal, 200, 202, 212, 221, 222-25
Washington Post (newspaper), 171
Watson, James, 2, 14, 17, 152-54, 155, 166, 168, 169, 222
Weber, Robert, 61
Wegener, Alfred, 156, 273n50
Weismann, August, 57
Weldon, W. F. R., 117, 118
Wells, H. G., 90, 91, 92-93, 161
Wenz, Charles, 207, 233
Weyl, Hermann, 129
Whewell, William, 5
Wiedemann, Gustav, 120
Wilhelm, Kaiser, II, 124
Wilkins, Maurice, 152-54
Williamson, Alexander W., 62-63
Wilson, J. Tuzo, 162, 164
Woltereck, R., 141
Wood, Robert W., 202
Wordsworth, William, 33, 77
World War I (Great War), 15, 17, 95, 96, 119, 121, 122, 123-24, 125-28, 129, 131, 133, 134, 141, 146
World War II, 18, 145-46, 147-48, 150, 186, 189, 192, 193
Worthington, A. M., 87-88
Wright, C. Hagberg, 135
Würtzberg, University of, 140

X Club, 15, 25, 26, 29, 31, 39-46, 47, 49-50, 51, 53, 54, 60, 63, 77, 114
X-rays (Röntgen rays), 86, 101-2, 103-5, 106, 112

Yale University, 100, 112, 216

Zemlicka, Jiri, 191

NOV 0 3 2015